HEINZ GÖTZE

MISCELLANEA MATHEMATICA

M. Atiyah
F.L. Bauer
H. Cartan
S.-S. Chern
J.H. Conway
B. Eckmann
L.D. Faddeev
H. Grauert
P. Hilton
H. Hironaka
F. Hirzebruch
L. Hörmander
F. John
M. Koecher
R. Narasimhan
C. Reid
R. Remmert
J-P. Serre
N. J. A. Sloane
J. Tits
A. Weil
D. Zagier

Edited by
Peter Hilton
Friedrich Hirzebruch
Reinhold Remmert

SPRINGER-VERLAG
Berlin Heidelberg New York
London Paris Tokyo
Hong Kong Barcelona
Budapest

ISBN-13:978-3-642-76711-1   e-ISBN-13:978-3-642-76709-8

DOI: 10.1007/978-3-642-76709-8

© Springer-Verlag Berlin Heidelberg 1991
Softcover reprint of the hardcover 1st edition 1991

40/3114-543210 – Printed on acid-free paper

QUAPROPTER BONO CHRISTIANO,
SIVE MATHEMATICI, SIVE QUILIBET IMPIE DIVINANTIUM,
MAXIME DICENTES VERA, CAVENDI SUNT,
NE CONSORTIO DAEMONIORUM ANIMAM DECEPTAM,
PACTO QUODAM SOCIETATIS IRRETIANT.

Augustinus: De genesis ad literam
(Liber 2, Caput XVII, Nr. 37)

Deshalb soll sich
der gute Christ hüten
vor Mathematikern
und allen,
die in gottloser Weise
prophezeien,
so sehr sie auch
die Wahrheit sagen,
damit diese nicht, mit
den Dämonen vereint,
die irregegangene Seele
durch einen
Gemeinschaftspakt
täuschen.

Thus the good
Christian should
beware of
mathematicians and
all those who make
false prophesies,
however much
they may in fact
speak the truth;
lest, being in league
with the devil,
they may deceive
errant souls
into making
common cause.

Voilà pourquoi un
bon chrétien doit
se garder des
mathématiciens ou de
n'importe quels devins
usant de sortilèges,
surtout lorsqu'ils
disent vrai, de peur
d'entrer en rapport
avec les démons et,
en pactisant avec eux,
de se laisser captiver
dans leurs filets.

Contributions to
the Culture of Mathematics

It is difficult to pinpoint a particular moment when the idea of a Festschrift for Heinz Götze was first conceived. But, once formulated, the concept took shape quickly, with general agreement among those who undertook the agreeable task of editing the volume, together with their fellow-conspirators at Springer Verlag, that a number of distinguished mathematicians should be approached to contribute articles, with the single proviso that the Festschrift would not be an appropriate vehicle for the publication of original research.

An inevitable result of this editorial policy – indeed, a result which the editors anticipated with some excitement – is that the essays which follow range over a broad spectrum of topics of mathematical interest, with perhaps a predominant flavor of mathematical history. The articles are all infused with the conviction of the importance of mathematics in our culture, as an integral component of man's quest for knowledge and understanding.

It is thus especially appropriate that this volume be dedicated to our esteemed friend and colleague Dr. Heinz Götze, who has made so unique and invaluable a contribution to the spread of the mathematical culture that is reported in these pages.

Oberwolfach Peter Hilton
April 14, 1991 Friedrich Hirzebruch
  Reinhold Remmert

Contents

M. Atiyah The European Mathematical Society · 1

F. L. Bauer Sternpolygone und Hyperwürfel · 7

H. Cartan Sur quelques progrès dans la théorie des fonctions analytiques de variables complexes entre 1930 et 1950 · 45

S.-S. Chern Surface Theory with Darboux and Bianchi · 59

J. H. Conway The Cell Structures
N. J. A. Sloane of Certain Lattices · 71

B. Eckmann Von der Studierstube in die Öffentlichkeit · 109

L. D. Faddeev A Mathematician's View of the Development of Physics · 119

H. Grauert The Methods of the Theory of Functions of Several Complex Variables · 129

P. Hilton The Mathematical Component of a Good Education · 145

H. Hironaka Fame, Sweet and Bitter · 155

F. Hirzebruch Centennial of the German Mathematical Society · 177

L. Hörmander The First Woman Professor and Her Male Colleague · 195

F. John Memories of Student Days in Göttingen · 213

M. Koecher Castel del Monte und das Oktogon · 221

R. Narasimhan The Coming of Age of Mathematics in India · 235

C. Reid   Hans Lewy (1904–1988) · 259

R. Remmert   Inventiones mathematicae:
                         Die ersten Jahre · 269

J-P. Serre   Les petits cousins · 277

J. Tits   Symmetrie · 293

A. Weil   Sur quelques symétries dans l'Iliade · 305

D. Zagier   Lösungen von Gleichungen
                         in ganzen Zahlen · 311

Michael Atiyah   The Master's Lodge · Trinity College
                           Cambridge CB2 1TQ · United Kingdom

Friedrich L. Bauer   Villenstraße 19
                           W-8081 Kottgeisering
                           Federal Republic of Germany

Henri Cartan   95, Boulevard Jourdan
                           F-75014 Paris · France

Shiing-Shen Chern   Department of Mathematics
                           University of California at Berkeley
                           Berkeley, CA 94720 · USA

John H. Conway   Department of Mathematics
                           Princeton University
                           Princeton NJ 08544 · USA

Beno Eckmann   ETH Zentrum,
                           Lehrstuhl für Mathematik
                           CH-8092 Zürich · Switzerland

Ludvig D. Faddeev   LOMI
                           Steklov Mathematical Institute
                           Fontanka 27
                           Leningrad 19001 · USSR

Hans Grauert   Mathematisches Institut
                           Bunsenstraße 3–5
                           W-3400 Göttingen
                           Federal Republic of Germany

Peter Hilton   Department of Mathematical Sciences
                           State University of New York
                           at Binghamton
                           P.O. Box 6000
                           Binghamton, NY 13902 · USA

Heisuke Hironaka   RIMS
                          Kyoto University, Kitashirakawa
                          Sakyo-ku, Kyoto 606 · Japan

Friedrich Hirzebruch   Max-Planck-Institut für Mathematik
                          Gottfried-Claren-Straße 26
                          W-5300 Bonn 3
                          Federal Republic of Germany

Lars Hörmander   Department of Mathematics
                          University of Lund, Box 725,
                          S-22007 Lund · Sweden

Fritz John   66 Wellington Avenue
                          New Rochelle, NY 10804 · USA

Max Koecher   † February 7, 1990

Raghavan Narasimhan The University of Chicago
                          Department of Mathematics
                          5734 University Avenue
                          Chicago, IL 60637 · USA

Constance Reid   70 Piedmont Street
                          San Francisco, CA 94117 · USA

Reinhold Remmert   Mathematisches Institut
                          Westfälische Wilhelms-Universität
                          Einsteinstraße 62
                          W-4400 Münster
                          Federal Republic of Germany

Jean-Pierre Serre   Collège de France
                          11, Place Marcelin Berthelot
                          F-75005 Paris · France

Neil J. A. Sloane   AT&T Bell Laboratories
                          600 Mountain Avenue
                          Murray Hill, NJ 07974 · USA

Jacques Tits   Collège de France
                     11, Place Marcelin Berthelot
                     F-75005 Paris · France

André Weil   The Institute for Advanced Studies
                     School of Mathematics
                     Princeton, NJ 08540 · USA

Don Zagier   Max-Planck-Institut für Mathematik
                     Gottfried-Claren-Straße 26
                     W-5300 Bonn 3
                     Federal Republic of Germany

Michael Atiyah

The European Mathematical Society

1. THE FOUNDATION MEETING

A major new development on the European scene will be the establishment, later this year, of a European Mathematical Society (EMS). The foundation meeting will be held near Warsaw from October 26th to 30th, 1990, and will be attended by representatives from nearly all European countries*. Many years of preparation have been involved, but it is indeed fortunate that the EMS will be launched at a unique historical moment when so many barriers are breaking down.

The political changes now taking place in Europe have an immense potential, not least in the cultural and scientific area. The EMS is thus coming into existence at just the right moment to develop these new opportunities, and to see that mathematics plays its rightful role in the new Europe that we all hope is now being formed.

In principle the EMS should play the part on the European level that national mathematical societies play in their individual countries. It should concern itself with all aspects of mathematics including education, research, the applications of mathematics and its relation to other sciences. Exactly how it is to operate and what its detailed activities will be, will of course depend on the views and initiatives of the mathematical community. The first steps will be taken at the Warsaw meeting, but the subsequent development will require a wide base of support.

To provide an initial input on possible activities a number of preliminary committees have been established and will provide reports for discussion at the foundation meeting. These committees and their chairmen are as follows:

1. *Utilization of EEC Research Programme in Mathematics*,
   Chairman – Professor L. de Michele, Milan, Italy

---

* The manuscript of this article was written in summer 1990. In the meantime the EMS has been founded. A review of the Warsaw meeting is given in a postscript to this paper, see page 5.

2. *Publications,*
   Chairman – Professor S. Robertson, Southampton, England
3. *Education,*
   Chairman – Dr. Tibor Nemetz, Budapest, Hungary
4. *Industrial Mathematics,*
   Chairman – Professor H. Neunzert, Kaiserslautern,
   Federal Republic of Germany
5. *East/West Relations,*
   Chairman – Professor A. Kufner, Prague, Czechoslovakia.

In addition there is already a provisional proposal from the French
Mathematical Society for a European Mathematical Congress in
Paris 1992. It is hoped that this may be the first of a series of
major conferences sponsored by the EMS.

2. STRUCTURE OF THE EMS

The Statutes and By-laws of the EMS are now essentially complete
and have, except for minor details, been approved by national mathematical societies. In brief the EMS will have both individual
members and organizational members, although initially only the
latter are involved. The main body will be a council, with perhaps
50 members, which will meet every two years. A smaller Executive
Committee will deal directly with the regular affairs of the Society.
The president and other members of the Executive Committee will
be elected by the Council.

The legal base of the EMS will be in Helsinki and the Statutes
have been drafted to be in accordance with Finnish law. Professor
A. Lahtinen of Helsinki University has been responsible for the
legal negotiations.

The structure of the EMS differs, in important respects, from
that of other bodies, such as the International Mathematical Union.
In particular it can happen that, where appropriate, several different
societies in one country all belong to the EMS. This provides for
flexibility and avoids political controversy.

3. BACKGROUND

When the European Science Foundation was founded, around
15 years ago, it considered the existing European links in the various scientific disciplines. Noting that there was no European body

specifically concerned with mathematics, it decided to set up a committee to consider the situation, with a view to possibly setting up a European Mathematical Society. I was invited by Lord Flowers, the first President of the ESF, to chair this committee. The other members were mathematicians drawn from many European countries, and the meetings were hosted by the ESF in Strasbourg.

After many discussions our committee felt that there was indeed a role for a European body in mathematics. Several meetings were necessary to work out mutually acceptable statutes and in due course a final meeting was held during the International Congress in Helsinki in 1978. This meeting was intended to be the formal occasion at which national mathematical societies would agree to establish a Federation on the lines proposed. Unfortunately, as with so many international ventures, there were too many different points of view and it proved impossible to get the necessary agreement.

Although final agreement, dealing with statutes and other legalities, was not reached, the Helsinki meeting did agree to establish a European Mathematical Council (EMC). This Council was to act as a general forum for European mathematics, but it was to be a purely informal body with no statutes or authority. Financial contributions were made on a voluntary basis and national societies were responsible for financing the attendance of their representations. I agreed to continue as Chairman and the EMC was to meet once every two years. After a suitable trial period, the situation would be reviewed to determine its future.

The EMC met three times at Oberwolfach, by invitation of the Deutsche Mathematiker-Vereinigung (DMV) once at the Banach Centre, in connection with the Warsaw Congress and once in Prague by invitation of the Czechoslovak Union of Mathematicians and Physicists. At these meetings many topics of interest were discussed and some of these, concerned with new means of communication, led eventually to the Euromath project and the setting up of the European Mathematical Trust. I will deal with this item separately in a moment.

At the meeting in Prague in 1986 it was felt that the trial period, envisaged at Helsinki, was then over and the future of the EMC should be considered. It was decided that it was fulfilling a useful function but that it would be more effective if it were to become a formal body with appropriate authority and responsibility. In general, it was felt that the various changes taking place in Europe made this an appropriate time to launch a new European mathematical body.

A committee was set up, at the Prague meeting, to consider the future of the EMC and to make detailed recommendations. At the Oberwolfach meeting in 1988 the Committee's report was considered and a decision was taken to establish a European Mathematical Society. The detailed statutes have since undergone several revisions and have been approved in principle by most of the mathematical societies in Europe.

4. EUROMATH

As I have already mentioned one of the EMC's initiatives was to put forward a project to the European Commission concerned with the role of computers and electronic communication in European mathematics. It was clear to all concerned that the rapid development of computing was going to have a significant effect on mathematicians and their way of working. There were just two alternatives. Either the mathematical community would sit back and let other people make the relevant decisions or mathematicians would need to become involved and see that their particular needs and requirements were met. The EMC decided to play an active rather than a passive role and, after lengthy preparations, an ambitious proposal – the Euromath project – was submitted to the European Commission. This was supported by the Commission, but at a more modest financial level, and permitted the first preparatory phase of the project to proceed. A second application for funding the next stage was subsequently made, and support for this has also been given.

Because the EMC was a purely informal body, with no legal standing, it could not itself apply for funds to external bodies. For this reason we had to establish a separate legal entity, the European Mathematical Trust, which has the necessary powers. Had a formal European Mathematical Society been in existence, such a step would have been unnecessary. It is partly to avoid such situations in the future that a properly established EMS is desirable.

The European Mathematical Trust is run by a committee of management and the chairman is Flemming Topsoe from Denmark. Once the EMS is established its precise relation to the Trust will need to be clarified.

5. POSTSCRIPT (added November 11th, 1990)

As forecast the European Mathematical Society was indeed established at a meeting held in Madralin (near Warsaw) from October 28th to 30th, 1990. 33 Mathematical Societies participated covering the entire span of Europe, from Portugal to Georgia and from Finland to Italy. The declared aims of the Society are "to establish a sense of identity among European Mathematicians, to concern itself with the relations of mathematics to society, to be involved in mathematical education and to promote research in pure and applied mathematics."

The Society wishes to encourage a spirit of European community amongst young mathematicians and to co-ordinate postgraduate studies with the aim of facilitating student interchange; in this context the Society views with concern the "brain drain" out of, and across Europe, of young mathematicians. The Society will make a determined effort to explain the significance of mathematical research and its applications in the life of the modern world; communications between mathematicians, by electronic and other means, are to be developed; links between mathematicians working in similar areas are to be extended and activities of meetings are to be co-ordinated. The publication of a newsletter and of a journal for mathematics are under active consideration. Other future activities include the promotion of meetings and, in particular, the organisation of major European Congresses, the first in Paris in 1992.

The EMS Council elected as its first Executive Committee the following:

Prof. F. Hirzebruch	(Bonn)	President
Prof. C. Olech	(Warsaw)	Vice-President
Prof. A. Figa-Talamanca	(Rome)	Vice-President
Prof. C. Lance	(Leeds)	Secretary
Prof. A. Lahtinen	(Helsinki)	Treasurer
Prof. E. Bayer	(Besançon)	
Prof. A. Kufner	(Prague)	
Prof. P.-L. Lions	(Paris IX)	
Prof. L. Marki	(Budapest)	
Prof. A. St. Aubyn	(Lisbon)	

At my request I became the first individual member of the European Mathematical Society.

Friedrich L. Bauer

Sternpolygone und Hyperwürfel

VORREDE

Lieber Herr Dr. Götze,

sicher ist Ihnen geläufig, daß es Mathematiker schlechthin auszeichnet, auf die intellektuellen Gefühle ihrer Mitmenschen keine Rücksicht zu nehmen. Dazu verleitet den Mathematiker die Absolutheit der Mathematik; das hohe Niveau des technischen Stands der Mathematik erfordert es auch, daß er eine Fachsprache verwendet, die für Außenstehende eine Geheimsprache ist, und dazu eine Schreibweise, in der naive Menschen eine Hexenschrift sehen können. Die Überlieferung berichtet auch von einem Geheimbund der Mathematiker zu den Zeiten des Pythagoras, der das Pentagramm als Erkennungszeichen führte. Mathematiker in der Nähe der Mystik? Wir wissen, daß Kepler den faulen Zauber der Astrologie durchschaute. Aber kulturgeschichtlich interessant ist es schon, solche mathematisch orientierten Symbole zu untersuchen. Das soll mein Thema sein. Ich beschränke mich dabei auf Figuren, die bei der Kreisteilung entstehen: Polygone und Sternpolygone. Und es wird einen Ausblick geben auf Hyperwürfel, die viel leichter zu verstehen sind, als ihr Name sagt. (Die verwendete Mathematik ist elementar, die im Hintergrund stehende Galoissche Theorie der Substitutionen der Wurzeln wird nicht explizit benützt.)

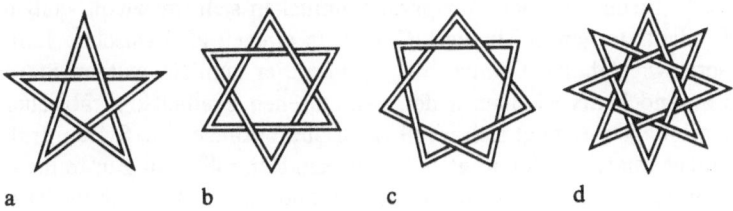

a   b   c   d

Abbildung 1. Geflochtene Sternpolygone

I. Blumen und Sternpolygone

Wer kennt sie nicht, die hübschen Kelchblüten der Tulpen! Die Fünfzähligkeit kommt bei Blüten häufiger als jede andere Symmetrie vor. Die blauen Sterne des Immergrüns (*Vinca minor*), die gelben Sterne des Johanniskrauts (*Hypericum perforatum*), die weißen Sterne des Studentenröschens (*Parnassia palustris*) sind mir eine liebe Erinnerung an meine Kindheit.

Fünfzählig ist neben dem Fünfeck, dem *Pentagon*, das *Pentagramm*, der *Drudenfuß*, das *Albkreuz*, ein uraltes magisches Zeichen sumerisch-babylonischen Ursprungs. Es ist *einzügig*, das heißt es kann ohne Absetzen gezeichnet werden, und ist das einfachste Beispiel eines *Sternpolygons*. Weil jeweils jede zweite Ecke benutzt wird, nennen wir es hier 2-Pentagramm. Das Pentagon ist dann als 1-Pentagramm zu bezeichnen.

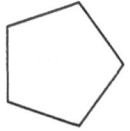  

1-Pentagramm   2-Pentagramm

Zwei Imperien, die Sowjetunion und die Vereinigten Staaten von Amerika, benutzen den Fünfstern als politisches Symbol: hier der Rote Stern, dort der Freiheitsstern, meist als weißer Stern auf blauem Grund. China und Vietnam wandelten den Roten Stern zum Gelben Stern um. Liberia, Panama, Samoa, aber auch Chile, Costa Rica, Honduras, Paraguay, die Niederländischen Antillen sowie Somalia, Kamerun, Senegal, Mauritius führen den Freiheitsstern. Birma und der Europarat schmücken sich mit zwölf weißen Freiheitssternen auf blauem Grund. In einigen afrikanischen Ländern, wie Ghana, Guinea-Bissau, wurde er zum Schwarzen Stern verwandelt als »Leitstern der afrikanischen Freiheit«. Arabischer Nationalismus zeigt sich im Grünen Stern von Jemen, Syrien, Irak und ehemals der Vereinigten Arabischen Republik. Die Zusammenstellung von Pentagramm und Halbmond markiert islamische Tradition: in der Türkei, in Mauretanien, Algerien, Tunesien, Pakistan, Komoren. In Singapur findet sich gar ein Kranz aus fünf Pentagrammen. Eine Besonderheit zeigt die Flagge von Marokko: das geflochtene Pentagramm (Abbildung 1a). Es ist das eigentliche magische Zeichen, das *Salomonssiegel.*

Sehr viele Blüten, insbesondere die der Kreuzblütler, sind vierzählig, von der Gänsekresse (*Arabis alpina*) und dem Wiesenschaumkraut (*Cardamine pratensis*) bis zum Kreuz-Enzian der Bergwiesen (*Gentiana cruciata*). Das Tetragon, das 1-Tetragramm, wird gemeinhin als Quadrat bezeichnet; dazu gehört das 2-Tetragramm, das Kreuzzeichen, das Symbol der christlichen Religion. Es ist ein ausgeartetes Polygon.

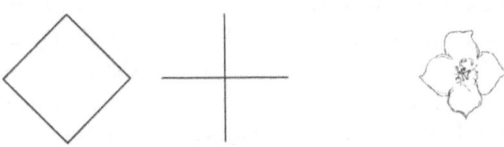

1-Tetragramm 2-Tetragramm

Einer Berufskrankheit der Mathematiker erliegend, gehe ich systematisch weiter, obschon es heimische dreizählige Blüten kaum gibt. Dreiblatt (*Trillium grandiflorum*) und Dreimasterblume (*Tradescantia virginiana*) sind wenigstens dem Blumenliebhaber bekannt. Das Trigon, das 1-Trigramm, wird landläufig als gleichseitiges Dreieck bezeichnet. Es symbolisiert in manchen Religionen die schöpferische Kraft des Gottes (mit der Spitze nach oben, das »Auge Gottes«) oder der Göttin (mit der Spitze nach unten, das »Delta der Venus«).

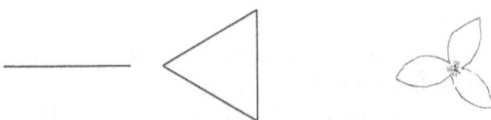

1-Digramm 1-Trigramm

Das Digon, das 1-Digramm, mag manchem nur noch als Spitzfindigkeit erscheinen. Es symbolisiert im Hinduismus (senkrecht stehend) den Lichtstrahl Shivas, das aktive Element. Im Jenischen wird der senkrechte Strich bis heute als »Zinken« verwendet mit der Bedeutung, durchzuhalten; der waagrechte jedoch mit der Bedeutung, sich zu ergeben. Das 2-Tetragramm stellt sich als Überlagerung zweier 1-Digramme dar.

Gehen wir auf die andere Seite. Als nächstes wäre die Sechszähligkeit zu nennen. Sechszählige Blüten sind nicht alltäglich. Sechszählig ist die Blüte einiger Frühlingsboten: des weißen Milchsterns
(*Ornithogalum umbellatum*), des blauen Leberblümchens (*Hepaticum
triloba*) und des gelben Safran (*Crocus aureus*). Sechszählig ist das
Hexagon (1-Hexagramm), dem Chemiker durch den Benzolring vertraut, und das klassische Hexagramm (2-Hexagramm); ein Stern-
polygon, das sich als Überlagerung aus zwei gegenseitig versetzten
1-Trigrammen darstellt. Es ist nicht mehr einzügig. Als *Davidstern*,
auch *Schild Davids*, hebräisch *Magen Dawid* ist es das Symbol der
mosaischen Religion; nach der Zerstreuung der Juden (70 n. Chr.)
wanderte es in die Welt als Symbol der Messiashoffnung. Die Prager
Judenschaft führte es 1356 in ihrem Banner, die Zionisten griffen
es auf, der Staat Israel zeigt es in der Fahne. Wie Burundi zu seinen
drei roten Sechsersternen in der Flagge kommt, ist nicht herauszu-
finden. Sechszählig ist aber auch der Schneestern, das ornamentale
Motiv der Rätoromanen Graubündens; das 3-Hexagramm, ebenfalls ein ausgeartetes Polygon, das eine Überlagerung dreier 1-
Digramme ist. Es kann von einem Kranz von sechs kleineren
Schneesternen umgeben sein. Fährt man so fort, gelangt man zu
Schneeflocken.

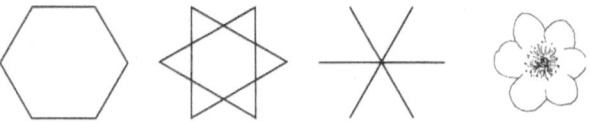

1-Hexagramm 2-Hexagramm 3-Hexagramm

Auch das 2-Hexagramm wird in geflochtener Form als das
eigentliche magische Zeichen gebraucht (Abbildung 1 b).
Sodann kommt die Siebenzähligkeit. Überraschenderweise liefert
uns die Flora auch hierfür Beispiele, wenn auch nur wenige: der
Siebenstern (*Trientalis europaea*) wäre zu nennen. Neben dem *Hep-*

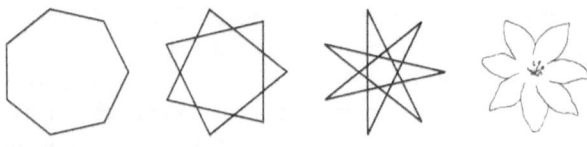

1-Heptagramm 2-Heptagramm 3-Heptagramm

tagon, dem 1-Heptagramm, haben wir zwei einzügige Sternpolygone, das 2-Heptagramm und das 3-Heptagramm. Australien und Jordanien führen Siebensterne im Wappen, in Australien bilden sie das Kreuz des Südens.

Das 3-Heptagramm spielt in der Kalenderwissenschaft eine Rolle: es gibt an, welcher Planet die erste Stunde des Tages regiert und damit dem Wochentag seinen Namen gibt (Arens, Zemanek, Abbildung 2). Die geflochtene Form der Dornenkrone zeigt Abbildung 1c.

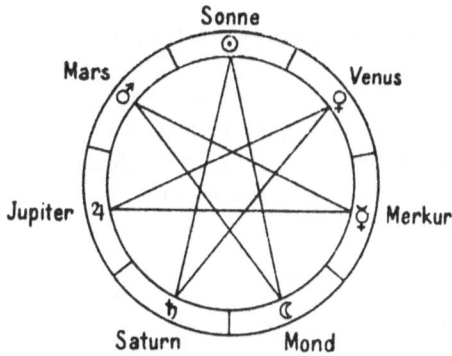

Abbildung 2. Stunden-Heptagramm, Ahrens, 1908

Und nun die Achtzähligkeit. Die Flora versteigt sich nicht gern zu dieser Vielfalt (wohl aber die Fauna: Spinnentiere (Arachnida) haben acht Beine). Immerhin: die achtblättrige Silberwurz (*Dryas octopetala*) steht als Beispiel. Für die politische Symbolistik geben Achtsterne jedoch etwas her, nämlich als Sonnenzeichen. Ein Achtstern findet sich auch in der Flagge der Philippinen, an die Befreiung von den Spaniern erinnernd. Die Freiheitssonne – sie wird häufig durch Unterteilung 16zählig geführt – trat vor allem in Südamerika als republikanisches Freiheitssymbol *El Sol de Mayo* auf, sie findet sich oder fand sich in den Wappen und Flaggen von Argentinien, Peru, Bolivien, Ekuador. Auch die stalinistische Sowjetunion und die meisten ihrer Gliedrepubliken sowie einige ihrer ehemaligen Satelliten verzichteten nicht auf den Symbolwert der aufgehenden (oder untergehenden, in jedem Fall halbierten) Freiheitssonne.

Das 2-Oktogramm stellt sich wieder als Überlagerung zweier 1-Tetragramme dar. Das einzügige 3-Oktogramm ist das Zeichen

der babylonisch-assyrischen Hauptgöttin Ischtar, es wird (auch in
der geflochtenen Form, Abbildung 1 d) ebenfalls als *Siegel Salomons*
bezeichnet. Das 4-Oktogramm – wieder ein ausgeartetes Polygon,
Überlagerung von vier 1-Digrammen – ist ein Zeichen der Windrose.

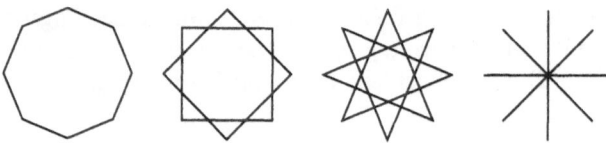

1-Oktogramm   2-Oktogramm   3-Oktogramm   4-Oktogramm

Für die Fälle der Neun-, Elf- und Dreizehnzähligkeit gibt es
weder in der Botanik noch in der Kabbalah Verwendung. Zehnzäh-
ligkeit und Zwölfzähligkeit bei Blüten ist oft nur durch tiefe Schlit-
zung von fünf oder sechs Blütenblättern vorgetäuscht.

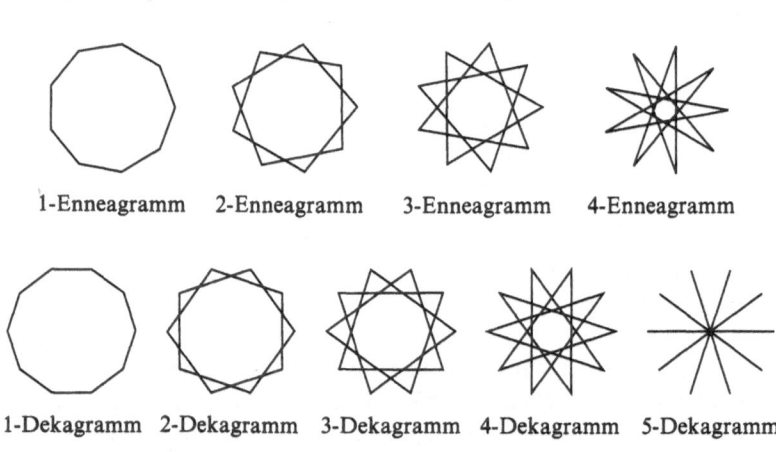

1-Enneagramm   2-Enneagramm   3-Enneagramm   4-Enneagramm

1-Dekagramm 2-Dekagramm 3-Dekagramm 4-Dekagramm 5-Dekagramm

Als Kuriosa wären ein weißer zwölfstrahliger Stern in der Flagge
von Nauru, ein vierzehnzipfliges Sonnensymbol in der Flagge von
Nepal, ein vierzehnstrahliger gelber Stern in der Flagge von Malay-
sia, ein fünfzehnfach unterteilter Schild im Wappen von Jordanien
zu nennen.

Die Sechzehnzähligkeit der durch acht eingefügte Zungen unter-
teilten Freiheitssonne wurde schon erwähnt. Ein sechzehnzähliges
Symbol besitzt auch das Wappen des japanischen Kaiserhauses,
eine goldene Chrysanthemenblüte, das *Sonnen-Mon*.

Abbildung 3. Sternpolygone bei Raimundus Lullus

9-, 14- und 16zählige überlagerte Sternpolygone finden sich in den mystisch-logischen kombinatorischen Diagrammen von Raimundus Lullus; auch ein 5-Pentadekagramm und ein 4-Hexadekagramm findet sich dort (Abbildung 3).

3-Dodekagramm 4-Dodekagramm 5-Dodekagramm

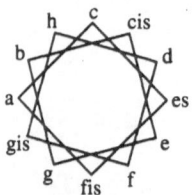

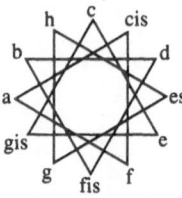

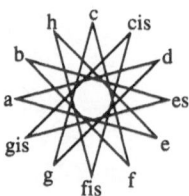

Kleinterzen/ Großterzen/ Quarten/
Großsextenzirkel Kleinsextenzirkel Quintenzirkel

Besonders hübsch »klingen« die zwölfzähligen Sternpolygone, die die Terz- und Quintverwandtschaft in der zwölfgeteilten gleichmäßig temperierten Stimmung der Oktave ausdrücken.

II. Das Oktogon in der Architektur

Beim Oktogon läßt sich der Bezug zum Grundriß des Castel del Monte, dem Dr. Götze viel Beachtung geschenkt hat, nicht übersehen. Das führt uns zu der Frage, welche Vielecke als Grundrißfiguren von Bauwerken auftreten. Beschränken wir uns auf Rundbauwerke, so ist in der Tat ein gleichseitig-achteckiger Grundriß von Türmen, Tempeln und Kirchen sehr häufig zu finden, neben dem trivialen gleichseitig-viereckigen (»quadratischen«) Grundriß.

Auf quadratische Türme setzt man, wenn nicht das zweizählige Satteldach ·(Steingaden), gern ein vierzähliges Zeltdach (Feste Marienberg ob Würzburg). Oft wird aber auch eine achtzählige Laterne mit Pyramidendach (Aschaffenburg, Schloß) oder eine achtzählige Haube (U.L. Frau, München) aufgesetzt.

Achteckige Türme, auch als schlanke Fortsetzungen viereckiger Grundtürme, finden sich in vielen bayrischen Barockkirchen (Kreuzkapelle bei Wessobrunn, St. Koloman bei Schwangau, Höglwörth im Chiemgau, Abbildung 4). Im Castel del Monte umgibt ein Kranz von acht achteckigen Türmen einen achteckigen Zentralbau.

Doch nun zu hallenartigen Rundbauten: bei ihnen sind achteckige Grundrisse häufig: in Castel del Monte ebenso wie in den Basilius-Kathedralen in Moskau und Kiew und in den Zisterzienserkirchen Frankreichs. Lange vor dem Hochmittelalter findet sich der achteckige Grundriß in den moslemischen Sakralbauten zwischen Buchara (um 900 das Mausoleum der Samaniden), Bagdad

Abbildung 4. Achteckig verjüngte Vierecktürme
(von links: Kreuzkapelle, St. Koloman, Höglwörth)

(um 860 das Qubbat as-Sulaibīya-Mausoleum) und Córdoba (um 960 die Omayyaden-Moschee). Auch in diesem Kulturkreis kennt man schon den Übergang von einem viereckigen Grundbau in eine achteckige Kuppelkonstruktion. In den Rippen der Kuppeln finden sich dann beispielsweise in Córdoba die beiden Achtstern-Figuren, das 2-Oktogramm und das 3-Oktogramm (Abbildung 5). Oktogonales findet sich weiterhin in der moslemischen Ornamentik, vor allem in Flecht-Figuren (Abbildung 6).

Auch in den Sakralbauten des Buddhismus in China und Japan, in südostasiatischen und indischen Tempeln kommt die achteckige

Abbildung 5. Achtzählige Gewölberippen
Abbildung 6. Moslemische Ornamentik

Grundfigur häufig vor, sie wird dort besonders oft durch Halbierung in 16zählige und 32zählige Grundrisse erweitert.

Gehen wir in der Geschichte noch weiter zurück, so finden wir um 100 n. Chr. beim römischen Pantheon eine achtzählige Nischenanordnung. Achteckige und sechzehneckige Grundrißfiguren sind in Rom und Byzanz häufig, letztere etwa um 320 n. Chr. im Mausoleum von Santa Constanza in Rom.

III. HEXAGON UND PENTAGON VERPÖNT?

Die eindrucksvolle Aufzählung der Familie von 2-, 4-, 8-, 16-, 32-, ... zähligen Grundrissen, die durch fortgesetzte Halbierung auseinander hervorgehen, läßt die Frage aufkommen: Kommen andere Grundfiguren tatsächlich seltener vor, und, wenn ja, warum?

»Warum hat keiner das Fünfeck zugrundegelegt? Warum hat aber ... einer das langweilige Quadrat als Basis gewählt?« fragt Helmuth Zebhauser hintersinnig[1]. Er gibt keine Erklärung, konstatiert nur einen Sachverhalt. Wenn man unvorsichtig genug ist, eine Deutung zu wagen, so wäre vielleicht darauf hinzuweisen, daß die Konstruktion des Fünf- oder des Zehnecks bereits etwas mathematische Kunstfertigkeit erfordert, während die des Quadrats und des Achtecks durch Falten von Papier bewerkstelligt werden kann und diese Manipulation sich auf dem Bau durch einen einfachen Riß nachvollziehen läßt. Die Konstruktion des Fünfecks hingegen, seit Fra Luca Pacioli (1445–1514) im Abendland wohlbekannt, ist nicht so, daß jeder Baumeister sie sich merken könnte; sie wird nicht so elegant von einer »anschaulichen Lösung« unterstützt. Ist das der Grund? Man könnte geneigt sein, es zu akzeptieren, wenn sich wenigstens in der Familie der 3-, 6-, 12-, 24-, 48-, ... zähligen Grundfiguren viele architektonische Beispiele finden ließen – denn die Konstruktion des Hexagons ist in einer Zirkelgeometrie sogar einfacher als die des Tetragons, und der Zirkel als Riß-Instrument ist auf dem Bau besonders angezeigt. Jeder, dem man es einmal gezeigt hat, wird sich merken, daß man mit fester Zirkelstellung nicht mehr als drei Abtragungen braucht, um ein Trigon zu konstruieren, und

---

[1] Helmuth Zebhauser, Hauben und Zwiebeln, Privatdruck 1989.

daß man durch sechsmalige Wiederholung dieser Konstruktion ein Hexagon mit der vorgegebenen Seite bekommt. Mein Vater hat es mir als Fünfjährigem gezeigt und mich damit entscheidend geprägt.

Käme es auf die Einfachheit der Konstruktion an, müßte es haufenweise sechszählige, zwölfzählige, etc. Grundfiguren geben. Allem Anschein nach ist dies nicht der Fall, und das, obwohl auch von der Bauausführung her das Sechseck nicht unmöglich ist (Abbildung 7).

»Ist die Zwiebel mit achteckigem Querschnitt besonders leicht zu bauen? Gewiß wäre sie sechseckig genauso leicht, müheloser und auch billiger herzustellen gewesen .... Warum wurde nie das mystische Pentagon gewählt?« (Helmuth Zebhauser). Das Pentagon in Washington, D.C. kann man in diesem Zusammenhang außer acht lassen.

Es verbleibt uns also nur die Erklärung, daß es nicht ein rationaler, sondern ein ästhetischer Grund ist, der zur Bevorzugung des Oktogons führt. Wenn auch möglicherweise in früher Zeit esoteri-

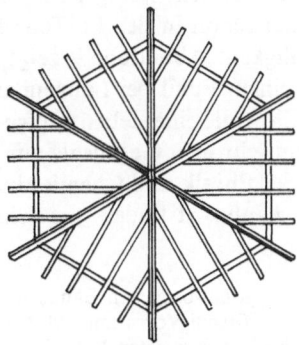

Abbildung 7. Sechszählige Sparrenkonstruktion

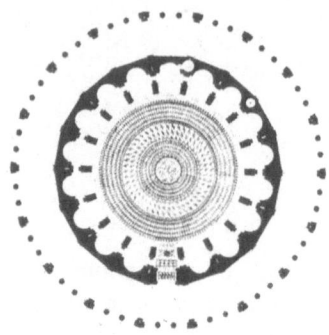

Abbildung 8. Grundriß der Befreiungshalle von Gärtner

sche Gründe die Oktogonal-Figur in den Vordergrund schoben,
so führten jedenfalls im Barock ästhetische Gründe dazu, den über-
kommenen Stil beizubehalten[2].

Um diese Erörterung abzuschließen, sei noch darauf hingewie-
sen, daß es in der Architektur Beispiele für die Familie der 9-,
18-, 36, ... zähligen Grundfiguren gibt, darunter (Abbildung 8) die
Befreiungshalle bei Kelheim, die ein kunstsinniger Bayernkönig
durch Friedrich von Gärtner entwerfen und bauen ließ. Für Fünf-
und für Siebenzähligkeit scheint nichts Klassisches bekannt zu sein,
wohl aber für Elfzähligkeit: die Kirche S. Stefano Rotondo, um
470 n. Chr. erbaut, hat 22 innere und 44 äußere Säulen. Sie ist dennoch wesentlich achtzählig: die 44 Säulen sind in viermal fünf und
viermal sechs gruppiert.

Im Kunsthandwerk herrscht weniger Einseitigkeit. Auf Gefäßen
aus Metall findet man schon in der La-Tène-Kultur Zirkelfiguren
verschiedenster Zähligkeit. Abbildung 9 zeigt eine durchbrochene
Bronze-Phalere aus einem Fund bei Langenhain in Hessen (datiert
auf ungefähr 500 v. Chr.), die eine siebenzählige Zirkelfigur aufweist.

Und nun zu einem schroffen Gegensatz zur Kunst: zur unbeleb-
ten Natur. Auch in der Physik und Chemie sind Zwei-, Drei-, Vier-
und Sechszähligkeit häufig zu finden. Sie schlagen sich nieder in

---

[2] Wenn man auf die hexagonal-basierten Bänke und Tische, die Professor
Johannes Ludwig für die Hörsaal-Vorräume auf dem Südgelände der Technischen Universität München entworfen hat, einen Blick wirft, sieht man,
daß auch Hexagonales ästhetisch ansprechend sein kann.

Abbildung 9. Siebenzählige Zirkelfiguren auf einer Bronze-Phalere

den Beugungsfiguren, die mittels Röntgenstrahlinterferenzen von Kristallen hervorgerufen werden (Max von Laue 1912). Die Kristalle, deren mathematische Klassifikation in 32 Klassen, die insgesamt 230 Raumgruppen umfassen, schon im 19. Jh. Schönflies und Fedorov gelang, kennen keine fünfzählige Symmetrie. Es war deshalb eine Sensation, als 1985 ein Röntgen-Beugungsbild mit fünfzähliger Symmetrie publiziert wurde (Abbildung 10). Es stammt von einem Quasikristall, einer quasiperiodischen Anordnung von Aluminium- und Manganatomen.

Fünfzähligkeit gibt es also auch bei Kristallen nur unter Komplikationen. Für Siebenzähligkeit oder Elfzähligkeit scheinen jedoch selbst Quasikristalle keine Gelegenheit mehr zu bieten.

Abbildung 10. Fünfzähliges Beugungsbild eines Quasi-Kristalls

IV. Vierzähligkeit im Himmel und auf Erden, und das Hexagesimalsystem

Die Bevorzugung der Vierzähligkeit in den ostasiatischen Kulturen hängt wohl mit der Einteilung des Himmels in vier Himmelsrichtungen

bei	»Rückseite«	Norden,
dong	»Sonne hinter Baum«	Osten,
xi	»Baum mit (Schlaf-)Nest«	Westen,
nan	»Pflanze«	Süden

zusammen. Der Hausherr oder Herrscher blickte stets nach Süden, der Haus- oder Palasteingang war stets im Süden. Die Vier-Zähligkeit des menschlichen Grundrisses mit ausgestreckten Armen und geradeaus gerichteten Augen spiegelt sich darin: Hinten, links, rechts, vorne. Auch die Sprachen unserer Hochkulturen haben nur für *diese* Begriffe urtümliche Ausdrücke verfügbar, nicht aber für »von der Richtung meiner Nase aus ein Fünftel meines Gesichtskreises nach links«.

Als artifizielles Hilfsmittel des terrestrischen und maritimen Zurechtfindens, zur Vermessung des gestirnten Himmels und der Erde führte der Mensch in den aufkeimenden Wissenschaften Astronomie, Chronologie, Geodäsie und Navigation die Unterteilung des vollen Kreisbogens in mehr als die vier urtümlichen Teile ein.

Bei dieser Aufgabe spielen ästhetische Fragen nur eine geringe Rolle, und esoterische auch keine große. Zwei Systeme haben sich entwickelt: Das eine geht von der Vierzähligkeit der Himmelsrichtungen aus und benutzt fortgesetzte Halbierung. Dies führt zu einer in der Seefahrt bis heute verwendeten 32er-Teilung, gelegentlich auch 64er-Teilung des Kreises, mit Bildern, die als *Kompaßrose* oder *Windrose* bezeichnet werden (Abbildung 11). Diese Teilung mag gerade fein genug sein, um einen Kurs zu bestimmen, den ein erfahrener Steuermann halten kann.

Für astronomische Zwecke reicht sie jedoch nicht aus. In Babylon entstand, unter dem Einfluß der Astronomie und Chronologie, eine ganz andere Art der Kreisteilung, die zu tiefen Rückwirkungen auf das verwendete Zahlsystem führte. Der Kreis wurde zunächst sechsgeteilt, der *Sextant* wurde dann in 60 Teile (*Grade*) unterteilt, jeder Grad wieder in 60 »winzige Teile« (*Minuten*), jede Minute in 60 Teile (*Sekunden*), und dies wurde notfalls fortgesetzt.

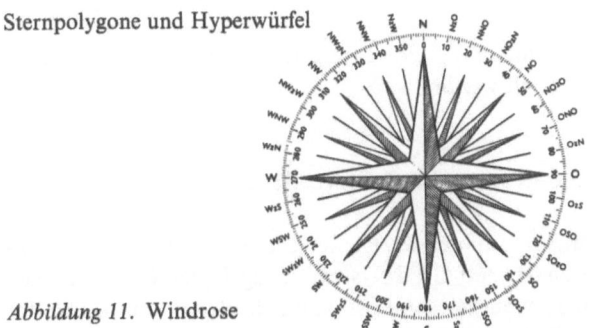

Abbildung 11. Windrose

Wieso eine 60er-Teilung? Sie entsprach dem Zahlsystem der Babylonier, einem Stellenwertsystem zur Basis 60 mit einer Unterteilung in 6×10; in Analogie zum Bi-Quinär-System der Römer müßte man von einem Sexien-Denär-System sprechen.

Von Babylon aus verbreitete sich die Astronomie im mediterranen und kleinasiatischen Raum; das babylonisch begründete hexagesimale Zahlsystem blieb in der abendländischen Wissenschaft bis Regiomontanus (1436–1476) unangefochten und wurde erst zu den Zeiten von Viète (1540–1603) endgültig vom dezimalen System verdrängt. Die Tafeln des Regiomontanus sind zwar in den trigonometrischen Funktionswerten dezimal, in den Winkeln aber noch hexagesimal, und daran hat sich nichts geändert, bis es vor einigen Jahren zum völligen Verschwinden des praktischen Gebrauchs der Tafeln kam. Heutige Taschenrechner erlauben zwar noch Gradangaben, aber keine Minutenangaben mehr.

An Versuchen, sich von der hexagesimalen Gradmessung zu lösen, hat es nicht gefehlt. Schon im Zuge der französischen Revolution wurde vorgeschlagen, von 90 Grad auf 100 Neugrad überzugehen. Die Umstellung, für die es noch in diesem Jahrhundert Bestrebungen gab, fand nur in die Geodäsie $\left(1 \text{ gon} = \dfrac{\pi}{200} \text{ rad}\right)$ Eingang.

Auch der Einfall des Militärs, den Kreis in 6400 »Striche« zu teilen (also eine dezimale Unterteilung der nautischen 64er-Teilung), blieb, so gut er auch gemeint war, unerheblich.

Mit der Computerisierung des Alltags, die in raschem Fortschreiten ist, kommt wenigstens rechnerisch (wenn auch nicht bei den Zeitangaben bis hinab zur Sekunde) das Hexagesimalsystem mehr und mehr gegenüber dem »natürlichen« System der fortgesetzten Halbierung, das dem Dualsystem entspricht, ins Hintertreffen. Welche besondere Vereinfachung das bringt, wird im letzten Abschnitt behandelt werden.

V. TAFELBERECHNUNG IM 360-GRAD-SYSTEM

Wäre die Antike nur an der Kreisteilung interessiert gewesen, müßten ihr die durch das 360-Grad-System nahegelegten Kreisteilungen besonders am Herzen gelegen sein, nämlich diejenigen n-Ecke, für die $n \geq 3$ ein Teiler von 360 ist:

$$n = 3, 4, 5, 6, 8, 9, 10, 12, 15, 18, 20, 24, 30, 36, 40, 45, 60, 90,$$
$$120, 180, 360.$$

Tatsächlich waren die Astronomen mehr an trigonometrischen Tafeln interessiert, und zwar nach der Vorliebe der Antike an »Sehnentafeln«, Tafeln, die für einen Winkelwert α die Sehne $y = \text{chord } \alpha$ des zugehörigen Kreisbogens (bei einem Radius 1) angaben:

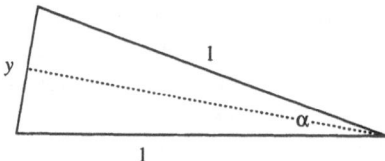

Mit Hilfe sogenannter »Additionstheoreme«[3] konnte man die *chorda* für die Summe oder Differenz zweier Winkel berechnen, wenn man nur die *chorda* für die Winkel selbst kannte. Die Antike wußte bereits, die Seitenlänge des Fünfecks mittels Quadratwurzeln[4] zu berechnen. Damit war die *chorda* von 72° $\left(= \sqrt{(5 - \sqrt{5})/2} \right)$ bestimmt, sowie selbstverständlich die von 60° $(=1)$ und die von 90° $\left(= \sqrt{2} \right)$ – die Quadratdiagonale. Nun ist aber der größte gemeinsame Teiler von 60, 72 und 90 gerade 6, das heißt, für alle Winkel, die Vielfache von 6° sind, läßt sich die *chorda* unter Verwendung der Quadratwurzeloperation so genau berechnen, wie eben nötig:

---

[3] Es gilt etwa
$\text{chord}(\alpha + \beta) = \left(\text{chord } \alpha \sqrt{4 - \text{chord}^2 \beta} + \text{chord } \beta \sqrt{4 - \text{chord}^2 \alpha} \right)/2.$
[4] $\sqrt{a}$ soll stets die nichtnegative Quadratwurzel aus einer nichtnegativen Zahl a bezeichnen.

$$18° = 90° - 72°$$
$$12° = 72° - 60°$$
$$6° = 18° - 12°.$$

Auch wußte man zu Zeiten von Ptolemaios, wie man die Sehne zum halben Winkel mittels Quadratwurzeln so genau wie nötig berechnet:

$$\text{chord } \frac{\alpha}{2} = \sqrt{2 - \sqrt{4 - \text{chord}^2 \, \alpha}}.$$

Damit konnte man zwar die *chorda* von 3°, $1\frac{1}{2}$°, $\frac{3}{4}$° berechnen, nicht aber von 1°. Ptolemaios half sich damit, chord 1° *näherungsweise* aus chord $\frac{3}{4}$° und chord $1\frac{1}{2}$° zu interpolieren. Die Muslims machten es nur ein geringes besser, indem sie chord $\frac{9}{8}$° und chord $\frac{15}{16}$° benutzten (Ibn-Yûnus, um 1000). Wonach man aber suchte, war eine Methode zur Konstruktion der Sehne des gedrittelten Winkels; damit hätte man die *chorda* von 2° und durch Halbierung auch von 1°, $\frac{1}{2}$°, usw. erhalten.

Das klassische Problem der Winkeldreiteilung mit Zirkel und Lineal war also für Ptolemaios ein sehr praktisches Problem. Daß es mit Zirkel und Lineal grundsätzlich nicht ging, wußte er zwar noch nicht. Jedoch kannte die Antike schon die Lösung der Winkeldreiteilungsaufgabe durch »unerlaubte« Konstruktionshilfsmittel, etwa durch Einschiebung der mittleren Proportionale oder durch Verwendung geeigneter Kurven wie der Conchoide, später auch zweier Ellipsen; was alles, wie wir heute wissen, auf die Lösung einer Gleichung dritten Grades hinausläuft. Die Gleichung dritten Grades zur Winkeldreiteilung benutzte 1595 explizit François Viète. Dies zeigt, daß das Problem der Winkeldreiteilung zu einer Haupttriebfeder für die Lehre von der Lösung von Gleichungen höheren Grades wurde.

Die Lösung der Neuneckaufgabe hätte übrigens Ptolemaios ebenfalls weitergeholfen: zum Neuneck gehört der Winkel 40°, und der größte gemeinsame Teiler von 72°, 60° und 40° ist 4°. Damit ließen sich mittels der Additionstheoreme bestimmen die *chorda* von

$$20° = 60° - 40°$$
$$12° = 72° - 60°$$
$$8° = 20° - 12°$$
$$4° = 12° - 8°.$$

VI. ALGEBRAISCHE GLEICHUNGEN FÜR DIE POLYGONSEITEN

1. Doch nun zu den algebraischen Gleichungen, denen die Längen der Seiten der gleichseitigen Polygone und Sternpolygone genügen. Aus technischen Gründen suchen wir zunächst statt einer Gleichung für die nichtausgeartete Sehne

$$\sigma_n^{(i)} = 2 \cdot \sin\left(i \cdot \frac{360°}{2n}\right)$$

zum Winkel $i \cdot \dfrac{360°}{2n}$, $\quad i = 1, 2, \ldots, \dfrac{n-1}{2}$ bzw. $\dfrac{n}{2} - 1$

eine Gleichung für den doppelten Cosinus zu diesem Winkel,

$\xi_n^{(i)} = 2 \cdot \cos\left(i \cdot \dfrac{360°}{n}\right)$. Elementare Trigonometrie ergibt, daß man

aus $\xi_n^{(i)}$ die n-Eck-Seite $\sigma_n^{(i)}$ erhält durch

$$(*) \qquad\qquad\qquad \sigma_n^{(i)} = \sqrt{2 - \xi_n^{(i)}}.$$

Umgekehrt gilt $\xi_n^{(i)} = 2 - (\sigma_n^{(i)})^2$.

Außerdem könnten wir uns eigentlich auf ungerades n beschränken, denn die Halbierung des Winkels und die Verdopplung der Eckenzahl kann durch

$$(**) \quad \begin{cases} \xi_{2n}^{(i)} = \sqrt{2 + \xi_n^{(i)}} \\ \xi_{2n}^{(n-i)} = -\sqrt{2 + \xi_n^{(i)}} \end{cases} \text{bzw.} \quad \begin{cases} \sigma_{2n}^{(i)} = \sqrt{2 - \sqrt{4 - (\sigma_n^{(i)})^2}} \\ \sigma_{2n}^{(n-i)} = \sqrt{2 + \sqrt{4 - (\sigma_n^{(i)})^2}} \end{cases}$$

bewerkstelligt werden. Dabei verdoppelt sich durch die Vorzeichenwahl bei jedem Schritt die Anzahl der Werte, der Verdoppelung der Anzahl der nichtausgearteten Polygone entsprechend.

2. Wir benötigen im weiteren einige formale Identitäten, die schon zu den Zeiten von Niels Henrik Abel (1802–1829) geläufig waren. Setzt man $z + \dfrac{1}{z} = x$, so ist $z^2 + 2 + \dfrac{1}{z^2} = x^2$, $z^3 + 3z + \dfrac{3}{z} + \dfrac{1}{z^3} = x^3$, usw. Der rationale Ausdruck $z^i + \dfrac{1}{z^i}$ ist also ein wohlbestimmtes Polynom in x.

Nennt man dieses Polynom $t_i(x)$ und definiert man es durch

$$z^i + \frac{1}{z^i} = t_i(x), \quad i \in \mathbf{N}$$

so ist[5]

$$t_0(x)=2, \quad t_1(x)=x, \quad \text{und} \quad t_{i+1}(x)=x\,t_i(x)-t_{i-1}(x)$$

eine Rekursionsbeziehung, die man in einer posthumen Ausgabe der *Sectiones angulares* von Viète (1615) bereits findet.

In Tabelle 1 findet man einige dieser Polynome explizit.

Definiert man weiterhin ein Polynom $u_i(x)$ durch

$$z^i+z^{i-2}+\ldots+\frac{1}{z^{i-2}}+\frac{1}{z^i}=u_i(x), \quad i\in\mathbb{N}$$

so gilt die gleiche Rekursion wie für die $t_i(x)$[6]

$$u_0(x)=1, \quad u_1(x)=x, \quad \text{und} \quad u_{i+1}(x)=x\,u_i(x)-u_{i-1}(x);$$

lediglich ein Startwert ist anders.

In Tabelle 2 findet man einige der u-Polynome explizit. Zwischen den beiden Sorten von Polynomen besteht aufgrund der Rekursion ein Zusammenhang[7]:

$$u_{i+1}(x)=t_{i+1}(x)+u_{i-1}(x),$$

überdies gilt die pythagoreische Beziehung

$$(t_i(x))^2+(4-x^2)(u_{i-1}(x))^2=4.$$

---

[5] Wegen

$$t_{i+1}(x)+t_{i-1}(x)=z^{i+1}+\frac{1}{z^{i+1}}+z^{i-1}+\frac{1}{z^{i-1}}=\left(z+\frac{1}{z}\right)\left(z^i+\frac{1}{z^i}\right)=x\cdot t_i(x)$$

[6] Wegen

$$u_{i+1}(x)+u_{i-1}(x)=\left(z^{i+1}+z^{i-1}+\ldots+\frac{1}{z^{i-1}}+\frac{1}{z^{i+1}}\right)$$
$$+\left(z^{i-1}+z^{i-3}+\ldots+\frac{1}{z^{i-3}}+\frac{1}{z^{i-1}}\right)=\left(z+\frac{1}{z}\right)\left(z^i+\ldots+\frac{1}{z^i}\right)$$
$$=x\cdot u_i(x)$$

[7] Wegen

$$u_{i+1}(x)=\left(z^{i+1}+z^{i-1}+\ldots+\frac{1}{z^{i-1}}+\frac{1}{z^{i+1}}\right)=\left(z^{i+1}+\frac{1}{z^{i+1}}\right)$$
$$+\left(z^{i-1}+z^{i-3}+\ldots+\frac{1}{z^{i-3}}+\frac{1}{z^{i-1}}\right)=t_{i+1}(x)+u_{i-1}(x)$$

Tabelle 1. t-Polynome

$$t_0(x) \triangleq 2$$
$$t_1(x) \triangleq x$$
$$t_2(x) \triangleq x^2 - 2$$
$$t_3(x) \triangleq x^3 - 3x$$
$$t_4(x) \triangleq x^4 - 4x^2 + 2$$
$$t_5(x) \triangleq x^5 - 5x^3 + 5x$$
$$t_6(x) \triangleq x^6 - 6x^4 + 9x^2 - 2$$
$$t_7(x) \triangleq x^7 - 7x^5 + 14x^3 - 7x$$
$$t_8(x) \triangleq x^8 - 8x^6 + 20x^4 - 16x^2 + 2$$
$$t_9(x) \triangleq x^9 - 9x^7 + 27x^5 - 30x^3 + 9x$$
$$t_{10}(x) \triangleq x^{10} - 10x^8 + 35x^6 - 50x^4 + 25x^2 - 2$$
$$t_{11}(x) \triangleq x^{11} - 11x^9 + 44x^7 - 77x^5 + 55x^3 - 11x$$
$$t_{12}(x) \triangleq x^{12} - 12x^{10} + 54x^8 - 112x^6 + 105x^4 - 36x^2 + 2$$
$$t_{13}(x) \triangleq x^{13} - 13x^{11} + 65x^9 - 156x^7 + 182x^5 - 91x^3 + 13x$$
$$t_{14}(x) \triangleq x^{14} - 14x^{12} + 77x^{10} - 210x^8 + 294x^6 - 196x^4 + 49x^2 - 2$$
$$t_{15}(x) \triangleq x^{15} - 15x^{13} + 90x^{11} - 275x^9 + 450x^7 - 378x^5 + 140x^3 - 15x$$
$$t_{16}(x) \triangleq x^{16} - 16x^{14} + 104x^{12} - 352x^{10} + 660x^8 - 672x^6 + 336x^4 - 64x^2 + 2$$

Tabelle 2. u-Polynome

$$u_0(x) \triangleq 1$$
$$u_1(x) \triangleq x$$
$$u_2(x) \triangleq x^2 - 1$$
$$u_3(x) \triangleq x^3 - 2x$$
$$u_4(x) \triangleq x^4 - 3x^2 + 1$$
$$u_5(x) \triangleq x^5 - 4x^3 + 3x$$
$$u_6(x) \triangleq x^6 - 5x^4 + 6x^2 - 1$$
$$u_7(x) \triangleq x^7 - 6x^5 + 10x^3 - 4x$$
$$u_8(x) \triangleq x^8 - 7x^6 + 15x^4 - 10x^2 + 1$$
$$u_9(x) \triangleq x^9 - 8x^7 + 21x^5 - 20x^3 + 5x$$
$$u_{10}(x) \triangleq x^{10} - 9x^8 + 28x^6 - 35x^4 + 15x^2 - 1$$
$$u_{11}(x) \triangleq x^{11} - 10x^9 + 36x^7 - 56x^5 + 35x^3 - 6x$$
$$u_{12}(x) \triangleq x^{12} - 11x^{10} + 45x^8 - 84x^6 + 70x^4 - 21x^2 + 1$$
$$u_{13}(x) \triangleq x^{13} - 12x^{11} + 55x^9 - 120x^7 + 126x^5 - 56x^3 + 7x$$
$$u_{14}(x) \triangleq x^{14} - 13x^{12} + 66x^{10} - 165x^8 + 210x^6 - 126x^4 + 28x^2 - 1$$
$$u_{15}(x) \triangleq x^{15} - 14x^{13} + 78x^{11} - 220x^9 + 330x^7 - 252x^5 + 84x^3 - 8x$$
$$u_{16}(x) \triangleq x^{16} - 15x^{14} + 91x^{12} - 286x^{10} + 495x^8 - 462x^6 + 210x^4 - 36x^2 + 1$$

VII. DIE FÄLLE UNGERADER UND GERADER ECKENZAHL

1. Nun hat, wie man spätestens seit Moivre (1667–1754) und Euler (1707–1783) weiß, die Gleichung (»Kreisteilungsgleichung«)

$$z^n - 1 = 0$$

neben der trivialen Wurzel 1 für $n \geq 2$ im Komplexen die $n-1$ verschiedenen Wurzeln

$$\zeta_n^{(i)} = \cos\left(i \cdot \frac{360°}{n}\right) + \sqrt{-1} \cdot \sin\left(i \cdot \frac{360°}{n}\right), \quad i = 1, 2, \ldots, n-1.$$

Beachte, daß $\zeta_n^{(i)} + \dfrac{1}{\zeta_n^{(i)}} = 2 \cdot \cos\left(i \cdot \dfrac{360°}{n}\right) = \zeta_n^{(n-i)} + \dfrac{1}{\zeta_n^{(n-i)}}$. Da 1 Wurzel ist, kann man faktorisieren $z^n - 1 = (z-1) \cdot (z^{n-1} + z^{n-2} + \ldots + z + 1)$, und die Gleichung $z^{n-1} + z^{n-2} + \ldots + z + 1 = 0$ hat die $n-1$ verschiedenen Wurzeln $\zeta_n^{(i)}$, $i = 1, 2, \ldots, n-1$.

2. Ist n *ungerade*, so wird durch $\zeta_n^{(i)}$, $i = 1, 2, \ldots, n-1$ auch

$$z^{\frac{n-1}{2}} + z^{\frac{n-3}{2}} + z^{\frac{n-5}{2}} + \ldots + \frac{1}{z^{\frac{n-5}{2}}} + \frac{1}{z^{\frac{n-3}{2}}} + \frac{1}{z^{\frac{n-1}{2}}} = 0$$

erfüllt, also hat die Gleichung vom Grad $\dfrac{n-1}{2}$

$$u_{\frac{n-1}{2}}(x) + u_{\frac{n-3}{2}}(x) = 0$$

die $\dfrac{n-1}{2}$ verschiedenen reellen und nicht-verschwindenden Wurzeln $\xi_n^{(i)}$, $\xi_n^{(i)} = 2 \cdot \cos\left(i \cdot \dfrac{360°}{n}\right)$, $i = 1, 2, \ldots, n-1$.

$\xi_n = \xi_n^{(1)} = 2 \cdot \cos\left(\dfrac{360°}{n}\right)$ ist die größte dieser Wurzeln. Aus ihr kann man mittels VI. (∗) die Seite des n-Polygons $\sigma_n = \sigma_n^{(1)}$ gewinnen, aus den anderen Wurzeln $\xi_n^{(i)}$ die Seitenlängen $\sigma_n^{(i)}$ der sämtlichen Stern-n-Polygone. Auch erhält man durch Einsetzen von $2 - y^2$ für x die Gleichung vom Grad $n-1$

$$\mu_{\frac{n-1}{2}}(2 - y^2) + u_{\frac{n-3}{2}}(2 - y^2) = 0,$$

deren $\dfrac{n-1}{2}$ *positive* Wurzeln die Seitenlängen $\sigma_n^{(i)}$, $i = 1, 2, \ldots, \dfrac{n-1}{2}$

des Polygons und der Sternpolygone mit n Ecken (n ungerade) sind.[8]

Es fällt auf, daß die Gleichungen für die Polygonseiten σ_n^i dieselben Koeffizienten aufweisen wie die Polynome t_n. In der Tat gilt für ungerades n

$$u_{\frac{n-1}{2}}(2-y^2)+u_{\frac{n-3}{2}}(2-y^2)=t_n(y)/y.$$

Ersetzt man nämlich z durch w^2, so geht die Kreisteilungsgleichung über in $w^{2n}-1=0$. Dies ist ebenfalls eine Kreisteilungsgleichung; sie besitzt die Wurzeln 1 und -1 sowie die $2n-2$ Wurzeln

$$\omega_n^{(i)}=\cos\left(i\cdot\frac{360°}{2n}\right)+\sqrt{-1}\cdot\sin\left(i\cdot\frac{360°}{2n}\right),$$

$$i=1,2,\ldots,n-1,n+1,\ldots,n-1.$$

Beachte, daß

$$\omega_n^{(i)}-\frac{1}{\omega_n^{(i)}}=\sqrt{-1}\cdot 2\sin\left(i\cdot\frac{360°}{2n}\right)=\sqrt{-1}\cdot\text{chord}\left(\frac{360°}{2n}\right).$$

Setzen wir also $w-\dfrac{1}{w}=\sqrt{-1}\cdot y$, so ist $w^2-2+\dfrac{1}{w^2}=-y^2$ oder $w^2+\dfrac{1}{w^2}=2-y^2$.

Nun wird mit $w^{2n}-1=0$ auch $w^n-\dfrac{1}{w^n}=0$ durch die Wurzeln ± 1 und $\pm w_n^{(i)}$ erfüllt. Wie man aber sofort nachrechnet, gilt für *ungerades n*

$$w^n-\frac{1}{w^n}=t_n(y).$$

$w=\pm 1$ entspricht $y=0$, das um den Faktor y reduzierte Polynom $t_n^*(y)=_{\text{def}}t_n(y)/y$ vom Grad $n-1$ hat also als Nullstellen

$$\pm 2\cdot\sin\left(i\cdot\frac{360°}{2n}\right)=\pm\text{chord}\left(i\cdot\frac{360°}{n}\right), i=1,2,\ldots,n-1; \text{ das heißt:}$$

---

[8] So ergeben sich

für $n=3$: $x+1=0$ und $y^2-3=0$
für $n=5$: $x^2+x-1=0$ und $y^4-5(y^2-1)=0$
für $n=7$: $x^3+x^2-2x-1=0$ und $y^6-7(y^4-2y^2+1)=0$
für $n=9$: $x^4+x^3-3x^2-2x+1=0$ und $y^8-9(y^6-3y^4+\frac{10}{3}y^2-1)=0$

Die $\frac{n-1}{2}$ verschiedenen Beträge der Wurzeln von $t_n^*(y)=0$ sind
die Seitenlängen $\sigma_n^{(i)}$ der n-Polygone und Stern-n-Polygone ungerader Eckenzahl n.

3. Ist aber n *gerade*, so ist neben $+1$ auch $-1=\zeta^{\left(\frac{n}{2}\right)}$ Wurzel der Kreisteilungsgleichung und es existiert die Faktorisierung

$$z^n-1=(z^2-1)\cdot(z^{n-2}+z^{n-4}+\ldots+z^2+1).$$

Damit wird auch

$$z^{\frac{n}{2}-1}+z^{\frac{n}{2}-3}+z^{\frac{n}{2}-5}+\ldots+\frac{1}{z^{\frac{n}{2}-5}}+\frac{1}{z^{\frac{n}{2}-3}}+\frac{1}{z^{\frac{n}{2}-1}}=0$$

durch $\zeta_n^{(i)}$, $i=1,2,\ldots,\frac{n}{2}-1,\frac{n}{2}+1,\ldots,n-1$ erfüllt. Also hat die Gleichung vom Grad $\frac{n}{2}-1$

$$u_{\frac{n}{2}-1}(x)=0$$

die $\frac{n}{2}-1$ verschiedenen reellen Wurzeln $\xi_n^{(i)}=2\cdot\cos\left(i\cdot\frac{360°}{n}\right)$, $i=1,2,\ldots,\frac{n}{2}-1$.

Erneut ist $\xi_n=\xi_n^{(1)}=2\cdot\cos\left(\frac{360°}{n}\right)$ die größte dieser Wurzeln; aus ihr gewinnt man mittels (*) die Seite des n-Polygons $\sigma_n=\sigma_n^{(1)}$; aus den übrigen die Seiten $\sigma_n^{(i)}$ der Stern-n-Polygone.

Wiederum durch Einsetzen von $2-y^2$ für x erhält man die Gleichung vom Grad $n-2$

$$u_{\frac{n}{2}-1}(2-y^2)=0,$$

deren $\frac{n}{2}-1$ *positive* Wurzeln die Seitenlängen $\sigma_n^{(i)}$, $i=1,2,\ldots,\frac{n}{2}-1$ des Polygons und der nicht-ausgearteten Sternpolygone mit n Ecken (n gerade) sind.[9]

---

[9] So ergeben sich
für $n=\ 4$: $x=0$ und $y^2-2=0$;
für $n=\ 6$: $x^2-1=0$ und $y^4-4y^2+3=0$;
für $n=\ 8$: $x^3-2x=0$ und $y^6-6y^4+10y^2-4=0$;
für $n=10$: $x^4-3x^2+1=0$ und $y^8-8y^6+21y^4-20y^2+5=0$;
für $n=12$: $x^5-4x^3+3x=0$ und $y^{10}-10y^8+36y^6-56y^4+35y^2-6=0$.

Trivialerweise ist $\sigma_n^{(n/2)} = 2$ die zu $\zeta_n^{(n/2)} = -1$ gehörige Seitenlänge des ausgearteten Stern-n-Polygons.

Wiederum fällt auf, daß die Gleichungen für die Polygonseiten dieselben Koeffizienten aufweisen wie die Polynome u_n. In der Tat gilt für gerades n

$$u_{\frac{n}{2}-1}(2-y^2) = u_{n-1}(y)/y,$$

was man ähnlich wie oben zeigt. Das um den Faktor y reduzierte Polynom $u_{n-1}^*(y) =_{\text{def}} u_{n-1}(y)/y$ (n gerade) vom Grad $n-2$ hat also als Nullstellen

$$\pm 2 \cdot \sin\left(i \cdot \frac{360°}{2n}\right) = \pm \operatorname{chord}\left(i \cdot \frac{360°}{n}\right), \quad i = 1, 2, \ldots, \frac{n}{2}-1;$$

das heißt:

Die $\frac{n}{2}-1$ verschiedenen Beträge der Wurzeln von $u_{n-1}^*(y) = 0$ sind die Seitenlängen $\sigma_n^{(i)}$ der n-Polygone und nicht-ausgearteten Stern-n-Polygone von gerader Eckenzahl n.

VIII. Faktorisierung der Polynome

1. Die nicht-einzügigen Stern-n-Polygone zerfallen in k einzügige Polygone mit n' Ecken, wenn n' ein echter Teiler von n ist: $n = k \cdot n'$. So zerfällt das 2-Hexagramm in zwei 1-Trigramme, das 2-Oktogon in zwei 1-Tetragramme. Aber auch das 3-Enneagramm zerfällt, und zwar in drei 1-Trigramme, während das 4-Dekagramm in zwei 2-Pentagramme zerfällt – man sehe sich dazu die Abbildungen in I. an. Die ausgearteten Sternpolygone vom geraden Grad n schließlich zerfallen in $\frac{n}{2}$ 1-Digramme.

Man sieht sofort ein, daß ein j-n-gramm (mit $2j \leq n$) genau dann einzügig ist, wenn j und n keinen Teiler außer 1 gemeinsam haben; und daß es in d j'-n'-gramme zerfällt, wenn $j = j' \cdot d$ und $n = n' \cdot d$ ist. Ist n prim, so sind alle j-n-gramme einzügig.

Die Anzahl einzügiger j-n-gramme ($n \geq 3$) ist gleich $\frac{1}{2}\phi(n)$, wo $\phi(n)$, die Eulersche Funktion, die Anzahl zu n teilerfremder Zahlen unter den Zahlen $1, 2, \ldots, n$ ist.

2. Die Bestimmung einer oder aller Wurzeln einer algebraischen Gleichung $P(x) = 0$ läuft auf die Abspaltung eines Linearfaktors

bzw. auf die Zerlegung in n Linearfaktoren hinaus. Verständlicherweise wird die damit verbundene Mühe reduziert, wenn es gelingt, das Polynom $P(x)$ von vornherein in gewisse Faktoren zu zerlegen.

Für die algebraischen Gleichungen, denen die Polygonseiten genügen

$$t_n^*(y) = 0 \ (n \text{ ungerade}) \quad \text{bzw.} \quad u_{n-1}^*(y) = 0 \ (n \text{ gerade})$$

bedeutet der Zerfall, daß ein Faktorpolynom von kleinerem Grad abgespalten wird, dem die Seiten einzügiger j'-n'-gramme mit $d \geq 2$ genügen. So ist zunächst für gerades n $u_{\frac{n}{2}-1}^*(y)$ bzw. $t_{\frac{n}{2}}^*(y)$ Faktor von $u_{n-1}^*(y)$, je nachdem, ob $\frac{n}{2}$ gerade oder ungerade ist.

Das bedeutet, daß es folgende Faktorisierungen gibt

für $n = 6$: $y^4 - 4y^2 + 3 = (y^2 - 3)(y^2 - 1)$

für $n = 8$: $y^6 - 6y^4 + 10y^2 - 4 = (y^2 - 2)(y^4 - 4y^2 + 2)$

für $n = 10$: $y^8 - 8y^6 + 21y^4 - 20y^2 + 5 = (y^4 - 5y^2 + 5)(y^4 - 3y^2 + 1)$

für $n = 12$: $y^{10} - 10y^8 + 36y^6 - 56y^4 + 35y^2 - 6$
$$= (y^4 - 4y^2 + 3)(y^6 - 6y^4 + 9y^2 - 2)$$

für $n = 18$: $y^{16} - 16y^{14} + 105y^{12} - 364y^{10} + 715y^8 - 792y^6$
$$+ 462y^4 - 120y^2 + 9 = (y^8 - 9y^6 + 27y^4 - 30y^2 + 9)$$
$$\cdot (y^8 - 7y^6 + 15y^4 - 10y^2 + 1)$$

Der restliche Faktor, der die Seitenlängen der verbleibenden einzügigen Polygone und Sternpolygone bestimmt, läßt sich auch wieder angeben: es gilt

für den Fall $n = 4m$: $u_{n-1}^*(y) = u_{\frac{n}{2}-1}^*(y) \cdot t_{\frac{n}{2}}(y)$;

für den Fall $n = 4m + 2$: $u_{n-1}^*(y) = t_{\frac{n}{2}}^*(y) \cdot u_{\frac{n}{2}-1}(y)$.

Es ist nämlich generell für gerades $n^{[10]}$ $u_{n-1}(y) = t_{\frac{n}{2}}(y) \cdot u_{\frac{n}{2}-1}(y)$.

Somit:

Die m verschiedenen Beträge der Wurzeln von $t_{2m}(y) = 0$ (für

---

[10] Wegen

$$z^{n-1} + z^{n-3} + \ldots + z + \frac{1}{z} + \ldots + \frac{1}{z^{n-3}} + \frac{1}{z^{n-1}}$$
$$= \left(z^{\frac{n}{2}-1} + z^{\frac{n}{2}-3} + \ldots + \frac{1}{z^{\frac{n}{2}-3}} + \frac{1}{z^{\frac{n}{2}-1}} \right) \cdot \left(z^{\frac{n}{2}} + \frac{1}{z^{\frac{n}{2}}} \right).$$

$n = 4m$) bzw. von $u_{2m}(y) = 0$ (für $n = 4m + 2$) liefern die Seitenlängen aller *einzügigen* n-Polygone und Stern-n-Polygone.[11]

Aber nicht alle Stern-n-Polygone, deren Seitenlängen dieser Gleichung genügen, sind einzügig:

Für $n = 12$ zerfällt das 3-Dodekagramm in drei 1-Tetragramme. Es gibt deshalb die weitergehende Faktorisierung

$$y^6 - 6y^4 + 9y^2 - 2 = (y^2 - 2)(y^4 - 4y^2 + 1)$$

(die Gleichung $y^4 - 4y^2 + 1 = 0$ liefert σ_{12}^1, σ_{12}^5 als Seiten einzügiger 18-Ecke).

Für $n = 18$ zerfällt das 3-Oktodekagramm in drei 1-Hexagramme. Es gibt deshalb die weitergehende Faktorisierung

$$y^8 - 7y^6 + 15y^4 - 10y^2 + 1 = (y^2 - 1)(y^6 - 6y^4 + 9y^2 - 1)$$

(die Gleichung $y^6 - 6y^4 + 9y^2 - 1 = 0$ liefert σ_{18}^1, σ_{18}^5, σ_{18}^7 als Seiten einzügiger 18-Ecke).

3. Für ungerade n, die nicht prim sind, ergeben sich ebenfalls Faktorisierungsmöglichkeiten. Ist etwa $n = k \cdot l$, so ist l ebenfalls ungerade und $t^*_{\frac{n}{k}}(y)$ ist Faktor von $t^*_n(y)$. Das bedeutet

für $n = 9$: $y^8 - 9y^6 + 27y^4 - 30y^2 + 9 = (y^2 - 3)(y^6 - 6y^4 + 9y^2 - 3)$

(die Gleichung $y^6 - 6y^4 + 9y^2 - 3 = 0$ liefert $\sigma_9^1, \sigma_9^2, \sigma_9^4$ als Seiten einzügiger 9-Ecke);

für $n = 15$: $y^{14} - 15y^{12} + 90y^{10} - 275y^8 + 450y^6 - 378y^4 + 140y^2$
$- 15 = (y^2 - 3)(y^4 - 5y^2 + 5)(y^8 - 7y^6 + 14y^4 - 8y^2 + 1)$

(die Gleichung $y^8 - 7y^6 + 14y^4 - 8y^2 + 1 = 0$ liefert $\sigma_{15}^1, \sigma_{15}^2, \sigma_{15}^4$, σ_{15}^7 als Seiten einzügiger 15-Ecke);

für $n = 27$:   aus $t^*_{27}(y)$ ergibt sich nach Abspaltung von $t^*_9(y)$ eine Gleichung $y^{18} - 18y^{16} + 135y^{14} - 546y^{12} + 1287y^{10} - 1782y^8$
$+ 1386y^6 - 540y^4 + 81y^2 - 3 = 0$;

---

[11] So ergeben sich

für $n = 6$:	$x - 1 = 0$	und $y^2 - 1 = 0$;
für $n = 8$:	$x^2 - 2 = 0$	und $y^4 - 4y^2 + 2 = 0$;
für $n = 10$:	$x^2 - x - 1 = 0$	und $y^4 - 3y^2 + 1 = 0$;
für $n = 12$:	$x^3 - 3x = 0$	und $y^6 - 6y^4 + 9y^2 - 2 = 0$;
für $n = 14$:	$x^3 - x^2 - 2x + 1 = 0$	und $y^6 - 5y^4 + 6y^2 - 1 = 0$;
für $n = 16$:	$x^4 - 4x^2 + 2 = 0$	und $y^8 - 8y^6 + 20y^4 - 16y^2 + 2 = 0$;
für $n = 18$:	$x^4 - x^3 - 3x^2 + 2x + 1 = 0$	und $y^8 - 7y^6 + 15y^4 - 10y^2 + 1 = 0$.

(sie liefert σ^1_{27}, σ^2_{27}, σ^4_{27}, σ^5_{27}, σ^7_{27}, σ^8_{27}, σ^{10}_{27}, σ^{11}_{27}, σ^{13}_{27} als Seiten einzügiger 27-Ecke);

für $n = 45$: aus $t^*_{45}(y)$ ergibt sich nach Abspaltung des kleinsten gemeinsamen Vielfachen von $t^*_{15}(y)$ und $t^*_{3}(y)$ eine Gleichung

$$y^{24} - 24y^{22} + 252y^{20} - 1519y^{18} + 5796y^{16} - 14553y^{14} + 24206y^{12}$$
$$- 26169y^{10} + 17253y^8 - 6623y^6 + 1182y^4 - 72y^2 + 1 = 0;$$

(sie liefert σ^1_{45}, σ^2_{45}, σ^4_{45}, σ^7_{45}, σ^8_{45}, σ^{11}_{45}, σ^{13}_{45}, σ^{14}_{45}, σ^{16}_{45}, σ^{17}_{45}, σ^{19}_{45}, σ^{22}_{45} als Seiten einzügiger 45-Ecke).

IX. Die geometrischen Hilfsfiguren der Muslims

1. Eine Kuriosität besteht noch im Fall $n = 4m + 2$: eine nochmalige allgemeine ganzzahlige Faktorisierungsmöglichkeit für $u_{2m}(y)$. Tatsächlich gilt

$$u_{2m}(y) = (u_m(y) + u_{m-1}(y))(u_m(y) - u_{m-1}(y)).$$

Ersetzt man y durch $-y$, so gehen die beiden Faktoren ineinander über. Jeder der beiden Faktoren enthält jedoch positive und negative Wurzeln. Um die m numerischen Werte der Seitenlängen zu erhalten, muß man nun schließlich den *Betrag* der Wurzeln von $(u_m(y) \pm u_{m-1}(y)) = 0$ nehmen.[12]

Gelegentlich kann man noch weiter faktorisieren. So hat man für $n = 18$

$$y^4 + y^3 - 3y^2 - 2y + 1 = (y+1)(y^3 - 3y + 1) \quad \text{und für } n = 30$$
$$y^7 - y^6 - 6y^5 + 5y^4 + 10y^3 - 6y^2 - 4y + 1$$
$$= (y-1)(y^2 - y - 1)(y^4 + y^3 - 4y^2 - 4y + 1).$$

---

[12] So bestimmen sich

für $n = 6$ aus der Gleichung $y - 1 = 0$: $\sigma^{(1)}_6 = 1$;

für $n = 10$ aus der Gleichung $y^2 + y - 1 = 0$:

$$\sigma^{(1)}_{10} = \frac{\sqrt{5} - 1}{2}, \quad -\sigma^{(3)}_{10} = \frac{\sqrt{5} - 1}{2};$$

für $n = 14$ aus der Gleichung $y^3 - y^2 - 2y + 1 = 0$:

$$\sigma^{(1)}_{14} \approx 0{,}445, \quad -\sigma^{(3)}_{14} \approx -1{,}247, \quad \sigma^{(5)}_{14} \approx 1{,}802;$$

für $n = 18$ aus der Gleichung $y^4 + y^3 - 3y^2 - 2y + 1 = 0$:

$$\sigma^{(1)}_{18} \approx 0{,}347, \quad -\sigma^{(3)}_{18} = -1, \quad \sigma^{(5)}_{18} \approx 1{,}532, \quad -\sigma^{(7)}_{18} \approx -1{,}879.$$

2. Nachdem so für $n = 10$, $n = 14$ und $n = 18$ auf algebraische Weise Gleichungen von Grad 2, 3 und 3 hergeleitet sind, deren Lösung die Länge der Seite des n-Ecks oder den Wert der *chorda* des Winkels $\dfrac{360°}{n}$ ergibt, sollen noch geometrische Figuren angegeben werden, die schon zu den Zeiten der Muslims bekannt waren und aus denen die betreffenden Gleichungen durch einfache Proportionalitäten gewonnen werden können.

(1) $\dfrac{1}{y} = \dfrac{y}{h}$ $\dfrac{1}{2-h} = \dfrac{y}{1}$ $5\gamma = 180°$

       $y^2 + 1 - 1 = 0$

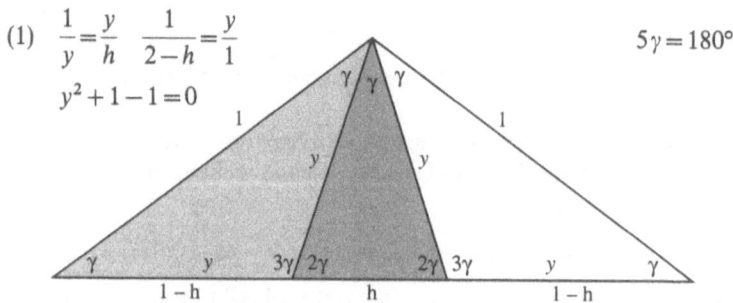

(2) $\dfrac{1}{y} = \dfrac{y}{h}$ $\dfrac{1}{2-h} = \dfrac{y}{1-h}$ $7\gamma = 180°$

       $y^3 - y^2 - 2y + 1 = 0$

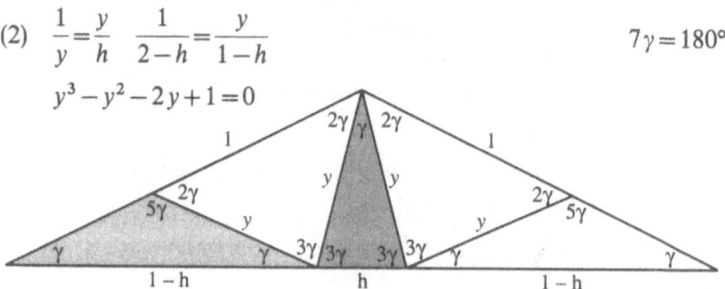

(3) $\dfrac{1}{y} = \dfrac{y}{h}$ $\dfrac{1}{2-h} = \dfrac{y}{1-y}$ $9\gamma = 180°$

       $y^3 - 3y + 1 = 0$

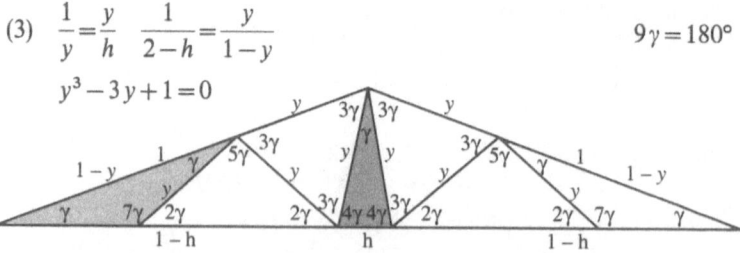

Für das Siebeneck gab schon Archimedes eine Konstruktion unter Verwendung der »unzulässigen« Einschiebung, und Ibn al Hai̮tam (um 965–nach 1039) eine Lösung durch Verwendung zweier Kegelschnitte, aus der sich durch Elimination das Polynom hätte gewinnen lassen. Al Bīrūni (973–1048) und Abū al-Ğūd (um 1000) gaben dann die Hilfsfigur (3) zur Bestimmung der Neuneck-Seite an. Die Hilfsfigur (1) zur Bestimmung der Fünfeck-Seite entspricht der Konstruktion bei Euklid.

X. Das Theorem der Indexmultiplikation

1. Bei der Aufstellung der Polynome $t_n(y)$ kann man sich auf den Fall beschränken, wo n prim ist, denn es gilt das Theorem der Indexmultiplikation

$$t_{ab}(y) = t_b(t_a(b)).$$

Zum Beweis muß man auf die Definition in VI. zurückgreifen: Es ist

$$t_{ab}(x) = t_{ab}\left(z + \frac{1}{z}\right) = z^{ab} + \frac{1}{z^{ab}} = (z^a)^b + \frac{1}{z^{(z^a)^b}} = t_b\left(z^a + \frac{1}{z^a}\right),$$

wobei $z^a + \dfrac{1}{z^a} = t_a\left(z + \dfrac{1}{z}\right) = t_a(x).$

Beispielsweise ist

$$t_9(y) = t_3(t_3(y)) = (y^3 - 3y)^3 - 3(y^3 - 3y) \quad \text{und}$$
$$t_{16}(y) = t_2(t_2(t_2(t_2(y)))) = (((y^2 - 2)^2 - 2)^2 - 2)^2 - 2.$$

Auf diese Weise lassen sich manche Polynome, wie

$$t_{15}(y) = t_5(t_3(y)) = t_3(t_5 y)) \quad \text{oder} \quad t_{30}(y) = t_2(t_3(t_5(y))) \quad \text{oder}$$
$$t_{45}(y) = t_3(t_3(t_5(y)))$$

sehr kompakt schreiben. Dies ist von Bedeutung für die Polynome $t_n(y)$ mit ungeradem n, die nach VII. 2 zu Polygonen mit ungerader Eckenzahl n gehören, sowie für die Polynome $t_{2m}(y)$, die nach VIII. 2 zu Polygonen mit gerader Eckenzahl $n = 4m$ gehören.

2. Für die Berechnung der Tafelwerte der trigonometrischen Funktionen ist es nicht unbedingt vorteilhaft, zuerst das betreffende Polynom aufzustellen und dann mit numerischen Methoden eine Lösung zu suchen, selbst wenn man das Polynom bereits faktorisieren kann.

Um etwa für das Neuneck $t_9(y) = 0$ zu lösen, löst man $t_3(y) = \eta$, wo $t_3(\eta) = 0$. Das ergibt zunächst $\eta_1 = 0$, $\eta_2 = \sqrt{3}$, $\eta_3 = -\sqrt{3}$ und damit die Gleichungen dritten Grades

$$y^3 - 3y = 0, \quad y^3 - 3y = \sqrt{3}, \quad y^3 - 3y = -\sqrt{3},$$

von denen die erste zerfällt und neben dem trivialen $y = 0$ eine quadratische Gleichung für die Lösungen zum 3-Enneagramm ergibt, während jede der beiden anderen betragsmäßig die Lösungen

$$\sigma_9^{(1)} \approx 0{,}684, \quad \sigma_9^{(2)} \approx 1{,}286, \quad \sigma_9^{(4)} \approx 1{,}970$$

zum 1-Enneagramm, 2-Enneagramm und 4-Enneagramm bringt.

Die Dreiteilung des Winkels läuft im übrigen auf die Lösung einer Gleichung $t_3(y) = \eta$ hinaus; somit, wie schon früher erwähnt, auf die Winkeldreiteilungsgleichung $y^3 - 3y = \eta$.

Hat man also für die Neuneckseite y das gestaffelte kubische Gleichungssystem

$$y^3 - 3y = v, \quad v^3 - 3v = 0,$$

so gibt es für die Fünfzehneckseite zwei Systeme:

$$y^3 - 3y = v, \quad v^5 - 5v^3 + 5v = 0 \quad \text{und}$$
$$y^5 - 5y^3 + 5y = v, \quad v^3 - 3v = 0.$$

Und um für das Sechzehneck $t_{16}(y) = 0$ zu lösen, löst man das System

$$y^2 - 2 = v_1, \quad v_1^2 - 2 = v_2, \quad v_2^2 - 2 = v_3, \quad v_3^2 - 2 = 0,$$

was sukzessive durch Quadratwurzeln geschehen kann, man erhält

$$\sigma_{16}^{(1)} = \sqrt{2 - \sqrt{2 + \sqrt{2}}}, \quad \sigma_{16}^{(3)} = \sqrt{2 - \sqrt{2 - \sqrt{2}}},$$
$$\sigma_{16}^{(5)} = \sqrt{2 + \sqrt{2 - \sqrt{2}}}, \quad \sigma_{16}^{(7)} = \sqrt{2 + \sqrt{2 + \sqrt{2}}}.$$

3. Auch für die u-Polynome gibt es eine, allerdings kompliziertere Indexmultiplikation

$$u_{ab-1}(y) = u_{b-1}(y) \cdot u_{a-1}(t_b(y)).$$

Sie paßt auf den Fall VII.3 der Polynome $u_{n-1}(y)$ mit ungeradem n, die nach VII.2 zu Polygonen mit gerader Eckenzahl n gehören, sowie für die Polynome $u_{2m}(y)$, die nach VIII.2 zu Polygonen mit gerader Eckenzahl $n = 4m + 2$ gehören.

Beispielsweise ist für $n = 12$ mit $a = 3$, $b = 4$ und mit $a = 4$, $b = 3$

$$u_{11}(y) = y^{11} - 10y^9 + 36y^7 - 56y^5 + 35y^3 - 6y$$
$$= u_3(y) \cdot u_2(t_4(y)) = (y^3 - 2y)((y^4 - 4y^2 + 2)^2 - 1)$$
$$= u_2(y) \cdot u_3(t_3(y)) = (y^2 - 1)((y^3 - 3y)^3 - (y^3 - 3y)).$$

Für den Spezialfall $a = 2$ ergibt sich mit $u_{2b-1}(y) = u_{b-1}(y) \cdot t_b(y)$, vgl. Fußnote 9, nichts Neues. Der Spezialfall $b = 2$ hingegen gibt $u_{2a-1}(y) = y \cdot u_{a-1}(y^2 - 2)$. Daraus gewinnt man

$$(***) \qquad\qquad u_{2n-1}^*(y) = u_{n-1}(y^2 - 2),$$

eine schon in VII.3 aufgetretene Beziehung, die zur fortgesetzten Verdopplung der Eckenzahl dienen kann und die Beziehung VI (**) ersetzt.

4. Die vielfachen algebraischen Beziehungen, die derart innerhalb der Menge der Wurzeln einer Kreisteilungsgleichung herrschen, veranlaßten übrigens zu Beginn des 19. Jh. Niels Henrik Abel und Évariste Galois dazu, Invarianten unter Substitutionen der Wurzeln zu betrachten, und führten damit sowohl zur Gruppentheorie wie zur Theorie der algebraischen Körpererweiterungen.

XI. Lösung durch Dekomposition in Gleichungssysteme

Diese verfeinerten Untersuchungen begannen wohl durch Lagrange um 1770, und zwar am Fünfzehneck bzw. am Dreißigeck. Da $\dfrac{360°}{30} = 12°$ durch Linearkombination aus $60°$ und $72°$ berechnet werden kann, sollte man nach dem Additionstheorem mit Quadratwurzeln auskommen, während doch in IX. 1 die Faktorisierung nur bis zu einer Gleichung 4. Grades gelang, und auch in X. das System

$$y^3 - 3y = v, \qquad (v^4 - 5v^2 + 5) \cdot v = 0$$

neben einer biquadratischen eine kubische Gleichung enthält.

Lagrange machte nun zur Faktorisierung von $y^4 + y^3 - 4y^2 - 4y + 1$ einen Ansatz wie

$$y^4 + y^3 - 4y^2 - 4y + 1 = \left(y^2 + \gamma y - \frac{1}{\gamma^2}\right)\left(y^2 - \frac{1}{\gamma}y - \gamma^2\right),$$

der durch $\gamma^2 - \gamma - 1 = 0$ erfüllt wird. Diese Gleichung hat die Wurzeln γ_1 und γ_2, wobei $\gamma_2 = -\dfrac{1}{\gamma_1}$ ist. Also lautet die Faktorisierung in symmetrischer Form

$$y^4 + y^3 - 4y^2 - 4y + 1 = \left(y^2 + \gamma_1\, y - \frac{1}{\gamma_1^2}\right)\left(y^2 + \gamma_2\, y - \frac{1}{\gamma_2^2}\right),$$

wobei γ_1 und γ_2 Wurzeln sind von $\gamma^2 - \gamma - 1 = 0$. Dies ist ein gestaffelt lösbares System von Gleichungen, wie es zuerst 1762 Étienne Bézout (1730–1783) betrachtet hat:

$$y^2 w^2 + y w^3 - 1 = 0, \quad w^2 - w - 1 = 0$$

und es erfordert zur Lösung nur Quadratwurzeln! Das Dreißigeck ist also mit Zirkel und Lineal konstruierbar, und somit erst recht das Fünfzehneck. Die Krönung dieses ersten geglückten Ansatzes gelang Gauß, als er zeigen konnte, daß auch das Vierunddreißigeck bzw. Siebzehneck eine Lösung lediglich mit Quadratwurzeln erlaubt.

Nehmen wir jedoch zunächst ein weit weniger berühmtes, aber auch nicht ganz uninteressantes Beispiel:

Zum Sechsundzwanzigeck (bzw. Dreizehneck) gehört das Polynom $y^6 + y^5 - 5y^4 - 4y^3 + 6y^2 + 3y - 1$; es erlaubt eine Faktorisierung in zwei Polynome dritten Grades: Sind ω_1 und ω_2 Wurzeln der quadratischen Hilfsgleichung $w^2 + w - 3 = 0$, so ist die Faktorisierung

$$(y^3 - \omega_1\, y^2 - y + (\omega_1 - 1))(y^3 - \omega_2\, y^2 - y + (\omega_2 - 1))$$

möglich, die Dekomposition lautet

$$y^3 - w y^2 - y + w - 1 = 0, \quad w^2 + w - 3 = 0.$$

Nimmt man dagegen $w^3 + w^2 - 4w + 1 = 0$ mit den drei Wurzeln ω_1, ω_2 und ω_3 als Hilfsgleichung, so ist die Faktorisierung

$$\left(y^2 - \omega_1\, y + \left(1 - \frac{1}{\omega_1}\right)\right)\left(y^2 - \omega_2\, y + \left(1 - \frac{1}{\omega_2}\right)\right)\left(y^2 - \omega_3\, y + \left(1 - \frac{1}{\omega_2}\right)\right)$$

möglich, die Dekomposition lautet

$$w y^2 - w^2 y + w - 1 = 0, \quad w^3 + w^2 - 4w + 1 = 0.$$

Das Beispiel zeigt, daß ganz verschiedenartige Zerlegungsmöglichkeiten bestehen mögen, insbesondere daß es nicht einfach sein kann, einen Überblick über alle zu gewinnen. Die hauptsächliche Bedeutung der Kreisteilungsaufgaben war denn auch, den Anstoß für eine allgemeinere Theorie gegeben zu haben.

Doch nun zu Gauß! Zum Vierunddreißigeck (bzw. Siebzehneck) gehört $y^8 + y^7 - 7y^6 - 6y^5 + 15y^4 + 10y^3 - 10y^2 - 4y + 1$; dieses Polynom erlaubt eine Faktorisierung in zwei Polynome vierten Grades: Sind ω_1 und ω_2 Wurzeln der quadratischen Hilfgleichung $w^2 + w - 4 = 0$, so ist die Faktorisierung

$$(y^4 - \omega_1 y^3 - (\omega_1 + 2) y^2 + (2\omega_1 + 3) y - 1)$$
$$\cdot (y^4 - \omega_2 y^3 - (\omega_2 + 2) y^2 + (2\omega_2 + 3) y - 1)$$

möglich.

Das linke Polynom $(y^4 - \omega_1 y^3 - (\omega_1 + 2) y^2 + (2\omega_1 + 3) y - 1)$ erlaubt weiterhin eine Faktorisierung in

$$(y^2 - \xi_1 y + \xi_2)(y^2 - \xi_3 y + \xi_4),$$

wo

ξ_1, ξ_3 Wurzeln der Hilfsgleichung $x^2 - \omega_1 x - 1 = 0$ und

ξ_2, ξ_4 Wurzeln der Hilfsgleichung $x^2 - \omega_2 x - 1 = 0$ sind.

Eine ähnliche Zerlegung gilt für das rechte Polynom. Damit sind durch Wahl aller Vorzeichenkombinationen die Seiten aller acht einzügigen Heptadekagramme lediglich durch Quadratwurzeln zu berechnen.

Die vorstehende Behandlung des Dreizehnecks findet sich als Schulbeispiel 1872 bei Paul Bachmann. Für das Elfeck hat schon 1771 Alexandre Théophile Vandermonde eine Lösung der zugehörigen Gleichung lediglich mit Quadratwurzeln und mit fünften Wurzeln angegeben.

Eine gewisse Belebung könnten diese ganz in Vergessenheit geratenen Überlegungen erfahren, wenn die beträchtliche kombinatorische Arbeit, die bei der gruppentheoretischen Behandlung algebraischer Gleichungen hohen Grades notwendig ist, durch neuerdings verfügbare algebraische Manipulationssysteme erleichtert wird. Kreisteilungspolynome werden heute herangezogen bei Verfahren zur Faktorisierung großer Zahlen, die vor allem für Anwendungen in der Kryptanalyse wichtig sind. Bedeutung haben solche Untersuchungen allerdings nicht nur in Körpern mit Charakteristik Null, sondern, etwa für Codierungsverfahren, auch in endlichen Körpern.

Natürlich trat auch bald die Frage auf, welche Kreisteilungsaufgaben nicht mit Quadratwurzeln lösbar sind. Daß dies für das

Siebeneck und das Neuneck so ist, daß nämlich wesentlich eine kubische Gleichung verbleibt, wird in elementaren Algebravorlesungen meist erwähnt; es nachzurechnen macht etwas mehr Mühe.

Den Fragenkreis der Nichtfaktorisierbarkeit hat zuerst Ferdinand Eisenstein (1823–1852) angegangen; der sehr weitreichende Eisensteinsche Irreduzibilitätssatz entstand zunächst beim Nachweis, daß die Polynome $t_p^*(y)$ irreduzibel sind, wenn p Primzahl ist.

XII. NUMERISCHE BERECHNUNG DER TAFELWERTE IM DUALSYSTEM

Die Seiten des 2^{m+1}-Ecks erhält man nach X.2 aus $t_2(t_2(t_2(\ldots t_2(y)\ldots)))=0$, also in der Form

$$\sigma_{2^{m+1}}^{(i)}=\sqrt{2+\sqrt{2+\sqrt{2+\ldots\sqrt{2+\sqrt{2}}}}}$$

mit m Quadratwurzeln. Welche Wahl der m Vorzeichen hat man vorzunehmen, um eine bestimmte Lösung, beispielsweise $\sigma_{2^{m+1}}^{(1)}$, zu bekommen? Die Halbierungsformel (∗∗) von VI. kann iteriert werden. Der erste Schritt liefert

$$\sigma_{4n}^{(i)}\quad=\sqrt{2-\sqrt{2+\sqrt{4-(\sigma_n^{(i)})^2}}}$$

$$\sigma_{4n}^{(n-i)}=\sqrt{2-\sqrt{2-\sqrt{4-(\sigma_n^{(i)})^2}}}$$

$$\sigma_{4n}^{(n+i)}=\sqrt{2+\sqrt{2-\sqrt{4-(\sigma_n^{(i)})^2}}}$$

$$\sigma_{4n}^{(2n-i)}=\sqrt{2+\sqrt{2+\sqrt{4-(\sigma_n^{(i)})^2}}}.$$

Aus $\sigma_{16}^{(1)}$ beispielsweise ergibt sich zunächst $\sigma_{32}^{(1)}$, $\sigma_{32}^{(15)}$ und dann $\sigma_{24}^{(1)}$, $\sigma_{64}^{(15)}$, $\sigma_{64}^{(17)}$, $\sigma_{64}^{(31)}$. Dies zeigt, daß bei der Halbierung ein fortgesetztes Spiegelungsverhalten bezüglich des oberen Indexes gilt.

Die 2^m Vorzeichenkombinationen bezeichnen in naheliegender Weise die Ecken eines m-dimensionalen Würfels. Wolfgang Händler hat 1958 eine Darstellung des m-Hyperwürfels gegeben (»Händler-Diagramm«), bei der die 2^m Ecken auf einem Kreis liegen und die die Ecken codierenden binären m-Tupel (in unserem Fall die Vorzeichen-Tupel) einen einschrittigen Code bilden – einen reflektierten Code, der durch fortgesetzte Spiegelung aufgebaut wird (»Gray-Code«). Verbindet man jeweils zwei Tupel, die sich nur in *einer*

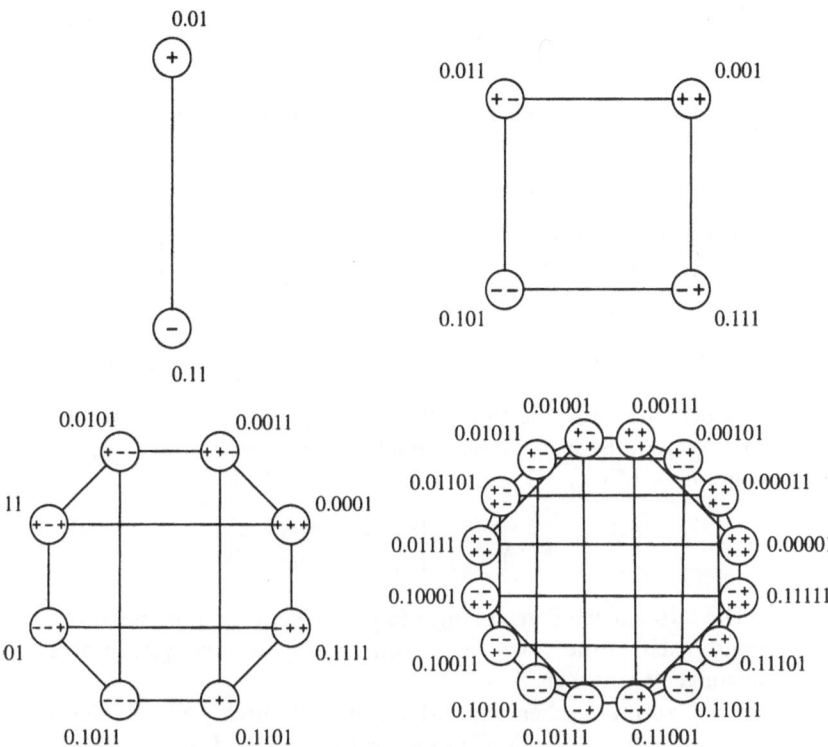

Abbildung 12. Händler-Diagramme und Pincherle-Formel

Stelle unterscheiden, durch eine Linie (Abbildung 12), so sind Tupel, die sich nur in der ersten Stelle unterscheiden, durch eine horizontale Linie verbunden; Tupel, die sich nur in der zweiten Stelle unterscheiden, durch eine horizontale Linie verbunden; Tupel, die sich nur in der zweiten Stelle unterscheiden, durch eine vertikale Linie. Tupel, die sich nur in der dritten Stelle unterscheiden, sind durch eine unter 45° oder 135° verlaufende Linie verbunden, Tupel, die sich nur in der vierten Stelle unterscheiden, durch eine unter $22\frac{1}{2}°$ oder $67\frac{1}{2}°$ oder $112\frac{1}{2}°$ oder $157\frac{1}{2}°$ verlaufende Linie, usw. Man kann im Händler-Diagramm direkt ablesen, auf welchen Wert der Polygonseite die Wahl eines bestimmten Vorzeichen-Tupels führt.

Dies führt uns zurück zur Berechnung von Tafelwerten trigonometrischer Funktionen.

Die reelle Zahl a, $0 \leq a < 1$, sei als echter Dualbruch vorgegeben:

$$a = 0.a_1, a_2, a_3, a_4, a_5, a_6, a_7, \ldots .$$

Gesucht ist der Wert der trigonomischen Funktion $\cos(a \cdot \pi)$.
Die zum abbrechenden Dualbruch

$$a^{(k)} = 0.a_1, a_2, a_3, \ldots, a_{k-1}, a_k, L$$

gehörigen Näherungen $\cos(a^{(k)} \cdot \pi)$ ergeben sich wegen
$\cos \alpha = \frac{1}{2}(2 - \text{chord}^2 \, \alpha)$ als

$$\cos(0.L \cdot \pi) = 0$$

$$\cos(a^{(k)} \cdot \pi) = \varepsilon_1 \sqrt{\frac{1}{2} + \varepsilon_2 \sqrt{\frac{1}{2} + \varepsilon_3 \sqrt{\frac{1}{2} + \ldots + \varepsilon_{k-1} \sqrt{\frac{1}{2} + \varepsilon_k \sqrt{\frac{1}{2}}}}}} \quad (k \geq 1).$$

Dabei erhält man die Vorzeichenwerte ε_1, ε_2, $\varepsilon_3 \ldots \varepsilon_k$ in anschaulicher Weise aus dem k-dimensionalen Hyperwürfel des Händler-Diagramms. Mit $a_0 = 0$ ergibt sich (Pincherle 1917)[13]

$$\varepsilon_i = \begin{cases} +\frac{1}{2}, & \text{falls } a_{i-1} = a_i \\ -\frac{1}{2}, & \text{falls } a_{i-1} \neq a_i . \end{cases}$$

Die tatsächliche Berechnung erfolgt, wie bei Kettenbrüchen, von hinten nach vorne; man muß von Anfang an die gewünschte Genauigkeit vorsehen.[14]

Doch zu praktischen Fragen! Wenn man durch geeignete *hardware*-Maßnahmen die Ausführungszeit einer Quadratwurzel-Operation in die Größenordnung der Multiplikationszeit bringt, kann man in einem modernen mikrominiaturisierten Rechner mit einem der verlangten Genauigkeit proportionalen Aufwand trigonometrische Funktionen stets neu berechnen und sich die Speicherung von

---

[13] Salvatore Pincherle (1853–1936). Siehe Pólya und Szegö, Bd. I, S. 191. Zum Problem der Vorzeichenwahl für die Quadratwurzeln beim Algorithmus des arithmetisch-geometrischen Mittels hat neuerdings besonders John Todd schöne Ergebnisse erzielt.

[14] Definiert man $b_i = \frac{1}{2} + \varepsilon_i$, so wird die Abbildung

$$(a_1, a_2, a_3, a_4, a_5, a_6, a_7, \ldots) \mapsto (b_1, b_2, b_3, b_4, b_5, b_6, b_7, \ldots)$$

als Bildung des abgeleiteten Codes bezeichnet. Die dadurch vermittelte involutorische Abbildung des Intervalls $[0, 1)$ der reellen Zahlen in sich:

$$a = 0.a_1, a_2, a_3, a_4, a_5, a_6, a_7, \ldots \mapsto b = 0.b_1, b_2, b_3, b_4, b_5, b_6, b_7, \ldots$$

ist zur Herstellung von Blätterteig nach E. Hopf (›On Causality, Statistics and Probability‹, J. Math. Physics vol. 13, pp. 51–102, 1934) geeignet.

Tabellen sparen, wobei man gegenüber der – ebenfalls der verlangten Genauigkeit proportionalen – Suchzeit nicht verliert. Für einige Standardaufgaben der Numerik, etwa die schnelle Fouriertransformation, wird dies mit steigenden Anforderungen an die Auflösungsgenauigkeit mit Erfolg auszunützen sein. Im übrigen kann man eine dem iterativen Algorithmus $\varphi := \varepsilon_i \sqrt{\frac{1}{2} + \varphi}$ entsprechende Formel auch ausgerollt auf einem Chip in *hardware* gießen und den Berechnungsstrom nach Art eines Fließbands hindurchführen.

NACHREDE

Lieber Herr Dr. Götze,

von den Sternpolygonen sind wir zu den Händler-Diagrammen der Hyperwürfel gelangt, die ihrerseits Bestandteile von Sternpolygonen in sich tragen. Die Ästhetik der Formen mag Sie über das Abracadabra der Formeln hinwegtrösten. Doch auch Formeln können schön sein, wie die Bleisetzer wußten.

Es mag Ihnen auch eine besondere Genugtuung sein, zu sehen, daß die Familie der 2-, 4-, 8-, 16-, 32-, ... zähligen Sternpolygone das besondere Interesse der Informatiker besitzt. Die Informatik an der Technischen Universität München hat nicht ohne Grund als ihr Logogramm (Abbildung 13) einen sechzehnzähligen Stern gewählt.

Abbildung 13. Logogramm INFORMATIK MÜNCHEN

Abbildung 14. Rosette des nördlichen Glasfensters im Querschiff von Notre-Dame in Paris

Und wieder einmal möchte ich zum Schluß ein Bild bringen, das mich stets aufs neue fasziniert: Die Rosette des nördlichen Glasfensters im Querschniff von Notre-Dame in Paris (Abbildung 14). Sie ist innen 16zählig, außen 32zählig. Wir dürfen sie als Wahrzeichen christlich-abendländischer Kultur auffassen.

Sur quelques progrès dans la théorie des fonctions analytiques de variables complexes entre 1930 et 1950*

Au Dr. Heinz Götze
en hommage reconnaissant pour tout ce qu'il a
fait en faveur de la diffusion des mathématiques
contemporaines

La théorie des fonctions analytiques de plusieurs variables complexes a connu, dans les années 1930–1934, un renouveau auquel l'école de Heinrich Behnke a beaucoup contribué. Peter Thullen, dont nous fêtons aujourd'hui le 80-ème anniversaire, fut l'un des membres les plus actifs de cette école à Münster. Le «Bericht» de Behnke-Thullen [2], paru en 1934, atteste ce renouveau; il donne une photographie de la situation à cette époque et indique une foule de problèmes ouverts. Aussi a-t-il exercé une influence décisive sur les développements ultérieurs de la théorie: par exemple, Oka s'y réfère constamment avant d'apporter la solution des principaux problèmes laissés ouverts dans le Behnke-Thullen.

Je me bornerai ici à évoquer trois sujets; à chacun d'eux Thullen a apporté une contribution qui fut ensuite la source d'importants développements.

I. Automorphismes des domaines bornés

1. Soit D un domaine *borné* de $\mathbb{C}^n$. On se propose d'étudier le groupe $G(D)$ de tous les automorphismes (bijections holomorphes) de D. Plus généralement, que peut-on dire de l'ensemble des isomorphismes $D_1 \to D_2$ lorsque D_1 et D_2 sont des domaines bornés de $\mathbb{C}^n$? (Cet ensemble peut être vide.)

Le cas $n = 1$ étant classique, on s'est d'abord attaqué au cas $n = 2$. Les exemples les plus simples étaient la boule $|x|^2 + |y|^2 < 1$ et le dicylindre $|x| < 1$, $|y| < 1$. Le groupe des automorphismes était connu pour chacun d'eux (groupe à 7 paramètres réels, resp. à

---

* Ce texte est celui d'une conférence prononcée le 13 novembre 1987 à l'Université de Fribourg (Suisse) lors d'un colloque en l'honneur du 80-ème anniversaire de Peter Thullen.

6 paramètres réels); ces groupes n'étant pas isomorphes, cela entraînait que la boule et le dicylindre ne sont pas isomorphes.

Ces deux exemples sont des cas particuliers des *domaines de Reinhardt*, c'est-à-dire des domaines D qui contiennent l'origine $(0, 0)$ et sont stables par le groupe à 2 paramètres réels

$$(*) \qquad (x, y) \mapsto (\lambda x, \mu y), \quad \text{où } |\lambda| = 1, |\mu| = 1.$$

Reinhardt [17] avait prouvé en 1921 que si un tel domaine D est *borné* et n'a pas la forme $A|x|^2 + B|y|^2 < 1$ $(A > 0, B > 0)$, le sous-groupe de $G(D)$ formé des automorphismes qui laissent fixe l'origine se réduit au groupe de définition (*), éventuellement combiné avec une transformation de la forme

$$(x, y) \mapsto (\alpha y, \alpha^{-1} x) \quad (\text{où } \alpha \in \mathbb{C} - \{0\}).$$

Une fois ce résultat acquis, que pouvait-on dire des automorphismes qui ne laissent pas fixe l'origine? On présumait qu'il n'y en a pas, «en général». En fait, Kritikos [13] avait prouvé en 1928 que tout automorphisme du domaine $|x| + |y| < 1$ laisse fixe l'origine. Dès lors était posé le problème de déterminer tous les domaines de Reinhardt bornés D dont le groupe $G(D)$ admet des automorphismes ne laissant pas fixe l'origine.

C'est ce problème que résout Peter Thullen dans sa dissertation de 1931 [21]: si un domaine de Reinhardt borné admet des automorphismes qui ne laissent pas fixe l'origine, il est «équivalent» (au moyen d'une transformation affine $x \mapsto ux$, $y \mapsto vy$ ou $x \mapsto uy$, $y \mapsto vx$) à l'un des domaines suivants:

– le dicylindre $|x| < 1$, $|y| < 1$,
– a désignant un nombre réel > 0, le domaine

$$(D_a) \qquad\qquad |x|^2 + |y|^a < 1.$$

(D_2 est la boule-unité). Thullen prouve en outre que, pour $a \neq b$, D_a et D_b ne sont pas isomorphes, et il détermine tous les automorphismes de D_a pour $a \neq 2$:

$$x \mapsto \lambda f(x), \qquad y \mapsto \mu y (f'(x))^{1/a}$$

où

$$|\lambda| = 1, \quad |\mu| = 1, \quad f(x) = (x + x_0)/(1 + \bar{x}_0 x) \quad (|x_0| < 1).$$

2. Une classe plus générale que celle des domaines de Reinhardt avait été introduite en 1928 par Carathéodory [3]: la classe des domaines D qui contiennent l'origine $(0, 0)$ et sont stables par le

groupe à un paramètre réel

$$(x, y) \mapsto (\lambda x, \lambda y), \quad (|\lambda| = 1).$$

On les appelle domaines *cerclés* (Kreiskörper), car la trajectoire de tout point $\neq (0, 0)$ est un cercle. L'origine s'appelle le *centre* du domaine. Cette définition s'étend aussitôt à un nombre quelconque n de variables $x_1, \ldots, x_n$, le groupe étant alors

$$(x_1, \ldots, x_n) \mapsto (\lambda x_1, \ldots, \lambda x_n) \quad (|\lambda| = 1).$$

En 1930, dans le cas $n = 2$, Behnke [1] démontrait, sous quelques hypothèses restrictives pour les frontières, que si un isomorphisme $f: D_1 \to D_2$ de domaines cerclés bornés envoie le centre de D_1 sur le centre de D_2, f est $\mathbb{C}$-linéaire. Ceci conduisit H. Cartan [6] à donner de ce théorème une démonstration simple, valable pour tout n (et même pour le cas des espaces de Banach complexes) et ne nécessitant aucune hypothèse restrictive sur les frontières. Cette démonstration tient en quelques lignes: soit $f: D_1 \to D_2$ un isomorphisme tel que $f(0) = 0$; pour θ réel, soit T_θ la multiplication par $e^{i\theta}$. Alors $f^{-1} \circ T_{-\theta} \circ f \circ T_\theta$ est un automorphisme de D_1 qui laisse fixe le centre et est tangent à l'identité en ce point. En considérant les itérées de cette transformation, et tenant compte du fait que D_1 est borné, on conclut que cette transformation est l'identité, et donc que $f(x e^{i\theta}) = e^{i\theta} f(x)$. Le développement de f en série de polynômes homogènes montre alors que f est un polynôme homogène de degré un.   C.Q.F.D.

Ce résultat valut à H. Cartan d'être invité en 1931 par H. Behnke à l'Université de Münster et d'y faire la connaissance du jeune Peter Thullen. Ce fut le début d'une collaboration dont je parlerai tout à l'heure.

Le théorème de Thullen relatif aux automorphismes des domaines de Reinhardt bornés laissait ouvert le problème suivant: déterminer, pour $n = 2$, tous les domaines *cerclés* bornés qui admettent des automorphismes ne laissant pas fixe le centre. On considère alors comme «équivalents» deux domaines qui se déduisent l'un de l'autre par une transformation linéaire $(x, y) \mapsto (ax + by, cx + dy)$, $ad - bc \neq 0$. Le résultat de E. et H. Cartan (1931, [5]) est le suivant: outre les D cerclés qui sont équivalents à l'un des domaines trouvés par Thullen, il n'y a que les domaines équivalents aux domaines suivants:

$$(\Delta_\alpha) \qquad |x| < 1, \quad |y| < 1, \quad \left| \frac{y - x}{1 - \bar{x} y} \right| < \alpha \quad (0 < \alpha < 1).$$

Pour $\alpha \neq \beta$, $\varDelta_\alpha$ n'est pas isomorphe à $\varDelta_\beta$; le groupe des automorphismes de $\varDelta_\alpha$ (qui est indépendant de α) se compose des transformations

$$x \mapsto f(x), \quad y \mapsto f(y)$$

$$(\text{où } f(x) = \lambda(x + x_0)/(1 + \bar{x}_0 x), \ |\lambda| = 1, \ |x_0| < 1)$$

et de leurs composés avec $(x, y) \mapsto (y, x)$.

En conclusion, on voit que, pour $n = 2$, le groupe $G(D)$ des automorphismes d'un domaine cerclé borné D est un groupe qui dépend de k paramètres réels, où k a l'une des valeurs suivantes:

$k = 7$ (boule),
$k = 6$ (dicylindre)
$k = 4$ (domaine D_a de Thullen pour $a \neq 2$),
$k = 3$ (domaine $\varDelta_\alpha$),
$k = 2$ (domaine de Reinhardt général),
$k = 1$ (domaine cerclé général).

3. *Résultats généraux sur le groupe $G(D)$.* – Soit D un domaine borné *quelconque* de $\mathbb{C}^n$. En 1936 H. Cartan [8] prouve que $G(D)$ est un *groupe de Lie réel*. D'une façon précise: le groupe $G(D)$, muni de la topologie de la convergence uniforme sur les compacts de D, est un groupe *localement compact*; le sous-groupe formé des automorphismes qui laissent fixe un point $a \in D$ est *compact* et isomorphe au groupe linéaire-complexe des transformations linéaires tangentes; de plus la composante connexe de l'élément neutre $G_0(D)$ possède une unique structure de variété analytique réelle qui en fait un groupe de Lie réel tel que l'action $G_0(D) \times D \to D$ soit analytique-réelle. Tout sous-groupe à un paramètre réel de $G_0(D)$ est engendré par une transformation infinitésimale, c'est-à-dire un champ de vecteurs $x \mapsto \xi(x)$, où ξ est holomorphe. On montre que si D est cerclé, $\xi(x)$ est un polynôme de degré ≤ 2.

Les domaines bornés symétriques. Dès 1935, E. Cartan [4], s'appuyant sur les résultats précédents, se propose de déterminer tous les domaines bornés *homogènes* de $\mathbb{C}^n$, c'est-à-dire tels que le groupe $G(D)$ opère transitivement dans D. Donc D est quotient d'un groupe de Lie réel par un sous-groupe compact. Parmi les D homogènes il y a la classe des D *symétriques*: D est symétrique si pour chaque point $a \in D$ il existe un automorphisme involutif σ_a de D admettant a comme point fixe isolé (un tel automorphisme est alors unique, et la transformation linéaire tangente en a est la symétrie par rapport à l'origine). E. Cartan démontre que pour $n \leq 3$ tous les domaines bornés homogènes sont symétriques (la réciproque étant vraie pour tout n); pour $n \geq 4$ il doit renoncer à déterminer tous les domaines bornés homogènes, mais détermine effectivement tous les domaines bornés symétriques.

Voici, très brièvement, le résultat obtenu par E. Cartan: tout domaine borné symétrique est, d'une seule manière, produit de domaines symétriques *irréductibles*. Il existe quatre grandes classes de domaines symétriques irréductibles, et en outre deux domaines symétriques exceptionnels (pour $n = 16$ et $n = 27$). Voici en quoi consistent les 4 grandes classes: pour la classe (I), la donnée de 2 entiers p et q $(p \geq q \geq 1)$ détermine un domaine borné de $\mathbb{C}^{pq}$. Pour la classe (II), la donnée d'un entier $p \geq 2$ definit un domaine borné de $\mathbb{C}^n$, avec $n = p(p-1)/2$. Pour la classe (III), la donnée de $p \geq 1$ définit un domaine borné de $\mathbb{C}^p$. Enfin, pour la classe (IV), la donnée de $p \geq 1$ définit un domaine borné de $\mathbb{C}^n$, avec $n = p(p+1)/2$. Tous ces domaines sont symétriques et irréductibles, et tout domaine borné symétrique et irréductible est isomorphe à l'un d'eux ou à l'un des 2 domaines exceptionnels. En outre, les domaines des 4 grandes classes sont deux à deux non isomorphes, sauf pour les basses dimensions.

Il se trouve que tous les domaines ainsi obtenus sont *cerclés*. Réciproquement d'ailleurs, tout domaine cerclé homogène est évidemment symétrique. Le fait que tout domaine borné symétrique est isomorphe à un domaine cerclé était une constatation expérimentale de E. Cartan; il fallut attendre 1976 pour avoir de ce résultat une démonstration a priori: c'est un résultat général que J.-P. Vigué [23] a démontré pour les domaines bornés symétriques d'un espace de Banach complexe.

E. Cartan avait dû laisser ouverte la question de savoir s'il existe des domaines bornés homogènes qui ne soient pas symétriques. Il en existe effectivement, déjà pour $n = 4$, comme l'a montré Piatetskii-Shapiro en 1959 [15]. Cela l'a conduit à développer toute une théorie des «domaines de Siegel». Mais ceci sort du cadre du présent exposé.

II. Théorie des domaines d'holomorphie

1. La notion de *variété analytique complexe* n'avait pas encore été dégagée en 1931. On travaillait alors uniquement sur des ouverts de l'espace $\mathbb{C}^n$.

Dès 1906, Hartogs [12] avait donné l'exemple, surprenant pour l'époque, d'un couple (U, V) de domaines de $\mathbb{C}^n$, avec $U \subset V$, $U \neq V$, tel que toute fonction holomorphe dans U se prolonge en une fonction holomorphe dans V. Voici cet exemple: V est le polydisque

$$|x_i| < 1, \quad 1 \leq i \leq n,$$

et U est la réunion des deux ouverts

$$1 - \varepsilon < |x_1| < 1, \quad |x_i| < 1 \text{ pour } 2 \leq i \leq n,$$

et

$$|x_1| < 1, \quad |x_i| < \varepsilon \quad \text{pour } 2 \leq i \leq n$$

(le nombre $\varepsilon > 0$ étant arbitrairement petit). U est souvent désigné sous le nom de «marmite de Hartogs».

Cet exemple suggère de formuler la définition suivante (qui va s'avérer incorrecte): un domaine U est un *domaine d'holomorphie* s'il existe une fonction holomorphe f dans U qui ne se prolonge à aucun ouvert V contenant U et distinct de U. En fait, l'étude du prolongement analytique d'une fonction uniforme peut conduire à des fonctions non uniformes, et c'est pourquoi la bonne définition consiste à exiger que f ne soit prolongeable en aucun point frontière de U, ce qui est une condition plus forte que la précédente.

L'exemple de $\sqrt{z}$ fera comprendre la chose: dans le domaine U formé du plan $\mathbb{C}$ privé du demi-axe réel ≥ 0, il existe une fonction holomorphe $f(z)$ égale à une détermination de $\sqrt{z}$; cette $f(z)$ ne peut pas se prolonger dans un ouvert strictement plus grand que U, et pourtant U n'est pas le domaine total d'existence de $f(z)$. Ce domaine d'existence est le revêtement à 2 feuillets de $\mathbb{C} - \{0\}$.

L'étude du prolongement analytique amène ainsi à la notion de *domaine étalé dans* $\mathbb{C}^n$. Rappelons d'abord la définition d'une application étale: X et Y étant deux espaces topologiques séparés, une application $f: X \rightarrow Y$ est dite *étale* si f est continue et si tout point $x \in X$ possède un voisinage ouvert U tel que la restriction $f | U$ soit un homéomorphisme de U sur un ouvert de Y. Ceci posé, nous appellerons «domaine étalé dans $\mathbb{C}^n$» la donnée d'un espace topologique X séparé et *connexe* et d'une application étale

$p: X \to \mathbb{C}^n$. Alors X hérite une structure de variété analytique complexe au moyen de p, qui fournit en chaque point de X un système de coordonnées locales.

Par définition, un morphisme $(X, p) \to (Y, q)$ de domaines étalés est une application continue $\varphi: X \to Y$ qui rend commutatif le diagramme

$$X \xrightarrow{\quad \varphi \quad} Y$$
$$p \searrow \qquad \swarrow q$$
$$\mathbb{C}^n$$

(φ est nécessairement étale et analytique). Nous avons ainsi défini la *catégorie $\mathcal{E}$* des domaines étalés dans $\mathbb{C}^n$. Les (X, p) tels que p soit un homéomorphisme de X sur l'ouvert $p(X)$ sont identifiés aux domaines de $\mathbb{C}^n$; la notion de morphisme généralise la notion d'inclusion pour les domaines.

Soit $\mathcal{H}(X)$ l'algèbre des fonctions holomorphes dans X. Un morphisme $\varphi: (X, p) \to (Y, q)$ induit un homomorphisme $\varphi^*: \mathcal{H}(Y) \to \mathcal{H}(X)$ (à savoir $g \mapsto g \circ \varphi$) qui est *injectif* (principe du prolongement analytique). Dire que φ^* est *bijectif*, c'est dire que pour toute $f \in \mathcal{H}(X)$ existe une $g \in \mathcal{H}(Y)$, nécessairement unique, telle que $g \circ \varphi = f$. Nous dirons alors que g est le *prolongement analytique* de f selon le morphisme $\varphi: X \to Y$. Un morphisme φ sera dit *privilégié* si φ^* est bijectif.

Fixons maintenant un domaine étalé (X, p). Les morphismes privilégiés de source (X, p) sont les objets d'une nouvelle catégorie, notée $\mathcal{E}(X, p)$: par définition, un morphisme de $(X, p) \xrightarrow{\varphi} (Y, q)$ dans $(X, p) \xrightarrow{\psi} (Z, r)$ est un $\mathcal{E}$-morphisme $\chi: (Y, q) \to (Z, r)$ tel que $\psi = \chi \circ \varphi$. Un tel morphisme χ est unique, et il est privilégié.

Ces définitions un peu pédantes nous permettent de formuler correctement un résultat fondamental concernant le prolongement analytique: *la catégorie $\mathcal{E}(X, p)$ possède un objet final.* Cela signifie qu'il existe un domaine étalé $(\tilde{X}, \tilde{p})$ et un morphisme privilégié

$$\alpha: (X, p) \to (\tilde{X}, \tilde{p})$$

jouissant de la propriété suivante: pour tout morphisme privilégié $\varphi: (X, p) \to (Y, q)$, il existe un morphisme privilégié $\psi: (Y, q) \to (\tilde{X}, \tilde{p})$ nécessairememt unique, tel que $\alpha = \psi \circ \varphi$. Un tel objet final $(\tilde{X}, \tilde{p})$ est unique à isomorphisme près. On l'appelle *l'enveloppe d'holomorphie* de (X, p).

On peut généraliser. Donnons-nous un sous-ensemble I de $\mathcal{H}(X)$. Les morphismes $\varphi\colon (X, p) \to (Y, q)$ tels que I soit contenu dans l'image de $\varphi^*\colon \mathcal{H}(Y) \to \mathcal{H}(X)$ sont les objets d'une catégorie $\mathcal{E}_I(X, p)$ qui possède un objet final: c'est en quelque sorte «le plus grand» des domaines étalés (Y, q), munis d'un morphisme $(X, p) \to (Y, q)$, permettant le prolongement analytique à Y de toutes les fonctions de $I \in \mathcal{H}(X)$.

Si $I \subset J$, il est clair que $\mathcal{E}_J(X, p)$ est une sous-catégorie de $\mathcal{E}_I(X, p)$. On a donc un unique morphisme de l'objet final de $\mathcal{E}_J(X, p)$ dans l'objet final de $\mathcal{E}_I(X, p)$. Lorsque I se réduit à une seule fonction $f \in \mathcal{H}(X)$, l'objet final s'appelle le *domaine d'holomorphie de la fonction f*.

Soient maintenant (X, p) et (Y, q) deux domaines étalés, $(\tilde{X}, \tilde{p})$ et $(\tilde{Y}, \tilde{q})$ leurs enveloppes d'holomorphie, $\alpha\colon X \to \tilde{X}$ et $\beta\colon Y \to \tilde{Y}$ les applications canoniques. Si $\varphi\colon (X, p) \to (Y, q)$ est un morphisme de la catégorie $\mathcal{E}$, soit $I \in \mathcal{H}(X)$ l'image de $\varphi^*\colon \mathcal{H}(Y) \to \mathcal{H}(X)$. L'objet final de $\mathcal{E}_I(X, p)$ n'est autre que l'enveloppe d'holomorphie $(\tilde{Y}, \tilde{q})$. On a donc un morphisme $\tilde{\varphi}\colon (\tilde{X}, \tilde{p}) \to (\tilde{Y}, \tilde{q})$ qui rend commutatif le diagramme

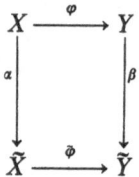

Ainsi l'enveloppe d'holomorphie (X, p), munie du morphisme $\alpha\colon (X, p) \to (\tilde{X}, \tilde{p})$, est un *foncteur contravariant* de (X, p).

On peut montrer (mais c'est un peu subtil) que le résultat précédent s'étend au cas d'une application holomorphe et étale $\varphi\colon X \to Y$. D'une façon précise:

THÉORÈME. Soient (X, p) et (Y, q) deux domaines étalés dans $\mathbb{C}^n$, soient $(\tilde{X}, \tilde{p})$ et $(\tilde{Y}, \tilde{q})$ leurs enveloppes d'holomorphie, et notons $\alpha\colon X \to \tilde{X}$ et $\beta\colon Y \to \tilde{Y}$ les applications canoniques. Si $\varphi\colon X \to Y$ est une application holomorphe et étale (ne satisfaisant pas nécessairement à $p = q \circ \varphi$), il existe une unique application $\tilde{\varphi}\colon \tilde{X} \to \tilde{Y}$ holomorphe et étale telle que le diagramme ci-dessus soit commutatif.

Il est immédiat que si φ est un isomorphisme, $\tilde{\varphi}$ est aussi un isomorphisme. Ainsi l'enveloppe d'holomorphie $\alpha: X \to \tilde{X}$ ne dépend pas (à isomorphisme près) du choix de l'application étale $X \to \mathbb{C}^n$. L'enveloppe d'holomorphie est donc définie pour toute variété analytique connexe X susceptible d'être étalée dans $\mathbb{C}^n$.

REMARQUE IMPORTANTE. Si (X, p) est le domaine d'holomorphie d'une $f \in \mathcal{H}(X)$, (X, p) est évidemment sa propre enveloppe d'holomorphie. Mais la réciproque pose problème: si (X, p) est sa propre enveloppe d'holomorphie, existe-t-il une $f \in \mathcal{H}(X)$ dont (X, p) soit le domaine d'holomorphie? On verra plus loin que la réponse est affirmative. On dira désormais que (X, p) est un *domaine d'holomorphie* s'il existe une $f \in \mathcal{H}(X)$ dont (X, p) est le domaine d'holomorphie.

2. *La notion de convexité holomorphe.* Soit X une variété analytique complexe supposée connexe. A tout compact $K \subset X$ associons l'ensemble $\tilde{K}$ formé des $x \in X$ tels que

$$|f(x)| \leqq \sup_{y \in K} |f(y)| \quad \text{pour toute } f \in \mathcal{H}(X).$$

DÉFINITION. On dit que X est une *variété de Stein* si elle satisfait à la condition (C) suivante:

(C) $\begin{cases} \text{(i)} & \text{les fonctions holomorphes sur } X \text{ séparent} \\ & \text{les points de } X; \\ \text{(ii)} & \text{quel que soit le compact } K \subset X,\ \tilde{K} \text{ est compact.} \end{cases}$

(La conditions (ii) s'exprime en disant que X est *holomorphiquement convexe*.)

Cette définition n'est pas exactement celle donnée par Karl Stein en 1951 [20], mais Grauert a démontré en 1955 [11] qu'elle lui est équivalente.

Il est clair que toute sous-variété analytique connexe et fermée de $\mathbb{C}^n$ est une variété de Stein. Inversement, Remmert [18] a annoncé en 1956 que toute variété de Stein de dimension n peut se plonger comme sous-variété fermée de $\mathbb{C}^N$, N grand. Nous n'aurons pas besoin de ce résultat.

La notion de variété de Stein s'applique à un domaine étalé (X, p) dans $\mathbb{C}^n$. Mais on peut aussi considérer d'autres conditions que la condition (C) ci-dessus. La condition (C') ci-dessous est due à Thullen:

(C') $\begin{cases} \text{(i)} & \text{les fonctions holomorphes dans } X \text{ séparent} \\ & \text{les points de } X; \\ \text{(ii)} & \text{pour tout compact } K \subset X, \text{ la distance de } \tilde{K} \\ & \text{à la frontière de } X \text{ est égale à la distance de } K \\ & \text{à la frontière de } X. \end{cases}$

Enfin, on peut envisager pour (X, p) une condition plus faible:

(C'') $\begin{cases} \text{(i)} & \text{les fonctions holomorphes dans } X \text{ séparent} \\ & \text{les points de } X; \\ \text{(ii)} & \text{pour tout compact } K \subset X, \text{ tout } x \in X \text{ est centre} \\ & \text{d'une boule de } X \text{ non contenue dans } \tilde{K}. \end{cases}$

(Dans les conditions ci-dessus, les notions de distance et de boule se rapportent à l'espace $\mathbb{C}^n$ ambiant au moyen de l'application étale p).

Comparons les conditions (C), (C') et (C'') pour un domaine étalé (X, p). On a vu que $(C') \Rightarrow (C'')$. De plus on voit facilement que $(C) \Rightarrow (C'')$. Les résultats de Thullen et de Cartan [10] peuvent alors se formuler comme suit:

THÉORÈME DE THULLEN. Si (X, p) est une enveloppe d'holomorphie, alors (X, p) satisfait à la condition (C').

THÉORÈME DE CARTAN. Si (X, p) satisfait à (C''), alors il existe une $f \in \mathscr{H}(X)$ dont (X, p) est le domaine d'holomorphie.

De là résultent d'importantes conséquences:

(a) toute enveloppe d'holomorphie est le domaine d'holomorphie d'une fonction.

(b) les conditions (C') et (C'') sont équivalentes.

Ceci nous amène à dire qu'un domaine étalé satisfait à la *condition de convexité de Cartan-Thullen* s'il satisfait à (C') ou (C'').

Puisque $(C) \Rightarrow (C'')$, on voit que si un domaine étalé est une variété de Stein, il satisfait à la condition de Cartan-Thullen et est donc un domaine d'holomorphie.

Tout domaine d'holomorphie est-il réciproquement une variété de Stein? Cartan et Thullen n'avaient pu donner qu'une réponse partielle à cette question, à savoir pour les domaines d'holomorphie n'ayant qu'un nombre fini de feuillets. Il a fallu attendre K. Oka pour prouver ce résultat en toute généralité ([14], 1953), à savoir:

THÉORÈME D'OKA. *Tout domaine d'holomorphie est holomorphiquement convexe* (ceci entraîne l'equivalence des conditions (*C*), (*C'*) ou (*C''*)).

La démonstration d'Oka est un tour de force. Il résout en même temps le fameux «problème de Levi». Voici comment procède Oka :

Si *D* est un domaine étalé dans $\mathbb{C}^n$, on définit d'abord la notion de point-frontière de *D* et de voisinage d'un point-frontière. Disons alors que *D* est *localement domaine d'holomorphie* si tout point-frontière de *D* possède un voisinage ouvert *V* (dans *D*) tel que *V* soit un domaine d'holomorphie. Il est facile de voir que tout domaine d'holomorphie est localement domaine d'holomorphie.

On considère alors une *condition de Hartogs* qui s'exprime par une propriété des marmites de Hartogs au voisinage des points-frontière de *D*. Un domaine étalé qui satisfait à cette condition est dit *pseudo-convexe*. Tout domaine étalé qui est localement domaine d'holomorphie est pseudo-convexe.

Alors Oka prouve que *tout domaine pseudo-convexe est holomorphiquement convexe,* ce qui implique que *D* est *domaine d'holomorphie*. C'est là que réside le tour de force, et il n'est évidemment pas question de détailler ici la démonstration.

Finalement, on a équivalence des propriétés:

D est domaine d'holomorphie,

D est localement domaine d'holomorphie,

D est pseudo-convexe,

D est holomorphiquement convexe.

La «condition de Levi» entraîne aussitôt que *D* est pseudo-convexe, donc domaine d'holomorphie, ce qui résout un problème fameux.

III. LE THÉORÈME DE PROLONGEMENT DE THULLEN
    ET SES GÉNÉRALISATIONS

En 1935, Thullen [22] considère la situation suivante: *X* est une variété analytique complexe de dimension *n*, connexe pour fixer les idées. Soit *Y* une hypersurface analytique de *X* (fermée dans *X*); *Y* est donc définie localement par l'annulation d'une fonction holomorphe non identiquement nulle. Soit maintenant, dans le complémentaire *X − Y*, une hypersurface analytique *Z* (fermée dans *X − Y*), et soit $\bar{Z}$ son adhérence dans *X*. Un point $y \in Y$ sera dit *régulier* pour *Z* si *y* possède un voisinage ouvert *V* tel que $\bar{Z} \cap V$

soit une hypersurface analytique dans V; dans le cas contraire, y sera dit *singulier essentiel* pour Z.

Le résultat de Thullen s'énonce ainsi: si Y est (globalement) irréductible dans X, ou bien tous les points de Y sont réguliers pour Z, ou bien ils sont tous singuliers essentiels. La démonstration, assez subtile, utilise notamment une généralisation d'un théorème de prolongement de Radó, qu'on peut formuler comme suit ([9]): si, dans une variété analytique U, on a une application continue $f : U \to \mathbb{C}$ qui est holomorphe en tout point $x \in U$ tel que $f(x) \neq 0$, alors f est holomorphe en tout point de U sans exception.

En 1953, Remmert et Stein [19] ont réussi à étendre le théorème de Thullen au cas plus général que voici: Y est un sous-ensemble analytique de X (fermé dans X), et Z est un sous-ensemble analytique de $X - Y$ (fermé dans $X - Y$). On suppose que Y est de codimension $\geq p$ dans X en chacun de ses points, et que Z est de codimension p en chacun de ses points. Un point $y \in Y$ est dit régulier pour Z si y possède un voisinage ouvert V tel que $\bar{Z} \cap V$ soit un sous-ensemble analytique de V, singulier essentiel dans le cas contraire. Remmert et Stein prouvent que *l'ensemble des points $y \in Y$ qui sont singuliers essentiels est un sous-ensemble analytique de Y, de codimension p en chacun de ses points*. En particulier, si Y est de codimension $> p$ en chacun de ses points, tous les points de Y sont réguliers pour Z.

Enoncé sous cette forme, le théorème de Remmert-Stein a été étendu par Ramis ([16]) au cas où X est une variété analytique banachique de dimension infinie.

Signalons, pour terminer, comment le théorème de Remmert-Stein fournit une démonstration simple du *théorème de Chow*: tout sous-ensemble analytique (fermé) de l'espace projectif $P_n(\mathbb{C})$ est *algébrique*. Un tel sous-ensemble analytique est représenté par un cône analytique Z dans l'espace $\mathbb{C}^{n+1} - \{0\}$. Ici, $X = \mathbb{C}^{n+1}$ et $Y = \{0\}$. La codimension de Z en chacun de ses points est $\leq n$, tandis que celle de Y dans X est $n + 1$. Le théorème de Remmert-Stein permet de conclure: le cône Z est analytique en son sommet 0. Or un cône Z qui est analytique en son sommet est algébrique; en effet, si f holomorphe en 0 s'annule identiquement sur Z, chaque terme du développement de f en série de polynômes homogènes s'annule identiquement sur Z, qui est donc défini par l'annulation d'une famille de polynômes homogènes, et par suite d'une sous-famille finie. C.Q.F.D.

RÉFÉRENCES BIBLIOGRAPHIQUES

1. Behnke, H. (1930): Die Abbildungen der Kreiskörper. Abh. math. Sem. Hamburg 7:329–341
2. Behnke, H., Thullen, P. (1934): Theorie der Funktionen mehrerer komplexen Veränderlichen. Ergebnisse der Mathematik, vol. 3
3. Carathéodory, C. (1928): Über die Geometrie der analytischen Abbildungen. Abh. math. Sem. Hamburg 6:96–145
4. Cartan, E. (1935): Sur les domaines bornés homogènes de l'espace de n variables complexes. Abh. math. Sem. Hamburg 11:116–162
5. Cartan, E. et H. (1931): Les transformations des domaines cerclés bornés. Comptes Rendus Acad. Sci. Paris 192:709–712. Voir aussi [7]
6. Cartan, H. (1930): Les transformations analytiques des domaines cerclés les uns dans les autres. Comptes Rendus Acad. Sci. Paris 190:718–720
7. Cartan, H. (1932): Sur les transformations analytiques des domaines cerclés et semi-cerclés. Math. Annalen 106:540–573
8. Cartan, H. (1936): Sur les groupes de transformations analytiques. Collection J. Herbrand. Hermann, Paris
9. Cartan, H. (1952): Sur une extension d'un théorème de Radó. Math. Annalen 125:49–50
10. Cartan, H., Thullen, P. (1932): Zur Theorie der Singularitäten der Funktionen mehrerer Veränderlichen. Math. Annalen 106:617–647
11. Grauert, H. (1955): Charakterisierung der holomorph-vollständigen komplexen Räume. Math. Annalen 129:233–259
12. Hartogs, F. (1906): Über analytische Funktionen mehrerer unabhängigen Veränderlichen. Math. Annalen 62:1–88
13. Kritikos, N. (1928): Über analytische Abbildungen einer Klasse von vierdimensionalen Gebieten. Math. Annalen 99:321–341
14. Oka, K. (1953): Sur les fonctions analytiques de plusieurs variables IX. Domaines finis sans point critique intérieur. Jap. Journal Math. 27:97–155
15. Piatetskii-Shapiro, II. (1959): Sur un problème de E. Cartan. Dokl. Akad. Nauk SSSR 124:272–273
16. Ramis, J.-P. (1970): Sous-ensembles analytiques d'une variété banachique complexe. Ergebnisse der Mathematik, vol. 53
17. Reinhardt, K. (1921): Über Abbildungen durch analytische Funktionen zweier Veränderlichen. Math. Annalen 84:211–255
18. Remmert, R. (1956): Sur les espaces analytiques holomorphiquement séparables et holomorphiquement convexes. Comptes Rendus Acad. Sci. Paris 243:118–121
19. Remmert, R., Stein, K. (1953): Über die wesentlichen Singularitäten analytischer Mengen. Math. Annalen 126:263–306
20. Stein, K. (1951): Analytische Funktionen mehrerer komplexen Veränderlichen zu vorgegebenen Periodizitätsmoduln und der zweite Cousinsche Problem. Math. Annalen 123:201–222
21. Thullen, P. (1931): Zu den Abbildungen durch analytische Funktionen usw. Math. Annalen 104:244–259

22. Thullen, P. (1935): Über die wesentlichen Singularitäten analytischer Funktionen und Flächen. Math. Annalen 111:137–157

23. Vigué, J.-P. (1976): Le groupe des automorphismes analytiques d'un domaine borné d'un espace de Banach complexe, application aux domaines bornés symétriques. Ann. scientifiques de l'Ecole Normale Sup. 4e série 9: 203–282

Surface Theory with Darboux and Bianchi

The treatises of Darboux (1842–1917) and Bianchi (1856–1928) on surface theory are among the great works in the mathematical literature. They are:

G. Darboux, *Théorie générale des surfaces*, Tome 1 (1887), 2 (1888), 3 (1894), 4 (1896), and later editions and reprints.

L. Bianchi, *Lezioni di Geometria Differenziale*, Pisa 1894; German translation by Lukat, *Lehrbuch der Differentialgeometrie*, 1899.

The subject is basically local surface theory. There are beautiful spots and I wish to guide you through some of them. Needless to say, the corresponding global questions deserve study. They are interesting and are usually difficult.

1. ISOMETRY

Classically this is known as the form problem: Given

(1)
$$ds^2 = E\,du^2 + 2F\,du\,dv + G\,dv^2,$$
$$ds'^2 = E'\,du'^2 + 2F'\,du'\,dv' + G'\,dv'^2,$$

both positive definite, to decide whether there is a transformation

(2)
$$u' = u'(u, v),$$
$$v' = v'(u, v),$$

such that after substitution

(3)
$$ds'^2 = ds^2.$$

The fundamental invariant is the Gaussian curvature $K(u, v)$. They have to be equal at corresponding points:

(4)
$$K(u, v) = K'(u', v').$$

If one is a constant, the other must be the same constant. The surface then admits a three-parameter group of isometries. It is the euclidean plane, the hyperbolic plane, or the elliptic plane, according as $K = 0$, < 0, or > 0.

In the general case the main tool consists of the Beltrami differential parameters. For a function $F(u, v)$ on the surface the first Beltrami differential parameter ∇F is the square of the norm of its gradient. The second Beltrami differential parameter ΔF is its Laplacian. If $G(u, v)$ is another function, we have also the polarization $\nabla(F, G)$, so that $\nabla(F, F) = \nabla F$.

Given two invariant functions

(5)
$$\varphi(u, v) = \varphi'(u', v')$$
$$\psi(u, v) = \psi'(u', v'),$$

it follows that

(6) $\qquad \nabla \varphi = \nabla' \varphi', \quad \nabla \psi = \nabla' \psi', \quad \nabla(\varphi, \psi) = \nabla'(\varphi', \psi').$

If the functions φ, ψ are independent, so that it determines the transformation (2), equations (6) are necessary conditions for the isometry (3).

They are also sufficient. For, by the definition of the differential parameters, we have, by taking φ, ψ as parameters,

(7) $\qquad ds^2 = \dfrac{\nabla \psi \, d\varphi^2 - 2 \nabla(\varphi, \psi) \, d\varphi \, d\psi + \nabla(\varphi) \, d\psi^2}{\nabla \varphi \nabla \psi - (\nabla(\varphi, \psi))^2}.$

K being an invariant function, not a constant, we search the second invariant function from $\nabla K, \Delta K$. Hence the problem is solved, when there are two independent functions among $K, \nabla K, \Delta K$. The remaining case is when $\nabla K, \Delta K$ are functions of K, say

(8) $\qquad\qquad \nabla K = f(K), \quad \Delta K = g(K).$

For (3) to hold we must have

(9) $\qquad\qquad \nabla' K' = f(K'), \quad \Delta' K' = g(K'),$

with the same functions f, g.

This condition is sufficient. For with K and another parameter ψ we can write

(10) $\qquad f(K) \, ds^2 = dK^2 + \exp\left(2 \int \dfrac{g(K)}{f(K)} \, dK\right) d\psi^2.$

Such a ds^2 is isometric to that of a surface of revolution. We shall call it rotation-like. It admits a one-parameter group of isometries.

We notice the gap phenomenon: A ds^2 is generally rigid. It may admit a one-parameter group of isometries (rotation-like surfaces) or a three-parameter group of isometries ($K = \text{const}$), but not a two-parameter group. Such a property persists in high dimensions.

It should be interesting to study the global problem of complete rotation-like surfaces. Is it always a surface of revolution?

2. APPLICABLE SURFACES

Classically two surfaces with the same ds^2 are called applicable; in fact, one is also called a deformation of the other. Applicable surfaces may not be congruent. Their investigation is clearly an interesting and important problem.

Let the surface S be

$$(11) \qquad \vec{x}(u, v) = (x(u, v), y(u, v), z(u, v)),$$

with a ds^2 given by (1). Then the metric

$$dx^2 + dy^2 = ds^2 - dz^2$$

has Gaussian curvature zero. Expressing this fact, we get a long partial differential equation in the unknown function z, whose leading term is

$$(12) \qquad (EG - F^2)(z_{uu} z_{vv} - z_{uv}^2) + \ldots = 0.$$

Given a surface, to find another surface applicable to it thus becomes analytically the study of a Monge-Ampère equation. Its characteristics are the asymptotic curves of S. This fact is the basis of the following theorem: Let C be a curve on S. If C is not an asymptotic curve, a surface keeping C fixed and applicable to S must be S itself. If C is an asymptotic curve, there is an infinite number of surfaces through C and applicable to S.

More generally, given a curve C on S and a curve C' in space, one asks the question whether there is a surface S' through C' and applicable to S such that C goes into C'. For this to be true it is necessary that C and C' have the same geodesic curvature at corresponding points. For C this is equal to $\rho \sin \theta$, where ρ is the curvature of C (which we suppose to be ≥ 0) and θ is the angle between the principal normal of C and the surface normal of S; the same notation, with dashes, will be used for C'. It follows that C' must satisfy the condition $\rho' \geq |\rho \sin \theta|$. It can be proved that if $\rho' > |\rho \sin \theta|$ there are exactly two applicable surfaces S'

through C'. On the other hand, if $\rho' = |\rho \sin \theta|$, C' will be an asymptotic curve of S'. By a theorem of Beltrami-Enneper, its torsion is equal to $\pm \sqrt{-K}$, which is another condition to be fulfilled by C'. When C is given on S, the curve C' is then determined up to a rigid motion and there is an infinite number of surfaces S' applicable to S with C going to C'.

Bonnet, and others, studied applicable surfaces with further conditions imposed. Analytically such a problem leads to an over-determined system of partial differential equations. Bonnet proved the theorem: An isometry between two non-ruled surfaces which maps a family of asymptotic curves of one surface into the asymptotic curves of another is a rigid motion.

Two other problems of this nature are:

α) isometries preserving the lines of curvature;
β) isometries preserving the principal curvatures or the mean curvature, as the Gaussian curvature is always preserved.

The study of such problems leads to long calculations.

A more interesting question is the study of a family of applicable surfaces with the above properties (cf. [1, 2]). A family of ∞^1 surfaces is called non-trivial if it is not the orbit of one of them by a one-parameter group of rigid motions. We have the theorem: A non-trivial family of applicable surfaces preserving the lines of curvature is a family of cylindrical molding surfaces. We recall that a cylindrical molding surface is constructed as follows: Take a cylinder and a tangent plane π to it. On π take a curve C. A cylindrical molding surface is the locus of C as π rolls over the cylinder. When the cylinder is a line, the molding surface becomes a surface of revolution.

Concerning the property β) Bonnet observed that a surface of constant mean curvature can be deformed continuously in a non-trivial way. In general we have the theorem: There exist non-trivial families of applicable surfaces of non-constant mean curvature, depending on six constants, such that the mean curvature is preserved during the deformation. It should be remarked that the proofs of this theorem and the theorem in the last section involve the studies of the respective over-determined systems and their integrability conditions, which look complicated but lead to unexpected simple conclusions.

3. W-SURFACES

A Weingarten surface or a W-surface S is one which satisfies a relation between the principal curvatures:

$$(13) \qquad W(k_1, k_2) = 0.$$

Such surfaces include the minimal surfaces, the surfaces of constant mean curvature, the spherical and pseudospherical surfaces, etc.

The first properties come from the congruence of normals. The latter consists of the common tangent lines of the evolute or focal surfaces of S. In fact, let $x(u, v)$ be a point on S and let $v(u, v)$ be a unit normal vector at that point. Then the focal surface F_i, $i = 1, 2$, is the locus of the point

$$(14) \qquad y_i = x(u, v) + r_i(u, v)\, v(u, v),$$

where $r_i = 1/k_i$ is a principal radius of curvature. Weingarten proved the remarkable theorem: If S is a W-surface, F_i is applicable to a surface of revolution, whose form depends only on the relation (13). The converse of this is also true.

The normals establish a map between the two focal surfaces by mapping one focal point to the other. A congruence is called a W-congruence if this map preserves the asymptotic curves on the two focal surfaces. We have the following theorem of Ribaucour: A surface is a W-surface if and only if its normal congruence is a W-congruence.

Let K_i be the Gaussian curvature of F_i at y_i, $i = 1, 2$. For a W-surface we have the formula of Halphen:

$$(15) \qquad K_1 K_2 = (r_1 - r_2)^{-4}.$$

Sophus Lie proved another characterization of the W-surfaces: A surface is a W-surface if and only if the quadratic differential form

$$(16) \qquad \Psi = (v, dx, dv)$$

has Gaussian curvature zero. Since the equation $\Psi = 0$ defines the lines of curvature, it follows that on a W-surface these can be determined by quadratures. The same is true of the asymptotic curves. Such properties are of importance, but they have been neglected in modern works on differential geometry.

An example of a minimal surface is the catenoid. One of its focal surfaces is a surface of revolution obtained by rotating the evolute of the catenary. By Weingarten's theorem one of the focal surfaces of any minimal surface is applicable to it. Similarly, consider Beltrami's pseudosphere obtained by the rotation of a tractrix. Since the catenary is the evolute of the tractrix, it follows that one of the focal surfaces of a pseudospherical surface is applicable to a catenoid.

It should be remarked that the relation (13) is essentially a partial differential equation, generally non-linear, in two independent variables. For instance, if the surface is given as a graph $z = z(x, y)$, the condition for a minimal surface is

$$(17) \qquad (1 + z_y^2) z_{xx} - 2 z_x z_y z_{xy} + (1 + z_x^2) z_{yy} = 0$$

and the condition $K = -1$ becomes

$$(18) \qquad z_{xx} z_{yy} - z_{xy}^2 + (1 + z_x^2 + z_y^2)^2 = 0.$$

The latter is thus the equation of a pseudospherical surface. It can be put in a different form: It ψ denotes the angle between the asymptotic curves, there are asymptotic parameters u, v, such that

$$(19) \qquad \psi_{uv} = \sin \psi.$$

This is called the sine-Gordon equation. Thus the study of pseudospherical surfaces is equivalent to that of the sine-Gordon equation.

The above are remarks on some of the important local properties of W-surfaces. Their global properties, particularly those of minimal surfaces, have recently been exhaustively studied. For the global study of general W-surfaces I wish to refer to the works of H. Hopf; cf. [4].

4. W-CONGRUENCES

An important feature of euclidean geometry is the rôle played by the straight lines. Thus the study of a surface is intimately tied to that of its normal congruence. On the other hand, it is justified to study line congruences, i.e., a two-parameter family of lines, on their own right.

The first fundamental paper on line congruences was written

in 1860 by E. Kummer, the great algebraic number theorist. Let the lines be given by a point $x(u, v)$ and a direction $\xi(u, v)$, the latter being a unit vector. Kummer based his study on the two quadratic differential forms

$$(20) \qquad\qquad I' = (d\xi, d\xi), \quad II' = (dx, d\xi).$$

(We use the dashes to distinguish them from the forms in surface theory. Actually $\xi(u, v)$ defines an analogue of the Gauss map and I' is a generalization of the third fundamental form in surface theory.)

The line congruence is called isotropic, if the forms I' and II' are proportional. The notion is a generalization of the sphere (or plane) in surface theory. It has the following geometric interpretation: If the corresponding points of two applicable surfaces have a constant distance, the lines joining them form an isotropic congruence.

The line λ with the parameters u, v and a neighboring line $(u + du, v + dv)$ have a common perpendicular. Its foot as $du \to 0$, $dv \to 0$ gives a point on λ. All such points lie on a segment of λ, whose endpoints L_1, L_2 are called the limit points on λ. On the other hand, the equations $u = u(t)$, $v = v(t)$ define a ruled surface consisting of lines of the congruence. There are in general two directions when it becomes a developable, whose lines are the tangent lines of a curve. The points of contact give two points F_i, $i = 1$, 2, on λ. They are called the foci and their loci Φ_i, $i = 1$, 2, the focal surfaces. The congruence consists of the common tangent lines of its focal surfaces Φ_1 and Φ_2.

It can be proved that the foci F_i, $i = 1$, 2, belong to the segment $L_1 L_2$ and that their mid-point is the same as that of L_1 and L_2. A congruence is a normal congruence, i.e., consisting of the normal lines to a surface, if and only if the foci coincide with the respective limit points.

A congruence establishes a correspondence between the focal surfaces, sending one focal point to the other. The congruence is called a W-congruence, if the correspondence preserves the asymptotic curves, or, what is the same, the conjugate nets. An analogue of Weingarten's theorem on W-surfaces is the following: A focal surface of a W-congruence admits an infinitesimal deformation in a direction parallel to the normal at the corresponding point of the other focal surface.

Halphen's formula (15) can also be generalized: Let K_i be the Gaussian curvature of Φ_i at F_i, $i = 1, 2$. For a W-congruence we have

$$(21) \qquad\qquad K_1 K_2 = d^{-4},$$

where d is the distance between the limit points.

A W-congruence for which both distances $\overline{F_1 F_2}$ and $\overline{L_1 L_2}$ are constants is called pseudospherical. In this case Φ_i, $i = 1, 2$, is a pseudospherical surface of curvature $-d^{-2}$. The correspondence between Φ_i is called a Bäcklund transformation.

An important family of W-congruences was constructed by Darboux as follows: Consider a surface of translation

$$(22) \qquad\qquad x_i = f_i(u) + \varphi_i(v), \qquad i = 1, 2, 3.$$

The lines of intersection of the osculating planes of the generating curves form a W-congruence such that the generating curves correspond to the asymptotic curves of the focal surfaces. Moreover, if the generating curves have the constant torsions $+w$ and $-w$ respectively, the W-congruence is a normal congruence of a W-surface satisfying the relation

$$(23) \qquad\qquad k(r_2 - r_1) = \sin(k(r_1 + r_2)), \qquad k = \text{const.}$$

Its focal surfaces are applicable to a paraboloid of revolution. These Weingarten-Darboux surfaces have many interesting properties.

A W-congruence is a projective property. Its study is an important chapter in projective differential geometry.

5. Transformation of Surfaces

As remarked above, most properties of surfaces are described by partial differential equations. It is interesting, and mysterious, that the same property could be defined by equations which are very different in appearance. For example, the pseudospherical surfaces with $K = -1$ can be characterized either by Monge-Ampère equation (18) or the sine-Gordon equation (19). It is thus of clear interest to study the transformations of surfaces which preserve certain geometrical properties. We shall give some examples:

α) Bonnet's Transformation. Let S be a surface with the principal curvatures k_i and the radii of principal curvatures $r_i = 1/k_i$,

$i = 1, 2$. The principal curvatures of its parallel surface S_h at a distance h are given by

$$(24) \qquad\qquad k_i' = \frac{k_i}{1 - h k_i}, \qquad i = 1, 2.$$

From this Bonnet made the following observation:

If S has a constant Gaussian curvature $1/a^2$, its parallel surface at a distance $\pm a$ has a constant mean curvature $\mp 1/2a$. If S has a constant mean curvature $1/a$, its parallel surfaces at the distances a and $a/2$ have respectively the constant mean curvature $-1/a$ and the constant Gaussian curvature $1/a^2$. (Note: mean curvature $= (k_1 + k_2)/2$.)

Thus surfaces of constant mean curvature $\neq 0$ and surfaces of constant positive Gaussian curvature are in a sense equivalent problems.

β) ∂-TRANSFORM. When the surface S is oriented, it has a complex structure defined by the rotation of a tangent vector by 90° (multiplication by i!). This leads to the definition of the operators ∂ and $\bar{\partial}$, which are respectively the exterior differentiations with respect to the holomorphic and anti-holomorphic coordinates. In particular, ∂x, where $x(u, v)$ is the position vector, is a vectorial form of bidegree $(1, 0)$. The ratios of its components define a new complex surface, called the ∂-transform of S.

A theorem on minimal surfaces says that they can be characterized by the condition

$$(25) \qquad\qquad \partial \bar{\partial} x = 0,$$

i.e., the coordinate functions are harmonic. This is equivalent to saying that the ∂-transform is a holomorphic curve. The latter property is the main reason for the Weierstrass formulas of a minimal surface.

The notion of a ∂-transform is playing an important rôle in the study of minimal surfaces in other spaces; cf. [3].

γ) BÄCKLUND TRANSFORMATIONS. In 1883 Bäcklund proved the remarkable theorem:

Let S and S' be the focal surfaces of a pseudospherical congruence (for which the distances between the foci and the limit points are both constant). Then S and S' have the same Gaussian curvature $-d^{-2}$, where d is the distance between the limit points.

The transformation so defined between S and S' is called a Bäcklund transformation. Given a pseudospherical surface S, to construct S' it suffices to construct a vector field on S such that the tangent lines to S along the vector field form a pseudospherical congruence. This leads to a completely integrable total differential equation whose solution depends on the solution of a Riccati equation. Since a pseudospherical surface corresponds to a solution of the sine-Gordon equation (19), a Bäcklund transformation can be interpreted as transforming one solution of (19) into another. In this way new solutions of (19) are produced. The method plays an important rôle in the theory of solitons in mathematical physics.

δ) Laplace Transform ($\neq$ Laplace Transform in Harmonic Analysis). A net of curves N on a surface is called conjugate, if at every point the tangent directions to the curves of the net separate harmonically the asymptotic directions. Taking the net to be the parametric net with parameters u, v, we have, by one of the Gauss equations

(26) $x_{uv} = \Gamma_{12}^1 x_u + \Gamma_{12}^2 x_v$.

A conjugate net has the following geometrical interpretation: Take a v-curve C_v. The tangent lines of the u-curves at the points of C_v form a developable surface. On such a tangent line there is thus a point x_1 where it is tangent to the edge of regression. Reversing the rôle of u, v, we get a point x_{-1} on the tangent line of the v-curve C_v. As u, v vary, $x_1(u, v)$ and $x_{-1}(u, v)$ generally describe surfaces, which are called the Laplace transforms of the net N. The remarkable fact is that the u- and v-curves also form conjugate nets on $x_{-1}(u, v)$ and $x_1(u, v)$; we will denote them by N_{-1} and N_1 respectively. Moreover, the positive (resp. negative) Laplace transform of N_{-1} (resp. N_1) is N itself. Continuing this process, we get a Laplace sequence of conjugate nets

(27) $..., N_{-2}, N_{-1}, N_0(=N), N_1, N_2, ...,$

such that each one is the Laplace transform of the one to the left and is the negative Laplace transform of the one to the right.

A conjugate net is a projective property. For its treatment it is advantageous to use homogeneous coordinates in the three-dimensional ambient space. The homogeneous coordinates of a surface $x(u, v)$ satisfy an equation of the form

$$(28) \qquad x_{uv} + a x_u + b x_v + c x = 0,$$

if and only if the parametric net is a conjugate net. Equation (28) is called a Laplace equation. To every conjugate net is associated a Laplace equation, and vice versa. If $x(u, v)$ defines N, its Laplace transforms are given by

$$(29) \qquad x_1 = x_u + b x, \quad x_{-1} = x_v + a x.$$

To the Laplace sequence (27) corresponds a sequence of Laplace equations, and the solution of a Laplace equation reduces to the solution of one of the equations in the sequence. In particular, the last problem could become a simple one, when the corresponding surface degenerates to a curve.

REFERENCES

1. Bryant, R., Chern, S., Griffiths, P.A. (1990): Exterior differential systems. Proc. of 1980 Beijing DD-Symposium (1980), 219–338 or Bryant-Chern-Gardner-Goldschmidt-Griffiths, Exterior Differential Systems, Springer
2. Chern, S. (1989): Deformation of surfaces preserving principal curvatures. Differential geometry and complex analysis, volume in memory of H. Rauch. Springer, 1984, pp. 155–163, or Chern, selected papers, vol. 4, pp. 95–103, Springer
3. Chern, S., Wolfson, J. (1989): Harmonic maps of the two-sphere into a complex Grassmann manifold, II. Annals of Mathematics 125 (1987) 301–335 or S. Chern, selected papers, vol. 4, p. 189–223, Springer
4. Hopf, H. (1983): Differential geometry in the large, part. II. Lecture notes in mathematics, vol. 1000. Springer

Work done under partial support of NSF Grant DMS-87-01609. Research at MSRI supported in part by NSF Grant DMS-8505550.

John H. Conway · Neil J.A. Sloane

The Cell Structures of Certain Lattices

And out of the ground the Lord God formed
every beast of the field, and every fowl of the
air; and brought them into Adam to see what
he would call them: and whatsoever Adam
called every living creature, that was the name
thereof.   Genesis 2:19

Kaum nennt man die Dinge beim richtigen Namen, so verlieren sie ihren gefährlichen Zauber.
Der primitive Mensch benannte alles und jedes
falsch. Ein einziger furchtbarer Zauberbann um-
gab ihn, wo und wann war er nicht gefährdet?
Die Wissenschaft hat uns von Aberglauben und
Glauben befreit. Sie gebraucht immer die gleichen Namen, mit Vorliebe griechisch-lateinische,
und meint damit die wirklichen Dinge. Miß-
verständnisse sind unmöglich.*

Elias Canetti, Die Blendung, Hanser, Munich,
1963, p. 425

ABSTRACT. The most important lattices in Euclidean space of
dimension $n \leq 8$ are the lattices A_n ($n \geq 2$), D_n ($n \geq 4$), E_n ($n = 6$, 7,
8) and their duals. In this paper we determine the cell structures
of all these lattices and their Voronoi and Delaunay polytopes in
a uniform manner. The results for E_6^* and E_7^* simplify recent work
of Worley, and also provide what may be new space-filling polytopes in dimensions 6 and 7.

1. INTRODUCTION

The Coxeter-Dynkin diagrams of types A_n, D_n, E_6, E_7 and E_8 arise
in surprisingly different parts of mathematics – see the discussions

---

* From the English version *Auto-da-Fé* (Continuum, New York, p. 385)
as translated by C.V. Wedgwood: "You have but to know an object by
its proper name for it to lose its dangerous magic. Primitive man called
each and all by the wrong name. One single and terrible web of magic
surrounded him; where and when did he not feel threatened? Knowledge
has freed us from superstitions and beliefs. Knowledge makes use always
of the same names, preferably Graeco-Latin, and indicates by these names
actual things. Misunderstandings are impossible."

by Arnold [1] and Hazewinkel et al. [30]. In the present paper
we study the lattices associated with these diagrams (the so-called
root lattices) and their duals (the weight lattices), as well as some
of the polytopes arising from consideration of the cell structure
of these lattices. In accordance with our epigraphs we provide
names for many of these polytopes.

   The Coxeter-Dynkin diagrams provide an astonishing amount
of information about these lattices, much of which can be found
in Coxeter's book "Regular Polytopes" [20]. The lattices are discussed from the point of view of Lie algebras in references such
as Bourbaki [5] and Humphreys [33], and are also extensively
analyzed in our book [13]. In recent years they have been used
to construct modulation schemes (called trellis codes) for high-speed
transmission of digital data [7-9, 22-24].

   However, in spite of their long history, there are still new things
to be said about these lattices and their polytopes. We recently
had occasion to determine the covering multiplicity of these lattices,
that is, the maximal number of times the interiors of the covering
spheres[1] can overlap [14]. We found that to do this we needed
to understand the Voronoi and Delaunay polytopes[1] of these lattices in considerable detail. Much to our surprise, this information
does not appear to be available anywhere in the literature, and
we have decided to present it in this paper. Partial information
is available in Coxeter's works (especially [15-20]) and in our book
[13], and in a sense the present paper can be regarded as an extension to Chapter 21 of [13]. We shall offer no proofs, since the
results, although sometimes hard to discover, are not difficult to
verify once found. The results for the two hardest cases, the lattices
E_6^* and E_7^*, depend on recent work of Worley [42, 43].

2. Lattices and Polytopes

We begin with an informal discussion of some of the geometric
notions associated with a lattice. (For further information see for
example [13, 20, 36].) By an n-dimensional *lattice* Λ we mean a
discrete abelian subgroup of $\mathbb{R}^n$, consisting of all integer combina-
tions of n linearly independent vectors $v_1, \ldots, v_n \in \mathbb{R}^n$. The squared

---

[1] These terms are defined in the next section.

volume of the parallelepiped spanned by the v_i is called the *determinant* of Λ, denoted by det Λ. The *dual lattice* Λ^* is defined by

$$\Lambda^* = \{x \in \mathbb{R}^n : x \cdot v \in \mathbb{Z} \text{ for all } v \in \Lambda\},$$

and det $\Lambda^* = (\det \Lambda)^{-1}$. Many interesting lattices have the property that $\Lambda \subseteq \Lambda^*$; these are the *classically integral* lattices, and in this case the determinant of Λ equals the order of the quotient group Λ^*/Λ.

If the minimal distance between lattice points is $2r$, solid spheres of radius r drawn around the lattice points will just touch. The set of all such spheres forms an n-dimensional *sphere packing*, and r is called the *packing radius* of the lattice. The density Δ of this packing is the fraction of $\mathbb{R}^n$ occupied by these spheres, and is given by the formula

$$\Delta = \frac{V_n r^n}{\sqrt{\det \Lambda}},$$

where $V_n = \pi^{n/2}/(n/2)!$ is the volume of an n-dimensional sphere of radius 1. It is a classical problem to find the lattices that maximize Δ – we mention some of the results in Table 3 below.

If the radius of the spheres is increased (allowing them to overlap) until they just *cover* the whole space, we obtain a *sphere-covering* of $\mathbb{R}^n$. The radius R (say) of the spheres when this happens for the first time is called the *covering radius* of the lattice. The *covering density* or thickness Θ of Λ is

$$\Theta = \frac{V_n R^n}{\sqrt{\det \Lambda}}.$$

Another classical problem is to find the lattices that minimize Θ. (Except for dimensions 1, 2 and 24 the answers to the two problems

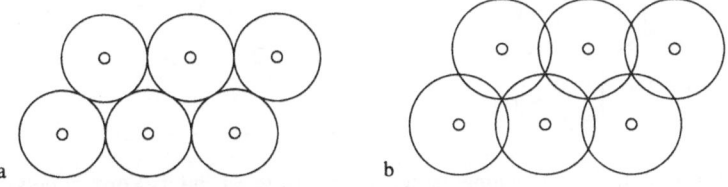

Figure 1. (a) The hexagonal lattice A_2 (small circles) and the associated packing of circles. (b) The corresponding covering

appear to be different – see [13] and Table 3.) Figure 1 shows the packing and covering obtained from the familiar planar hexagonal lattice.

The *Voronoi polytope* $V(u)$ centered at a lattice point $u \in \Lambda$ consists of the points

$$V(u) = \{x \in \mathbb{R}^n : N(x-u) \leq N(x-v), \text{ all } v \in \Lambda\},$$

where $N(x) = x \cdot x$ denotes the squared length or *norm* of $x \in \mathbb{R}^n$. The sphere of radius r centered at u (a packing sphere) is the inscribed sphere in $V(u)$, while the sphere of radius R at u (a covering sphere) is the circumsphere around $V(u)$. If the lattice points are used as codewords in a communication system, the polytopes $V(u)$ are the decoding regions: any point $x \in V(u)$ should be decoded as u.

The Voronoi polytopes $V(u)$, $u \in \Lambda$, are all congruent, and we shall usually just study $V(0)$, the polytope containing the origin. $V(0)$ is a convex centrally-symmetric polytope, and the $V(u)$, $u \in \Lambda$, form a tessellation of $\mathbb{R}^n$ by copies of $V(0)$.

The vertices of the Voronoi polytopes are especially interesting points of $\mathbb{R}^n$: they are the *holes* in the lattice, i.e. the points of $\mathbb{R}^n$ that are locally maximally distant from the lattice. In particular the vertices of $V(u)$ at distance R from u are the *deep holes* in Λ: they are the points that are globally maximally distant from the lattice.

Let $h \in \mathbb{R}^n$ be a hole in Λ. The convex hull of the lattice points closest to h is called the *Delaunay polytope* containing h. The Delaunay polytopes form a second tessellation of $\mathbb{R}^n$ into convex polytopes (dual to the Voronoi tessellation). Figure 2 shows the two tessellations in the case of the hexagonal lattice.

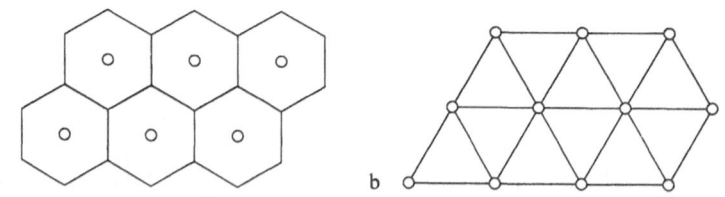

Figure 2. (a) The hexagonal lattice A_2 (small circles) and its Voronoi polygons (hexagons). (b) The corresponding Delaunay polygons (equilateral triangles in two orientations)

Another important polytope associated with a lattice is its *contact polytope:* take one of the spheres in the sphere packing, find all the points where neighboring spheres touch it, and form their convex hull.[2] The number of vertices of this contact polytope is thus the number of spheres that touch one sphere in the packing (the *kissing number* of the lattice). If the Voronoi polytope $V(0)$ has just τ walls (where τ is the kissing number), one wall bisecting the line from 0 to each of the neighboring lattice points, then the contact polytope and $V(0)$ are *dual* polytopes (cf. [29]).

The lattice itself determines an infinite polytope or *honeycomb*, in which the vertices are the lattice points, the edges join lattice points at distance $2r$ apart, etc., and the n-dimensional cells are the Delaunay polytopes.

There are many questions one can ask about lattices. In coding applications for example one wishes to maximize the probability of correct decoding, which is proportional to

$$\int_{V(0)} e^{-x \cdot x/2\sigma^2} dx$$

for a given $\sigma > 0$ (see [6]), while for applications to quantization or data compression one wishes to minimize the second moment of $V(0)$, which is proportional to

$$\int_{V(0)} x \cdot x \, dx.$$

The latter expression has been evaluated for many lattices in [12, 13, 42, 43].

A problem of recent interest (Sullivan [40]) is to find the lattices of smallest covering multiplicity. The *covering multiplicity* $CM(\Lambda)$ is the maximal number of times that the interiors of the covering spheres overlap. As illustrated in Figure 2, the covering multiplicity of the hexagonal lattice is 2. The problem is to find the minimal covering multiplicity of any n-dimensional lattice. In [14] we show that this is equal to n for all $n \leq 8$ – again see Table 3. We conjecture that it exceeds n in all other cases.

---

[2] By *the* contact n-polytope we mean the contact polytope for the densest lattice in $\mathbb{R}^n$. For $n = 1, \ldots, 8$ the contact n-polytope is respectively an interval, hexagon, cuboctahedron, regular polytope $\{3, 4, 3\}$, ambo-orthoplex, $1_{22}, 2_{31}$ and the Gosset polytope 4_{21}.

Table 1. Basic properties of root lattices and their duals

Λ	$g=g_0\cdot g_1$	$\det\Lambda$	r^2	R^2	Coset representatives for Λ^*/Λ
I_n	$2^n\cdot n!$	1	$\dfrac{1}{4}$	$\dfrac{n}{4}$	$[0]=(0^n)$
A_n	$(n+1)!\cdot 2$ $(n\geq 2)$	$n+1$	$\dfrac{1}{2}$	$\dfrac{a(n+1-a)}{n+1}$, $a=\left[\dfrac{n+1}{2}\right]$	$[i]=\left(\left(\dfrac{j}{n+1}\right)^i\left(\dfrac{-i}{n+1}\right)^j\right)$, $0\leq i\leq n,\ i+j=n+1$
A_n^*		$\dfrac{1}{n+1}$	$\dfrac{n}{4n+4}$	$\dfrac{n(n+2)}{12(n+1)}$	
D_n	$192\cdot 3!$ $(n=4)$ $2^{n-1}n!\cdot 2$ $(n\geq 5)$	4	$\dfrac{1}{2}$	$\dfrac{n}{4}\ (n\geq 4)$	$[0]=(0^n),\ [1]=\left(\dfrac{1}{2}^n\right)$ $[2]=(0^{n-1},1),$ $[3]=\left(\dfrac{1}{2}^{n-1},-\dfrac{1}{2}\right)$
D_n^*		$\dfrac{1}{4}$	1 $(n\geq 4)$	$\dfrac{n}{8}$ (n even ≥ 4) $\dfrac{2n-1}{16}$ (n odd ≥ 5)	
E_6	$2^7 3^4 5\cdot 2$	3	$\dfrac{1}{2}$	$\dfrac{4}{3}$	$[0]=(0^8),\ [2]=-[1],$ $[1]=\left(0;\dfrac{2}{3}^2,-\dfrac{1}{3}^4;0\right)$
E_6^*		$\dfrac{1}{3}$	$\dfrac{1}{3}$	$\dfrac{2}{3}$	
E_7	$2^{10} 3^4 5\,7\cdot 1$	2	$\dfrac{1}{2}$	$\dfrac{3}{2}$	$[0]=(0^8),$ $[1]=\left(\dfrac{3}{4}^2,-\dfrac{1}{4}^6\right)$
E_7^*		$\dfrac{1}{2}$	$\dfrac{3}{8}$	$\dfrac{7}{8}$	
E_8	$2^{14} 3^5 5^2 7\cdot 1$	1	$\dfrac{1}{2}$	1	$[0]=(0^8)$

In the present paper we are concerned with the lattices I_n, A_n, D_n, E_n and their duals. The lattice I_n (or $\mathbb{Z}^n$), $n \geq 1$, is the n-dimensional *cubic lattice*, and consists of all vectors $(u_1, \ldots, u_n)$ with integer coordinates u_i. This lattice is self-dual: $I_n^* = I_n$. The other lattices are defined in Sections 4–10, but for convenience we give a summary of their properties in Table 1. For each lattice Λ the table gives the order g of its point group (factorized as $g_0 \cdot g_1$, where g_0 is the order of the corresponding Weyl group), the determinant $\det \Lambda$, the squared packing and covering radii r^2 and R^2, and representative vectors for the cosets of Λ in Λ^*. All of I_n, A_n, D_n, E_n are classically integral, and we use $[i]$ to denote a minimal vector in the ith coset of Λ in Λ^*, for $i = 0, 1, \ldots, \det(\Lambda)$. Then the dual lattice is given by

$$\Lambda^* = \bigcup_{i=0}^{d-1} ([i] + \Lambda), \quad \text{where} \quad d = \det \Lambda.$$

Further information about a lattice Λ is provided by its *theta series*, given by

$$\theta_\Lambda(q) = \sum_{u \in \Lambda} q^{N(u)}, \quad |q| < 1,$$

$$= \sum_r M_r q^r,$$

where M_r is the number of lattice points of norm r. Analytical expressions (in terms of Jacobi theta functions) for the theta series of all these lattices can be found in Chapter 4 of [13]. Note that for I_n (the simplest of our lattices), the coefficient M_r is equal to the number of ways of writing r as a sum of n squares, a remark which hints at the enormous number-theoretic aspects of lattice theory. For further information see for example [10, 13, 27, 39].

For our present purposes the first few terms of the theta series will suffice: these are given in Table 2. Note in particular that the second term, $M_\mu q^\mu$ say, supplies both the minimal nonzero norm μ in the lattice, as well as the number of minimal vectors M_μ (the kissing number, i.e. the number of vertices of the contact polytope).

One of the chief reasons for our interest in these lattices is that they provide the best answers presently known (and often the best possible answers) to many common questions about lattices in low dimensions. To illustrate this Table 3 gives, in dimensions $n \leq 9$, the lattices which provide the densest packings, thinnest coverings, and have the smallest covering multiplicities. The question marks

Table 2. Theta series

$$I_n: \quad 1 + 2nq + 2^2\binom{n}{2}q^2 + 2^3\binom{n}{3}q^3 + \left(2^4\binom{n}{4} + 2n\right)q^4 + \ldots$$

$$A_n: \quad 1 + 2\binom{n+1}{2}q^2 + \binom{n+1}{4}\binom{4}{2}q^2 + \left(\binom{n+1}{6}\binom{6}{3} + 6\binom{n+1}{3}\right)q^6 + \ldots$$

$$A_n^*: \quad 1 + \sum_{i=1}^{n}\binom{n+1}{i}q^{i(n+1-i)/(n+1)} + 2\binom{n+1}{2}q^2 + \ldots$$

$$D_n: \quad 1 + 4\binom{n}{2}q^2 + \left(2^4\binom{n}{4} + 2n\right)q^4 + \left(2^6\binom{n}{6} + 3 \cdot 2^3\binom{n}{3}\right)q^6 + \ldots$$

$$D_n^*: \quad 1 + 2nq + 2^2\binom{n}{2}q^2 + 2^3\binom{n}{3}q^3 + \left(2^4\binom{n}{4} + 2n\right)q^4 + \ldots$$

$$+ 2^n q^{n/4} + 2^n n q^{(n+8)/4} + 2^n \binom{n}{2}q^{(n+16)/4} + \ldots$$

$$E_6: \quad 1 + 72q^2 + 270q^4 + 720q^6 + 936q^8 + \ldots$$
$$E_6^*: \quad 1 + 54q^{4/3} + 72q^2 + 432q^{10/3} + 270q^4 + \ldots$$
$$E_7: \quad 1 + 126q^2 + 756q^4 + 2072q^6 + 4158q^8 + \ldots$$
$$E_7^*: \quad 1 + 56q^{3/2} + 126q^2 + 576q^{7/2} + 756q^4 + \ldots$$
$$E_8: \quad 1 + 240q^2 + 2160q^4 + 6720q^6 + 17520q^8 + \ldots$$

Table 3. The optimal lattices

Dimension	1	2	3	4	5	6	7	8	9
Densest packing	I_1	A_2	A_3	D_4	D_5	E_6	E_7	E_8	Λ_9?
Thinnest covering	I_1	A_2	A_3^*	A_4^*	A_5^*	A_6^*?	A_7^*?	A_8^*?	A_9^*?
Smallest covering multiplicity and lattice	1 I_1	2 I_2, A_2	3 A_3^*	4 A_4^*, D_4	5 A_5^*	6 A_6^*, E_6^*	7 A_7^*	8 A_8^*	11? A_9^*?

indicate that these entries are only conjectured to be optimal. The results in the first row of Table 3 are primarily due to Blichfeldt (see [4, 41]), and in the second row to Ryskov and Baranovskii [37]. Λ_9 (in the first row) is a *laminated lattice* – see Chapter 6 of [13]. For the third row of Table 3 see [14]. The lattices in this row are not unique, since small perturbations usually do not change the covering multiplicity. For information about packings and coverings in higher dimensions, as well as what happens when nonlattice arrangements of spheres are permitted, the reader is referred to [13].

In later sections we shall encounter a number of different polytopes. To help distinguish them we have decided to give "English names" to many more polytopes than usual. What we call the Schläfli, Hesse and Gosset polytopes probably all first appeared (as polytopes) in Gosset [26]. However, we have used the names Schläfli and Hesse to recall well-known tactical configurations described by those authors (see Sections 9, 10). We use n-polytope to mean an n-dimensional polytope. The $(n-1)$-dimensional faces of an n-polytope are called its *cells*.

Table 4 summarizes the most important polytopes encountered.

Table 4. Summary of polytopes

Number of vertices	Dim.	Symbol	Name
$n+1$	n	n	simplex
$2n$	n	$(n-3)_{1,1}$	orthoplex
2^{n-1}	n	$1_{n-3,1}$	hemicube
16	7	–	diplo-simplex = Delaunay polytope for E_7^*
27	6	2_{21}	Schläfli polytope
			= Delaunay polytope for E_6
54	6	–	diplo-Schläfli polytope
			= Voronoi polytope for E_6
56	7	3_{21}	Hesse polytope = contact polytope of E_7^*
72	6	1_{22}	contact 6-polytope
126	7	2_{31}	contact 7-polytope
240	8	4_{21}	Gosset polytope = contact 8-polytope
576	7	1_{32}	Voronoi polytope of E_7^*
720	6	0_{221}	ambo contact 6-polytope
			= Voronoi polytope of E_6^*
2160	8	2_{41}	"deep hole polytope" of E_8
17280	8	1_{42}	"shallow hole polytope" of E_8

3. DIAGRAMS AND GRAPHS

In the 1930's Coxeter [15, 16] classified all irreducible discrete groups generated by reflections; the list is given on page 196 of [20]. (For the history of this theorem see [5, p. 237], [20, p. 209],

[21, p. 122], [34].) The groups are classified by diagrams introduced by Coxeter, now generally called Coxeter-Dynkin diagrams, which provide (among many other things) a set of defining relations for the groups.

For our application to the study of lattices we need only consider those groups on this list in which the angles between the reflecting hyperplanes are always 60° or 90°. The others either produce the same lattices or (the "non-crystallographic" groups) do not correspond to lattices at all. Thus we need only consider the groups defined by the diagrams [3] a_n $(n \geq 1)$, A_n $(n \geq 2)$, d_n $(n \geq 3)$, D_n $(n \geq 4)$, e_n $(n = 6, 7, 8)$ and E_n $(n = 6, 7, 8)$. These are shown in Figure 3.

The rules for constructing the reflection group from the Coxeter-Dynkin diagram can be found in [5, 11, 13, 20, 21, 28], etc., and we do not reproduce them here. The groups defined by the diagrams A_n, D_n, E_n are infinite and are called the *affine* (or *Euclidean*) *Weyl groups* $W(A_n)$, $W(D_n)$, $W(E_n)$. The a_n, d_n and e_n diagrams define the *finite* (or *spherical*) *Weyl groups* $W(a_n)$, $W(d_n)$, $W(e_n)$, of orders $(n+1)!$ (for a_n), $2^{n-1} n!$ (for d_n), $2^7 3^4 5$ (for e_6), $2^{10} 3^4 5 7$ (for e_7) and $2^{14} 3^5 5^3 7$ (for e_8).

The same Coxeter-Dynkin diagrams also represent simplexes (in Euclidean or spherical space) which are fundamental regions for these groups – see the references above, especially Chapters 4 and 21 of [13]. For example, Figures 5 and 13 below give fundamental simplexes for the groups $W(A_n)$ and $W(D_n)$.

By circling certain nodes, the diagrams are also used to represent polytopes (if the group is finite), or honeycombs and lattices (if the group is infinite). In this notation, used by Coxeter and others ([18; 20, p. 196; 44]) a diagram with a single circled node represents the set of images of the corresponding vertex of the fundamental simplex under the group. A diagram containing several circles represents the orbit of a suitably chosen point in the interior of the convex hull of the vertices represented by the individual circles. (In this paper we shall meet only a few instances when there is more than one circle.)

---

[3] The diagrams that we call a_n, d_n, e_n, A_n, D_n, E_n, are more usually called A_n, D_n, E_n, $\tilde{A}_n$, $\tilde{D}_n$, $\tilde{E}_n$, respectively. Since our chief interest is in the honeycombs we have preferred to adopt this slightly unusual terminology (used also in Chapter 23 of [13].)

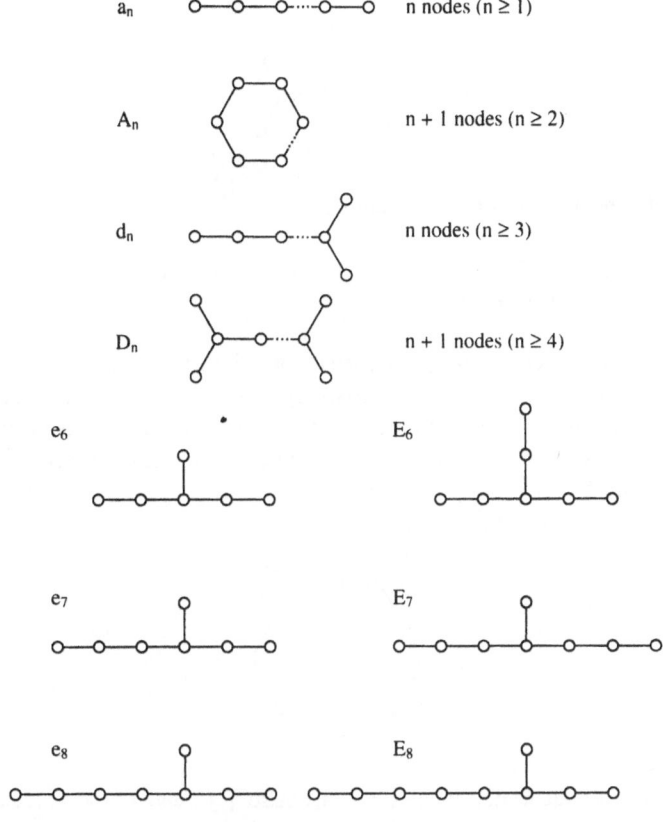

Figure 3. Coxeter-Dynkin diagrams

In particular, *the root lattices A_n, D_n, E_n are the vertices of the honeycombs obtained by circling (or starring) a single node in the A_n, D_n, E_n diagrams,* as shown in Figures 6, 14(M), 17(M), 19(M) and 23(M). The finite Weyl groups $W(a_n)$, $W(d_n)$, $W(e_n)$, when extended by the automorphism groups of the corresponding diagrams a_n, d_n, e_n, yield the point groups of the root lattices A_n, D_n, E_n, respectively (see column 2 of Table 1).

We follow Coxeter in using the symbol n_{ij} to denote the polytope or honeycomb specified by the diagram with a single circle shown in Figure 4. Then the root lattices E_6, E_7 and E_8 are the vertices of the honeycombs 5_{21}, 3_{31} and 2_{22} respectively.

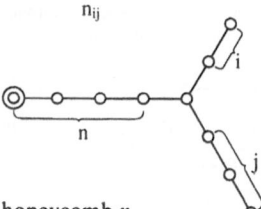

Fig. 4. The polytope or honeycomb n_{ij}

4. THE LATTICE A_n $(n \geq 2)$

The root lattice A_n $(n \geq 2)$ consists of all vectors $(u_0, u_1, \ldots, u_n)$ for which the u_i are integers satisfying $u_0 + u_1 + \ldots + u_n = 0$.[4] (Thus we are using $n + 1$ coordinates to define an n-dimensional lattice.) Coordinates for the vertices of a fundamental simplex are given in Figure 5.

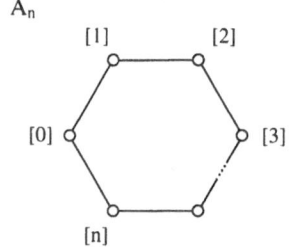

Figure 5. Fundamental simplex for A_n. Also $[i]$ denotes the particular vector

$$\left(\left(\frac{j}{n+1} \right)^i, \left(\frac{-i}{n+1} \right)^j \right), \quad \text{where } i+j=n+1$$

The lattice points are the vertices of the honeycomb represented in the circle notation by the diagram in Figure 6. That is, the lattice points consist of the set of images of the circled node of the fundamental simplex under the affine Weyl group $W(A_n)$.

By Coxeter's rule [26], [3, p. 197], the holes in this lattice, as well as the associated Delaunay polytopes, can be found as follows. The holes are the images under $W(A_n)$ of all the nodes of Figure 6

---

[4] Although this definition is also valid for $n=0$ and 1 (in particular $A_1 = \sqrt{2} I_1$), the diagrams would need special treatment in these cases.

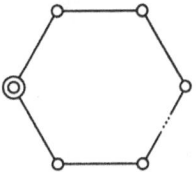

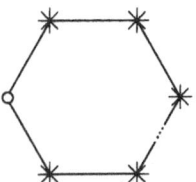

Figure 6.
Diagram for the root lattice A_n

Figure 7.
Holes in the root lattice A_n

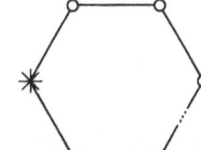

Figure 8. Star diagram
for the root lattice A_n

whose removal (together with their adjacent edges) does not disconnect any node from the circled node. All nodes except the circled one satisfy this condition, so the holes in A_n are the images under $W(A_n)$ of the vertices $[i]$, $1 \leq i \leq n$, of the fundamental simplex.

We describe the set of all these holes using a modification of the circle notation, shown in Figure 7. In this notation a diagram with a single star has the same meaning as that with a single circled node, and henceforth we shall prefer this notation. For example the points of A_n are now indicated as in Figure 8. A diagram with several stars denotes the union of the sets of points indicated by the individual stars.

The Delaunay polytope corresponding to a particular hole is obtained by deleting the corresponding node from Figure 6. So for A_n the Delaunay polytopes are as shown in Figure 9.

The vertices of a typical Delaunay polytope consist of the lattice vectors nearest to a starred vertex $[i]$ $(1 \leq i \leq n)$ in Figure 7. Equivalently, these are the points of the coset $A_n - [i]$ closest to the origin and consist of all $\binom{n+1}{i}$ permutations of

$$\left(\left(\frac{-j}{n+1} \right)^i, \left(\frac{i}{n+1} \right)^j \right), \quad j = n+1-i.$$

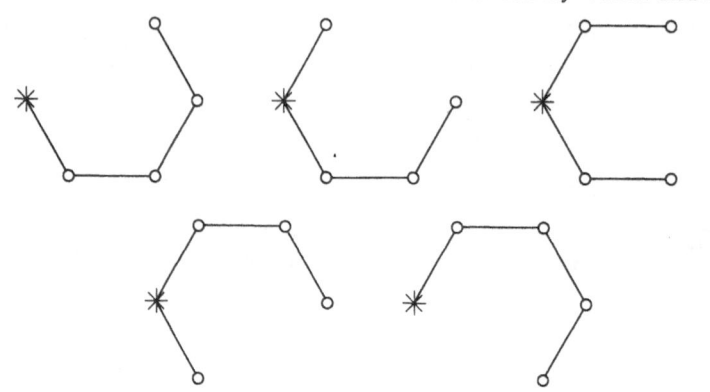

Figure 9. Delaunay polytopes of A_n

These are the midpoints of the $(i-1)$-dimensional faces of a regular simplex, so we call this Delaunay polytope an $(i-1)^{st}$-*order ambo-simplex.*[5] The 0th-order ambo-simplex is a simplex (whose vertices are the images of the vector [1] under $W(a_n)$). By *the* ambo-simplex we mean the first-order ambo-simplex, whose vertices are the midpoints of edges of a simplex (for example the vector [2] and its images). The second-order ambo-simplex is the convex hull of the midpoints of the two-dimensional faces of a simplex (for example [3] and its images), and so on. The *ambo-polytope* and *higher-order ambo-polytopes* for any sufficiently regular polytope are defined similarly. For example an ambo-tetrahedron is an octahedron and an ambo-cube is a cuboctahedron. The typical ambo-polytope of arbitrary order for a regular polytope is an intersection of suitably scaled versions of both that polytope and its dual. If we take a scaling factor between those for the tth and $(t+1)^{st}$ ambo-polytopes we obtain what we shall call the $(t+\frac{1}{2})^{th}$-order ambo-polytope.

Of course the holes (indicated in Figure 7) are also the vertices of the Voronoi tessellation corresponding to A_n. The vertices of the Voronoi cell $V(0)$ around the origin are the images of the starred nodes in Figure 7 under the finite Weyl group $W(a_n)$, consisting of all permutations of the $n+1$ coordinates. There are

---

[5] The prefix ambo- can mean either "edge" (or "rim"), as in the Greek αμβων, or "both", from the Latin *ambo*.

$$\binom{n+1}{1}+\binom{n+1}{2}+\ldots+\binom{n+1}{n}=2^{n+1}-2$$

vertices. Thus $V(0)$ is represented by the diagram shown in Figure 10. It is the convex hull of the union of ambo-simplexes of all orders.

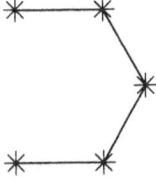

Figure 10. Voronoi cell $V(0)$ for A_n

EXAMPLES. The Delaunay polygons for the hexagonal lattice A_2 are triangles and ambo-triangles (which are inverted triangles) – see Figure 2b. The Voronoi polygon, a regular hexagon, is the convex hull of the union of two triangles – see Figure 2a. For A_3, the face-centered cubic lattice, the Delaunay polyhedra are tetrahedra, octahedra (ambo-tetrahedra), and inverted tetrahedra (second-order ambo-tetrahedra). The Voronoi polyhedron is the convex hull of the union of three such polyhedra, namely a rhombic dodecahedron with $4+6+4=14$ vertices.

The contact polytope of A_n is the convex hull of all points of the form $(1, -1, 0^{n-1})$. (This is an ambo-diplo-simplex in the notation introduced later.) This somewhat exceptional polytope can be represented by the diagram obtained by circling two ends of the a_n diagram (see Figure 11(C)).

THE CELL SCHEMATIC DIAGRAM. We summarize the analysis in this section in what we shall call the *Cell Schematic* (or CS) *diagram* for A_n, shown in Figure 11. (Later sections will be largely in the form of CS diagrams.) Such a schematic consists of a diagram for the vertices of the mother honeycomb (M), followed by a diagram (H) describing the holes and then diagrams (D) for the Delaunay polytopes corresponding to the various holes. In some cases we also give diagrams for the contact polytope (C) and the Voronoi polytope $V(0)$ (labeled V). We shall sometimes give cell schematics for certain polytopes, in which case the diagram (H) describes the points of the cells furthest from the vertices, and the diagrams labeled (D) describe the cells.

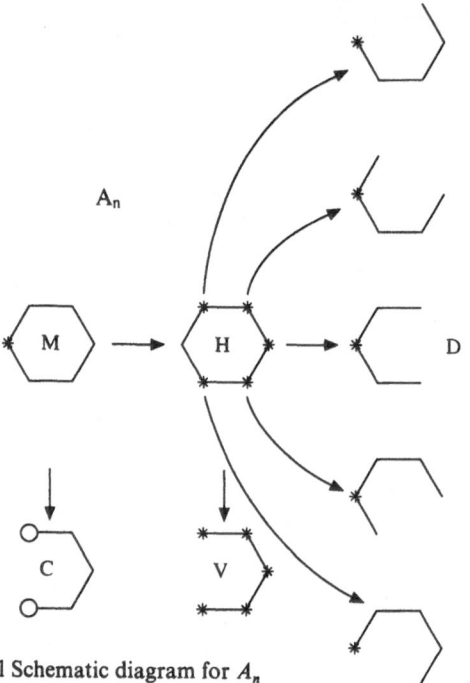

Figure 11. Cell Schematic diagram for A_n

Usually H is obtained from M by starring the nodes furthest in the graph from those starred in M. When this is the case the diagrams D are obtained from M by individually omitting the vertices that were starred in H. Again, C and V are usually obtained from M and H by omitting a vertex that was starred in M. However, these rules are only guidelines, not universal truths, and we occasionally see exceptions such as Figure 11(C). The true solution to any particular problem of this type involves finding the points of the fundamental simplex furthest from some given ones. Often these desired points will themselves be vertices of the fundamental simplex, but not always, as we shall see in the next section. We have found the computer programs MINOS [35] and AMPL [25] useful for solving particular instances of these problems (see [14]).

SUMMARY FOR A_n. *The Delaunay polytopes are ambo-simplexes of all orders, the Voronoi polytope $V(0)$ is the convex hull of the union of ambo-simplexes of all orders, and the contact polytope is an ambo-*

diplo-simplex. The squared packing and covering radii are respectively $\frac{1}{2}$ *and* $N([a]) = a(n+1-a)/(n+1)$, *where* $a = [(n+1)/2]$.

5. THE LATTICE A_n^*, $n \geq 2$

The dual lattice to A_n is the weight lattice A_n^*, consisting of all vectors $(u_0, u_1, \ldots, u_n)$ where $u_0 + u_1 + \ldots + u_n = 0$, $u_i \in \frac{1}{n+1} \mathbb{Z}$, and $u_0 \equiv u_1 \equiv \ldots \equiv u_n \pmod{1}$. A_n^* is the union of $n+1$ cosets of A_n, represented by the vectors $[i]$ shown in Figure 5. Since each of these cosets is one orbit under $W(A_n)$, A_n^* is represented by the first figure (M) in the CS diagram shown in Figure 12.

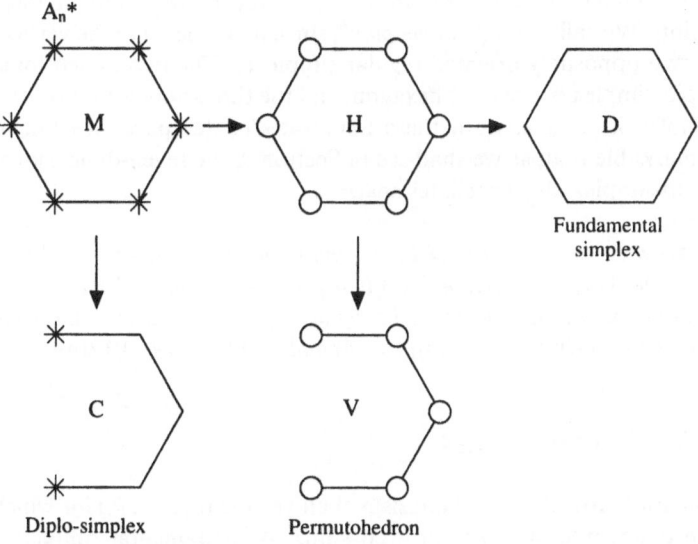

Figure 12. Cell Schematic diagram for A_n^*

In other words, A_n^* consists of the images of all vertices of the fundamental simplex for A_n under $W(A_n)$. Obviously the centroid of the fundamental simplex,

$$P = \frac{1}{n+1}\left(-\frac{n}{2}, 1-\frac{n}{2}, 2-\frac{n}{2}, \ldots, \frac{n}{2}-2, \frac{n}{2}-1, \frac{n}{2}\right),$$

is at the maximal distance from its vertices, and so this point and its images under $W(A_n)$ are the holes in A_n^*. In the circle notation this is represented by Figure 12(H).

The vertices of the Voronoi polytope $V(0)$ are the images of P under the finite group $W(a_n)$. Since this group consists of all permutations of the $n+1$ coordinates, this Voronoi polytope is sometimes called a *permutohedron* [13, p. 472].

The Delaunay polytope centered at P *is* the fundamental simplex. We adopt the convention that an uncircled and unstarred diagram (D) simply represents that simplex.

The minimal vectors of A_n^* are the $2n+2$ points of the form

$$\pm\left(\frac{n}{n+1}, \left(\frac{-1}{n+1}\right)^n\right).$$

The contact polytope (Figure 12(C)) is the convex hull of these points. We call this a *diplo-simplex*[6], since its vertices are the vertices of two oppositely oriented regular simplexes. The two-dimensional diplo-simplex is a regular hexagon, and the three-dimensional diplo-simplex is a cube. Both these figures tessellate space. We find it remarkable that, as we shall see in Section 9, the seven-dimensional diplo-simplex also tessellates space.

SUMMARY FOR A_n^*. *The Delaunay polytope is the fundamental simplex, the Voronoi polytope $V(0)$ is a permutohedron, and the contact polytope is a diplo-simplex. The squared packing and covering radii are respectively $\frac{1}{4}N([1])=n/(4n+4)$ and $N(P)=n(n+2)/12(n+1)$.*

6. THE LATTICE D_n, $n \geq 4$

The root lattice D_n, $n \geq 4$ consists of all vectors $(u_1, \ldots, u_n)$ for which the u_i are integers with an even sum.[7] A fundamental simplex is shown in Figure 13 and the CS diagram in Figure 14.

---

[6] The prefix diplo- means double. In general the vertices of the *diplo-polytope* of Π are the vertices of Π and its opposite polytope $-\Pi$.

[7] Although this definition is also valid for $n<4$ (in particular $D_1=2I_1$, $D_2\cong\sqrt{2}I_2$, $D_3\cong A_3$, where $\cong$ indicates congruent lattices, differing only by a rotation), the diagrams would need special treatment in these cases.

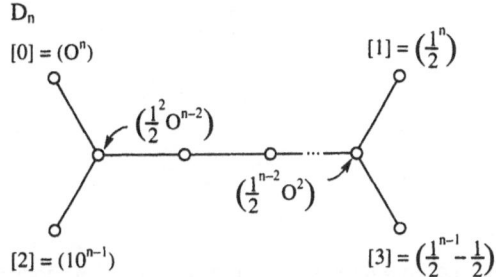

Figure 13. Fundamental simplex for D_n

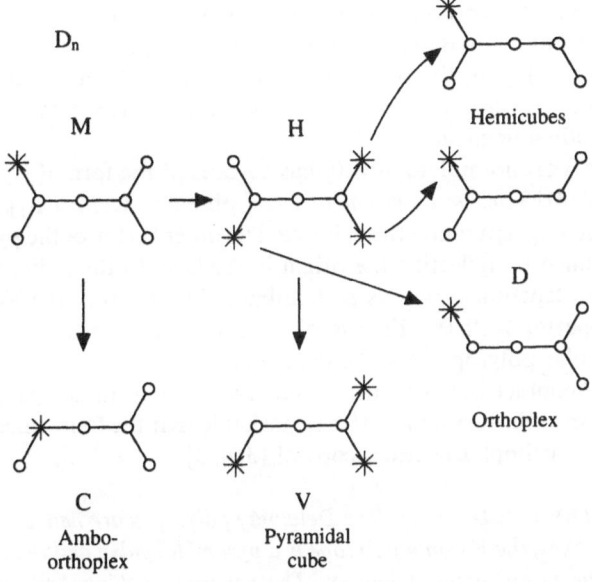

Figure 14. Cell Schematic diagram for D_n

Two of the Delaunay polytopes are *hemicubes*, consisting of alternate vertices of a cube. For example the vertices of the Delaunay polytope containing $[1] = \left(\frac{1}{2}^n\right)$ are the 2^{n-1} points of the form $(1^{2k}, 0^{n-2k})$, $0 \leq k \leq n/2$.

The vertices of the Delaunay polytope containing $[2] = (1, 0^{n-1})$ are all the points $[2] + v$, where v is one of the $2n$ vectors $(\pm 1, 0^{n-1})$. This is a regular polytope which we shall call an n-dimensional

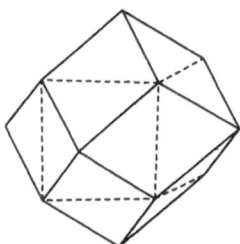

Figure 15. The Voronoi polytope for $A_3 \cong D_3$, a rhombic dodecahedron, may be obtained by attaching a pyramid to each face of a cube (dashed lines)

orthoplex[8] (abbreviating *orthant-complex*), since it has one cell for each orthant of n-dimensional space. It is represented by the diagram $(n-3)_{11}$, in the notation of Section 3. It is remarkable that the four-dimensional orthoplex is the same polytope as the four-dimensional hemicube.

The Voronoi polytope $V(0)$ has vertices of the form $((\pm \frac{1}{2})^n)$ and $(\pm 1, 0^{n-1})$. This is a *pyramidal cube*, obtained from a hypercube by attaching a pyramid to each face. The outer vertex of the pyramid is obtained by reflecting the origin in the face. In three dimensions this construction produces a rhombic dohecahedron, the Voronoi polytope for $A_3 \cong D_3$ (Figure 15). In four dimensions it produces the regular polytope $\{3, 4, 3\}$, the octaplex.

The contact polytope for D_n has $2n(n-1)$ vertices $(\pm 1^2, 0^{n-2})$, and is an ambo-orthoplex. It is remarkable that the four-dimensional ambo-orthoplex is also a copy of $\{3, 4, 3\}$.

SUMMARY FOR D_n, $n \geq 4$. *The Delaunay polytopes are hemicubes and orthoplexes, the Voronoi polytope is a pyramidal cube, and the contact polytope is an ambo-orthoplex. The squared packing and covering radii are respectively $\frac{1}{2}$ and $N([1]) = n/4$.*

7. THE LATTICE D_n^*, $n \geq 5$

The dual to the root lattice D_n is the weight lattice D_n^*, which is the union of the four cosets $D_n + [i]$, $0 \leq i \leq 3$. Since D_4^* is geometrically similar to D_4, we can suppose $n \geq 5$, and so avoid certain

---

[8] Other names are *generalized octahedron* or *cross-polytope*.

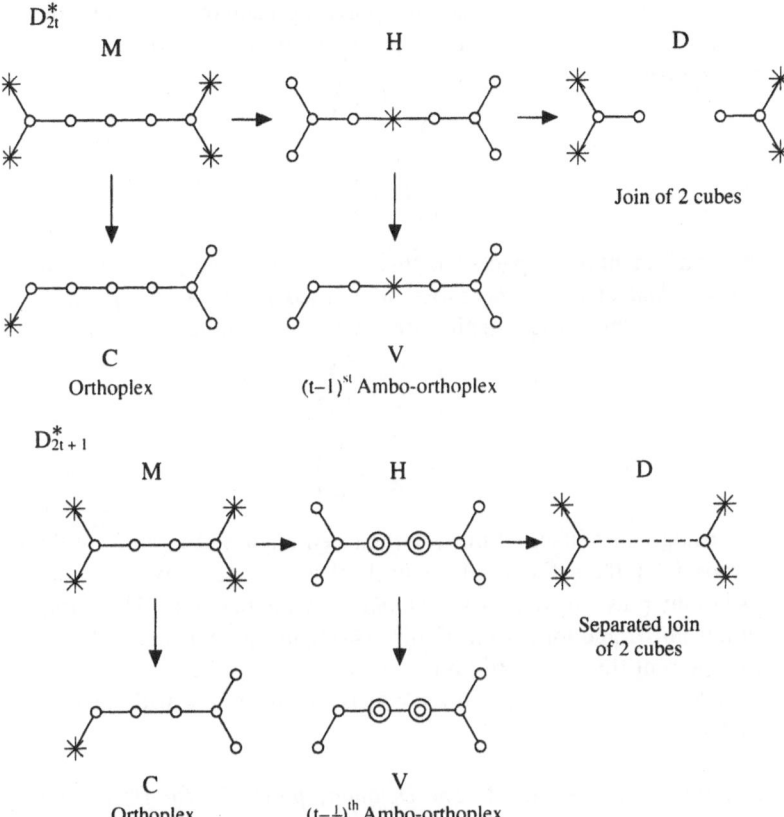

Figure 16. Cell Schematic diagram for D_n^*, $n \geq 5$. (Note that $D_4^* \cong D_4$)

modifications that would be required for $n \leq 4$. The CS diagram is given in Figure 16.

The holes in D_n^* are the images under $W(D_n)$ of that point P in the fundamental simplex which is at the maximal distance from the four vertices [0], ..., [3]. If $n = 2t$ is even, $P = \left(\dfrac{1^t}{2}, 0^t\right)$ is itself a vertex corresponding to the central node of the diagram. But if $n = 2t + 1$ is odd, $P = \left(\dfrac{1^t}{2}, \dfrac{1}{4}, 0^t\right)$ is the point midway between the two vertices represented by the middle nodes of the diagram. In either case this is a $\frac{1}{2}(n-2)^{\text{th}}$ order ambo-orthoplex.

The vertices of the Delaunay polytope centered at P are the lattice points nearest P. If $n = 2t$ these form two t-dimensional hypercubes

$$P + \left(\pm \frac{1^t}{2}, 0^t \right),$$

$$P + \left(0^t, \pm \frac{1^t}{2} \right),$$

in complementary t-spaces, so this Delaunay polytope is what we call the *join* of two hypercubes, illustrated for $t = 4$ in Figure 16. If $n = 2t + 1$ the vertices again form two t-dimensional hypercubes

$$P + \left(\pm \frac{1^t}{2}, 0, 0^t \right),$$

$$P + \left(0^t, \frac{1}{4}, \pm \frac{1^t}{2} \right),$$

in orthogonal t-spaces, but now their centers are separated by the vector $(0^t, \frac{1}{4}, 0^t)$ orthogonal to both these t-spaces. We call this Delaunay polytope the *separated join* of two hypercubes. This situation is illustrated for $t = 3$ in Figure 16(D), using an *ad hoc* notation to represent the separated join.

The contact polytope (for $n \geq 5$) is an orthoplex with vertices $(\pm 1, 0^{n-1})$.

SUMMARY FOR D_n^*, $n > 4$. *The Delaunay polytopes for D_n^* are the joins or separated joins of two $\left[\frac{n}{2} \right]$-dimensional hypercubes, according as n is even or odd. The contact polytope is an orthoplex, and the Voronoi polytope is a $\frac{1}{2}(n-2)^{\text{th}}$ order ambo-simplex. The squared packing and covering radii are respectively $\frac{1}{4} N([2]) = 1$ and $N(P) = \frac{n}{8}$ (n even) or $\frac{2n-1}{16}$ (n odd).*

8. THE LATTICE E_8 AND THE GOSSET POLYTOPE 4_{21}

The 8-dimensional root lattice E_8 consists of all vectors $(u_0, \ldots, u_7)$ for which the u_i are all in $\mathbb{Z}$ or all in $\mathbb{Z} + \frac{1}{2}$ and are such that $u_0 + \ldots + u_7$ is even. E_8 is self-dual, so that the weight lattice of

E_8

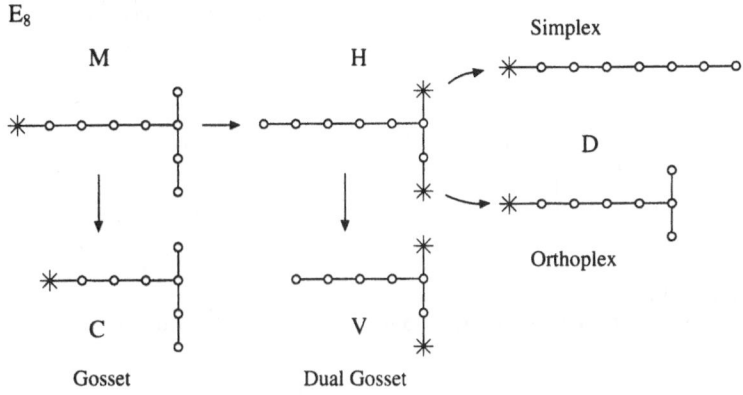

Figure 17. Cell Schematic diagram for E_8

E_8 is the same lattice. To save space we shall not give coordinates for the fundamental simplexes for E_6, E_7 or E_8; they can be found on page 461 of [13].

The CS diagram is given in Figure 17. There are two types of holes in E_8, deep holes such as $(0^7, 1)$ and shallow holes such as $((\frac{1}{3})^7, -\frac{1}{3})$.

The vertices of the Delaunay polytope centered at the hole $P = (0^7, 1)$ are all the points $P + (\pm 1, 0^7)$, and form an orthoplex. The other type of Delaunay polytope is a simplex, which is more easily described in an alternative coordinate system for E_8 obtained by negating the final coordinate.[9] In these coordinates the hole is $Q = \left(\frac{1}{3}^8\right)$, and the nearest lattice points are the origin and the eight vertices of shape $(-\frac{1}{3}, (\frac{1}{3})^7)$; the convex hull of these nine points is a regular simplex.

Note that P is $\frac{1}{2}$ of a vector of norm 4 in E_8 and Q is $\frac{1}{3}$ of a primitive[10] vector of norm 8. The Weyl group $W(e_8)$ is transitive on these two types of vectors. Thus the vertices of the Voronoi polytope $V(0)$ are $\frac{1}{2}v_4$ and $\frac{1}{3}v_8$, where $v_4 \in E_8$ is any norm 4 vector and v_8 is any primitive norm 8 vector. The images of these vectors

---

[9] This is the *odd coordinate system* for E_8 [13, p. 120].
[10] I.e. not the double of a norm 2 vector.

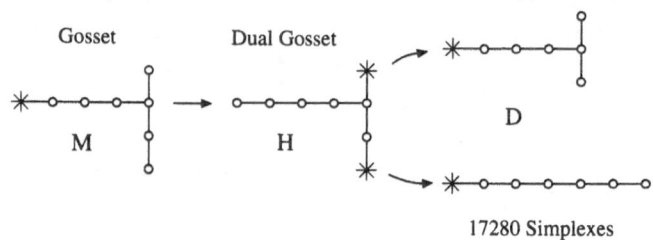

Figure 18.
Cell Schematic diagram for the 8-dimensional Gosset polytope 4_{21}

under $W(E_8)$ are all the vertices of the Voronoi tessellation, i.e. the holes in the lattice.

The contact 8-polytope (the contact polytope of E_8) is the vertex figure[11] 4_{21} of the honeycomb 5_{21} (see Figure 17(C)). We call this the *Gosset polytope*, after Gosset [29]. Apart from a scale factor it is the convex hull of the 240 minimal vectors of E_8. Its CS diagram is given in Figure 18. The numbers of cells of each type are easily found by the method given in [3, § 11.8]. Further calculations of the same type (see [3, p. 204]) give the numbers of faces of every dimension; these are shown in Table 5.

Table 5. d-dimensional faces of Gosset polytope 4_{21}

d	Face	Number
8	Gosset 4_{21}	1
7	7-orthoplex	2160
	7-simplex	17280
6	6-simplex	$69120 + 138240$
5	5-simplex	483840
4	4-simplex	483840
3	tetrahedron	241920
2	triangle	60480
1	edge	6720
0	vertex	240

[11] For definition see [3, p. 128].

Table 6. Sections of Gosset polytope 4_{21}

$\dfrac{1^8}{2}$	1	zenith
$\dfrac{1^6}{2} - \dfrac{1^2}{2}$ and $1^2 0^6$	56	Hesse polytope
$\dfrac{1^4}{2} - \dfrac{1^4}{2}$ and $1 - 1 0^6$	126	contact 7-polytope
$\dfrac{1^2}{2} - \dfrac{1^6}{2}$ and $-1^2 0^6$	56	Hesse polytope
$-\dfrac{1^8}{2}$	1	nadir
$\dfrac{1^7}{2} - \dfrac{1}{2}$	8	simplex
$1^2 0^6$	28	ambo-simplex
$\dfrac{1^5}{2} - \dfrac{1^3}{2}$	56	2nd ambo-simplex
$1 - 1 0^6$	56	ambo-diplo-simplex
$\dfrac{1^3}{2} - \dfrac{1^5}{2}$	56	2nd ambo-simplex
$-1^2 0^6$	28	ambo-simplex
$\dfrac{1}{2} - \dfrac{1^7}{2}$	8	simplex
$1; \pm 1 0^6$	14	orthoplex
$\dfrac{1}{2}; \dfrac{1^{7+}}{2}$	64	hemicube
$0; \pm 1^2 0^5$	84	ambo-orthoplex
$-\dfrac{1}{2}; \dfrac{1^{7-}}{2}$	64	hemicube
$-1; \pm 1 0^6$	14	orthoplex

Table 6 shows how the 240 vertices are distributed among the sections, starting at a vertex, simplex or orthoplex. It is convenient to use different coordinate systems for the different sections. In this table $\left(\dfrac{1^{n+}}{2}\right)$ indicates a vector $(\pm\frac{1}{2}, \ldots, \pm\frac{1}{2})$, with n components, the product of whose signs is positive. Similarly for $\left(\dfrac{1^{n-}}{2}\right)$, but now the product of the signs must be negative.

SUMMARY FOR E_8. *The holes are the images of $\frac{1}{2}v_4$ and $\frac{1}{3}v_8$ under
$W(E_8)$, where $v_4 \in E_8$ and $v_8 \in E_8$ are vectors of norm 4 and primitive
vectors of norm 8 respectively. The vertices of the Voronoi polytope
$V(0)$ are the images of $\frac{1}{2}v_4$ and $\frac{1}{3}v_8$ under $W(e_8)$ and form a polytope
dual to the Gosset polytope. The corresponding Delaunay polytopes
are an orthoplex and an 8-simplex. The contact 8-polytope is the
Gosset polytope 4_{21}. The squared packing and covering radii are
respectively $\frac{1}{2}$ and 1.*

9. THE LATTICES E_7, E_7^* AND THE HESSE POLYTOPE 3_{21}

The 7-dimensional root lattice E_7 is best defined to consist of those
vectors of E_8 that are perpendicular to a given minimal vector
$v \in E_8$. According as we take v to be

$$\left(\frac{1}{2}\right)^8, \quad (1, -1, 0^6), \quad \text{or} \quad (1, 1, 0^6),$$

we get three different coordinate systems for E_7: E_7 consists of
those vectors $(u_0, \ldots, u_7) \in E_8$ that satisfy

$$\Sigma u_i = 0, \quad u_1 = u_2, \quad \text{or} \quad u_1 + u_2 = 0,$$

respectively. The CS diagram is given in Figure 19.

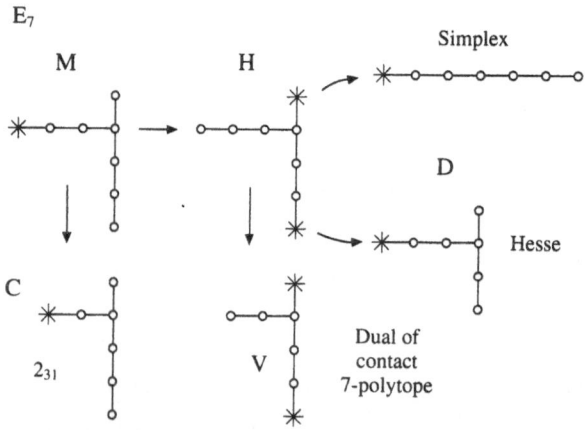

Figure 19. Cell Schematic diagram for E_7

There are two types of holes in E_7, deep holes such as $P = \left(\frac{3^2}{4}, -\frac{1^6}{4}\right)$ and shallow holes such as $Q = \left(\frac{7}{8}, -\frac{1^7}{8}\right)$, in the first coordinate system. The Delaunay polytope centered at Q has eight vertices, (0^8) and $(1; -1\,0^6)$, and is a simplex.

The Delaunay polytope 3_{21} centered at P has 56 vertices, consisting of all points of the form

$$P \pm \left(\frac{3^2}{4}, -\frac{1^6}{4}\right).$$

We call this polytope 3_{21} (discussed by Gosset) the *Hesse polytope*, because its 28 diameters have the same symmetries as the configuration of 28 bitangents to the general plane quartic curve studied by Hesse [32] in 1855. Since Hesse and Cayley these bitangents have been labeled by unordered pairs from a set of eight objects, and the Hesse group $(Sp_6(2) \cong O_7(2))$ consists of all permutations of the 28 bitangents that preserve the 315 quadruples such as $\{ab, bc, cd, da\}$ and $\{ab, cd, ef, gh\}$, where $a, \ldots, h$ are distinct. The bitangent ij in this notation corresponds to the diameter

$$\pm \left(\frac{3^2}{4}, -\frac{1^6}{4}\right)$$

with the $\frac{3}{4}$ in the ith and j^{th} coordinates. Since the 28 points of shape $\left(\frac{3^2}{4}, -\frac{1^6}{4}\right)$ form an ambo-simplex, the Hesse polytope might also be called a diplo-ambo-simplex, but this would be misleading, since the polytope has much more symmetry than this name would suggest. The Hesse group is described for example in [11, p. 46].

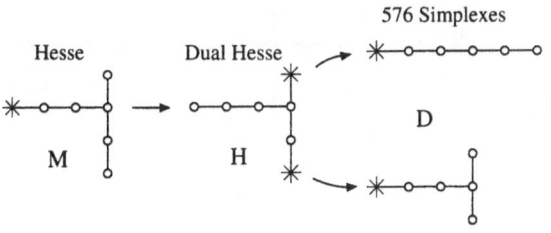

Figure 20. Cell Schematic diagram for Hesse polytope 3_{21}

See also [2, 15, 26]. The CS diagram for the Hesse polytope is given in Figure 20, and the numbers of faces of each dimension are given in Table 7 (after Coxeter [15, p. 7]).

The contact polytope for E_7 (i.e. the contact 7-polytope, see Figure 19(C)) is the vertex figure 2_{31} of the honeycomb 3_{31}, and is described in Figure 21. Its 126 vertices are the centers of the orthoplex cells of the Hesse polytope.

Tables 8 and 9 show the vertices of the Hesse and contact 7-polytope distributed by sections in the three appropriate directions.

The weight lattice E_7^* is the union of the two cosets E_7 and $E_7 + [1]$, where (in the first coordinate system) $[1] = \left(\dfrac{3^2}{4}, -\dfrac{1^6}{4}\right)$. Its CS diagram is shown in Figure 22, and E_7^* itself is shown in Figure 22(M).

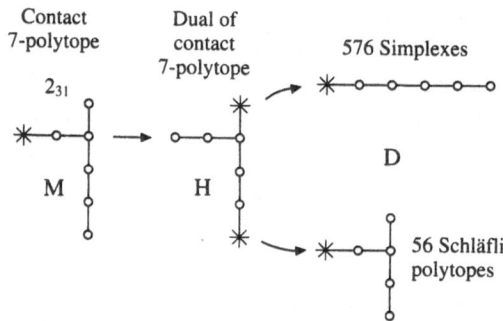

Figure 21. Cell Schematic diagram for contact 7-polytope

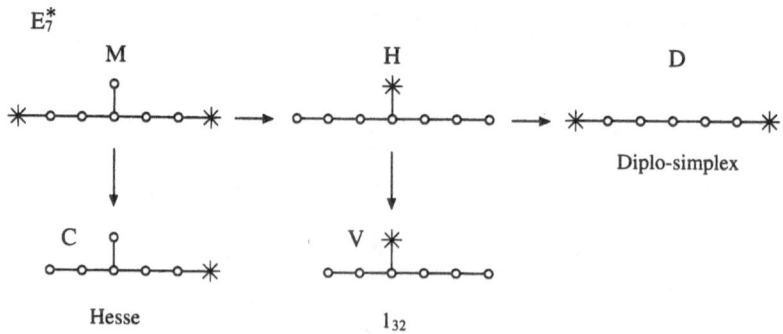

Figure 22. Cell Schematic diagram for E_7^*

Table 7. d-dimensional faces of Hesse polytope 3_{21}

d	Face	Number
7	Hesse 3_{21}	1
6	6-orthoplex	126
	6-simplex	576
5	5-simplex	$2016 + 4032$
4	4-simplex	12096
3	tetrahedron	10080
2	triangle	4032
1	edge	756
0	vertex	56

Table 8. Sections of Hesse polytope 3_{21}

Coordinates	Number	Shape
$\frac{3}{4}; \frac{3}{4} -\frac{1}{4}^{6}$	7	simplex
$\frac{1}{4}; -\frac{3}{4}^{2} \frac{1}{4}^{5}$	21	ambo-simplex
$-\frac{1}{4}; \frac{3}{4}^{2} -\frac{1}{4}^{5}$	21	ambo-simplex
$-\frac{3}{4}; -\frac{3}{4} \frac{1}{4}^{6}$	7	simplex
$\frac{1}{2}^{2}; \pm 1\,0^{5}$	12	orthoplex
$0^{2}; \frac{1}{2}^{6-}$	32	hemicube
$-\frac{1}{2}^{2}; \pm 1\,0^{5}$	12	orthoplex
$0^{2}; \frac{1}{2}^{6}$	1	zenith
$\frac{1}{2}^{2+}; 1\,0^{5}$ and $0^{2}; \frac{1}{2}^{4} -\frac{1}{2}^{2}$	27	Schäfli polytope
$\frac{1}{2}^{2+}; -1\,0^{5}$ and $0^{2}; -\frac{1}{2}^{4} \frac{1}{2}^{2}$	27	Schläfli polytope
$0^{2}; -\frac{1}{2}^{6}$	1	nadir

Table 9. Sections of contact 7-polytope

Coordinates	Number	Shape
$1; -1\,0^6$	7	simplex
$\dfrac{1}{2}; \dfrac{1}{2}^3 -\dfrac{1}{2}^4$	35	2nd ambo-simplex
$0; 1-1\,0^5$	42	ambo-diplo-simplex
$-\dfrac{1}{2}; -\dfrac{1}{2}^3 \dfrac{1}{2}^4$	35	2nd ambo-simplex
$-1; 1\,0^6$	7	simplex
$1^2; 0^6$	1	zenith
$\dfrac{1}{2}^2; \dfrac{1}{2}^{6+}$	32	hemicube
$0^2; \pm 1^2\,0^4$	60	ambo-orthoplex
$-\dfrac{1}{2}^2; \dfrac{1}{2}^{6+}$	32	hemicube
$-1^2; 0^6$	1	nadir
$\dfrac{1}{2}^{2-}; \dfrac{1}{2}^5 -\dfrac{1}{2}$ and $0^2; 1^2\,0^4$	27	Schläfli polytope
$1^{2-}; 0^6$ and $\dfrac{1}{2}^{2-}; \dfrac{1}{2}^3 -\dfrac{1}{2}^3$ and $0^2; 1-1\,0^4$	72	contact 6-polytope
$\dfrac{1}{2}^{2-}; -\dfrac{1}{2}^5 \dfrac{1}{2}$ and $0^2; -1^2\,0^4$	27	Schläfli polytope

Worley [43] showed that the unique point R of the fundamental simplex most distant from the starred vertices in Figure 22(M) is the starred vertex in Figure 22(H), so that the honeycomb 1_{33} symbolized by this figure is actually the Voronoi tessellation for E_7^*. The Delaunay polytope, Figure 22(D), is a diplo-simplex.

SUMMARY FOR E_7. *The holes are the images of vectors $v_{3/2}$ and $\frac{1}{2} v_{7/2}$ under $W(E_7)$, where $v_{3/2}$, $v_{7/2}$ are vectors of norms $\frac{3}{2}$, $\frac{7}{2}$ in E_7^*. The vertices of the Voronoi polytope $V(0)$ are the images of $v_{3/2}$ and $\frac{1}{2} v_{7/2}$ under $W(e_7)$. The corresponding Delaunay polytopes are the Hesse polytope 3_{21} and a 7-simplex. The contact 7-polytope* is a 2_{31}. *The squared packing and covering radii are respectively $\frac{1}{2}$ and $N(P)=\frac{3}{2}$.*

SUMMARY FOR E_7^*. *The holes are the images of $\frac{1}{2}v_{7/2}$ under $W(E_7)$, and consist of the vertices of the honeycomb 1_{33}. The vertices of the Voronoi polytope $V(0)$ are the images of $\frac{1}{2}v_{7/2}$ under $W(e_7)$ and form the polytope 1_{32}. The Delaunay polytope is a diplo-simplex and the contact polytope for E_7^* is a Hesse polytope 3_{21}. The squared packing and covering radii are respectively $\frac{1}{4}N([1])=\frac{3}{8}$ and $N(R)=\frac{7}{8}$.*

10. THE LATTICES E_6, E_6^* AND THE SCHLÄFLI POLYTOPE 2_{21}

The 6-dimensional root lattice E_6 is best defined to consist of those vectors of E_8 that are perpendicular to a given A_2 sublattice, spanned by a pair of minimal vectors $v, w \in E_8$ with $v \cdot w = -1$. According as we take v and w to be

$$(1,0^6,1) \text{ and } \left(-\frac{1^8}{2}\right), \quad \text{or} \quad (1,-1,0^6) \text{ and } (0,1,-1,0^5)$$

we get two different coordinate systems for E_6: E_6 consists of the vectors $(u_0, \ldots, u_7) \in E_8$ that satisfy

$$u_0 + u_7 = u_1 + \ldots + u_6 = 0, \quad \text{or} \quad u_0 = u_1 = u_2,$$

respectively. The CS diagram is given in Figure 23.

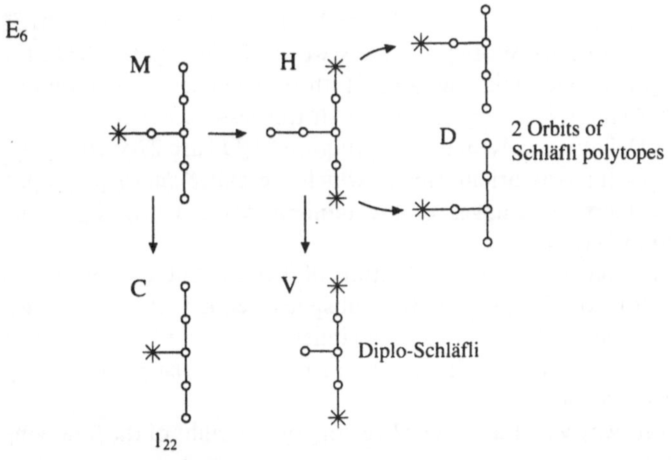

Figure 23. Cell Schematic diagram for E_6

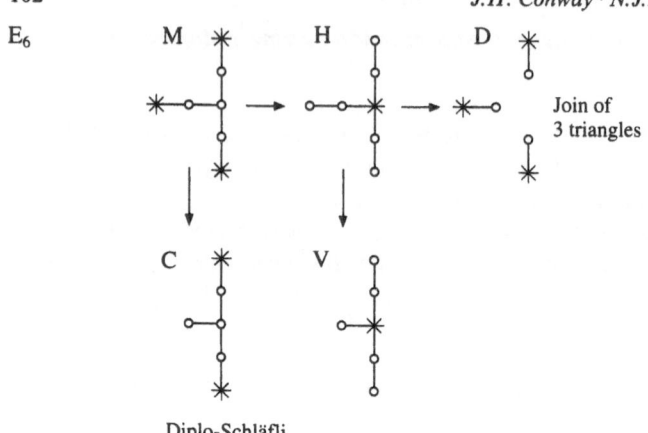

Figure 24. Cell Schematic diagram for E_6^*

The weight lattice E_6^* is the union of the three cosets E_6, $E_6 + [1]$, $E_6 + [2]$, where (in the first coordinate system)

$$[1] = \left(0; \frac{2^2}{3}, -\frac{1^4}{3}; 0\right), \quad [2] = -[1].$$

The CS diagram is given in Figure 24. The covering radius of E_6^* was first found by Worley [42]. As we see from Figure 24(D), the Delaunay polytope is the join of three triangles in orthogonal planes. Again it is a surprise to us that this tessellates space.

The Delaunay polytopes of E_6 (Figure 23(D)) are 27-vertex polytopes 2_{21} (in two orientations), which we call *Schläfli polytopes*. Their centers are images of the minimal vectors $v_{4/3} \in E_6^*$ under the group $W(E_6)$.

If we take the second definition of E_6, so that E_6 consists of the vectors of E_8 lying in the subspace defined by $u_0 = u_1 = u_2$, then any vector of E_8 has the same inner product with these vectors as its projections onto this space. In particular, *the projections of E_8 vectors lie in E_6^*.*

In this way we obtain the 27 vectors on the right of the following display:

(1)
$$(1\ 1\ 0; 0^5) \rightarrow \left(\frac{2}{3}\ \frac{2}{3}\ \frac{2}{3}; 0^5\right)$$

(16)
$$\left(\frac{1}{2}\ \frac{1}{2}\ -\frac{1}{2}; \frac{1}{2}^{5-}\right) \rightarrow \left(\frac{1}{6}\ \frac{1}{6}\ \frac{1}{6}; \frac{1}{2}^{5-}\right)$$

(10)
$$(-1\ 0\ 0; \pm1\ 0^4) \rightarrow \left(-\frac{1}{3}\ -\frac{1}{3}\ -\frac{1}{3}; \pm1\ 0^4\right).$$

These are all the shortest vectors from one coset of E_6 in E_6^*, and form a Schläfli polytope. The contact polytope for E_6^* consists of these vectors and their negatives, and so is a diplo-Schläfli polytope.

The reason for calling 2_{21} the Schläfli polytope is best seen using instead the first definition of E_6. The 27 vectors become six vectors

$$a_i = \left(\frac{1}{2}; \frac{5}{6}, -\frac{1}{6}^5; -\frac{1}{2}\right),$$

with the $\frac{5}{6}$ in position i, $1 \le i \le 6$; six vectors

$$b_j = \left(-\frac{1}{2}; -\frac{1}{6}^5, \frac{5}{6}; \frac{1}{2}\right),$$

with the $\frac{5}{6}$ in position j, $1 \le j \le 6$; and fifteen vectors

$$c_{ij} = \left(0; -\frac{2}{3}\ \frac{1}{3}^4\ -\frac{2}{3}; 0\right),$$

with the $-\frac{2}{3}$ in positions i and j, $1 \le i < j \le 6$.

These are 45 triples of these vectors that add to 0:

$$a_i + b_j + c_{ij} = 0, \quad c_{ij} + c_{kl} + c_{mn} = 0,$$

where $\{i, j, k, l, m, n\} = \{1, 2, \ldots, 6\}$, and the Weyl group $W(e_6)$ achieves all permutations of the 27 vectors that preserve this set of 45 triples. Now there are 27 lines in the general cubic surface in projective 3-space, that lie by 3's in 45 tritangent planes, for which Schläfli's notation is as above. It follows that $W(e_6)$ is isomorphic to the famous group of permutations of the 27 lines that preserve the 45 tritangent planes. (For additional information see [2, 3], [6, Appendix H], [11, p. 26], [15, 19, 31, 38, 43]).

The Schläfli polytope and the contact 6-polytope are further described in Figures 25, 26 and Tables 10, 11.

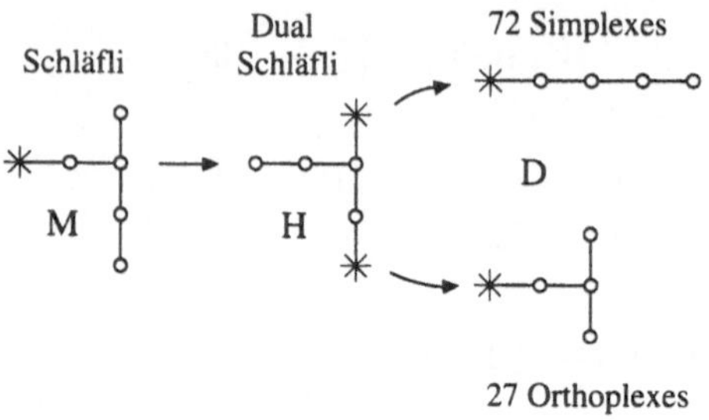

Dual   72 Simplexes
Schläfli   Schläfli

M   H

D

27 Orthoplexes

Figure 25. Cell Schematic diagram for Schläfli polytope 2_{21}

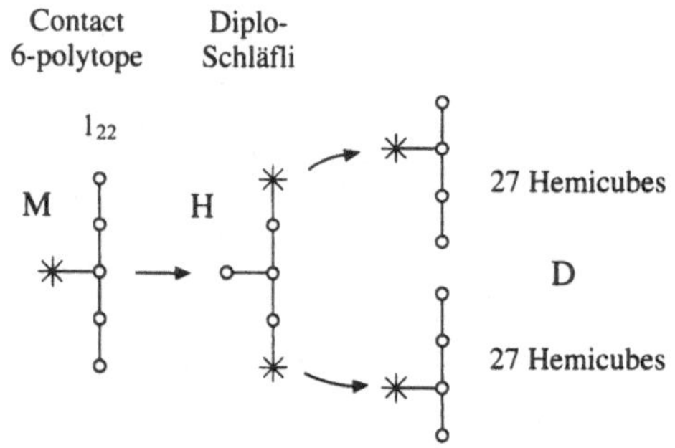

Contact   Diplo-
6-polytope   Schläfli

1_{22}

M   H   27 Hemicubes

D

27 Hemicubes

Figure 26. Cell Schematic diagram for contact 6-polytope

SUMMARY FOR E_6. *The holes are the images of* $v_{4/3} \in E_6^*$ *under* $W(E_6)$. *The vertices of the Voronoi polytope* $V(0)$ *are the images of* $v_{4/3}$ *under* $W(e_6)$, *and form a diplo-Schläfli polytope. The Delaunay polytopes are Schläfli polytopes* 2_{21}, *and the contact 6-polytope is a* 1_{22}. *The squared packing and covering radii are respectively* $\frac{1}{2}$ *and* $\frac{4}{3}$.

SUMMARY FOR E_6^*. *The holes are the images of* $\frac{1}{3}v_6$, $v_6 \in E_6$, *under* $W(E_6)$. *The vertices of the Voronoi polytope* $V(0)$ *are the images of* $\frac{1}{3}v_6$ *under* $W(e_6)$ *and form an ambo-1_{22}. The Delaunay polytope is the join of three triangles in orthogonal planes, and the contact polytope for* E_6^* *is a diplo-Schläfli polytope. The squared packing and covering radii are respectively* $\frac{1}{3}$ *and* $\frac{2}{3}$.

Table 10. d-dimensional faces of Schläfli polytope 2_{21}

d	Face	Number
6	Schläfli 2_{21}	1
5	5-orthoplex	27
	5-simplex	72
4	4-simplex	$216+432$
3	tetrahedron	1080
2	triangle	720
1	edge	216
0	vertex	27

Table 11. Sections of Schläfli polytope 2_{21}

Coordinates	Number	Shape
$\frac{2^3}{3}; 0^5$	1	apex
$\frac{1^3}{6}; \frac{1^{5-}}{2}$	16	hemicube
$-\frac{1^3}{3}; \pm 1\,0^4$	10	orthoplex
$a_i = \frac{1}{2}; \frac{5}{6}, -\frac{1^5}{6}; -\frac{1}{2}$	6	simplex $(1 \leq i \leq 6)$
$c_{ij} = 0; -\frac{2^2}{3}\frac{1^4}{3}; 0$	15	ambo-simplex $(1 \leq i < j \leq 6)$
$b_i = -\frac{1}{2}; \frac{5}{6}, -\frac{1^5}{6}; \frac{1}{2}$	6	simplex $(1 \leq i \leq 6)$

REFERENCES

1. Arnold, V.I. (1976): The $A-D-E$ classifications, Problems of present day mathematics no VIII. In: Browder, F.E. (ed.) Mathematical developments arising from Hilbert's problems (Proc. Symp. Pure. Math., vol. 28. Amer. Math. Soc., p. 46)
2. Baker, H.F. (1940): Principles of geometry, vol. IV. Cambridge Univ. Press, p. 105
3. Baker, H.F. (1946): A locus with 25920 linear self-transformations, Cambridge tracts in mathematics 39. Cambridge Univ. Press

4. Blichfeldt, H.F. (1935): The minimum values of positive quadratic forms in six, seven and eight variables. Math. Zeit. 39:1–15

5. Bourbaki, N. (1968): Groupes et algèbres de Lie, chapitres 4, 5 et 6. Hermann, Paris

6. Burnside, W. (1955): Theory of groups of finite order, 2nd edn. Cambridge Univ. Press, 1911. Dover, NY

7. Calderbank, A.R., Sloane, N.J.A. (1985): Four-dimensional modulation with an eight-state trellis code. AT&T Tech. J. 64:1005–1018

8. Calderbank, A.R., Sloane, N.J.A. (1986): An eight-dimensional trellis code. Proc. IEEE 74:757–759

9. Calderbank, A.R., Sloane, N.J.A. (1987): New trellis codes based on lattices and cosets. IEEE Trans. Information Theory 33:177–195

10. Cassels, J.W.S. (1978): Rational quadratic forms. Academic Press, NY

11. Conway, J.H., Curtis, R.T., Norton, S.P., Parker, R.A., Wilson, R.A. (1985): An ATLAS of finite groups. Oxford Univ. Press

12. Conway, J.H., Sloane, N.J.A. (1984): On the Voronoi regions of certain lattices. SIAM J. Algebraic and Discrete Methods 5:294–305

13. Conway, J.H., Sloane, N.J.A. (1988): Sphere packings, lattices and groups. Springer, NY

14. Conway, J.H., Sloane, N.J.A. (1991): On the covering multiplicity of lattices. Discrete and Computational Geometry, to appear.

15. Coxeter, H.S.M. (1928): The pure archimedean polytopes in six and seven dimensions. Proc. Camb. Phil. Soc. 24:1–9

16. Coxeter, H.S.M. (1931): Groups whose fundamental regions are simplexes. J. Lond. Math. Soc. 6:132–136

17. Coxeter, H.S.M. (1934): Discrete groups generated by reflections. Ann. Math. 35:588–621

18. Coxeter, H.S.M. (1968): Wythoff's construction for uniform polytopes. In: Coxeter, H.S.M. (ed.) Twelve Geometric Essays. Southern Univ. Illinois Press, Carbondale I.L., pp. 40–53 (Proc. London Math. Soc. 38 (1935) 327–339)

19. Coxeter, H.S.M. (1940): The polytope 2_{21}, whose 27 vertices correspond to the lines of the general cubic surface. Am. J. Math. 62:457–486

20. Coxeter, H.S.M. (1973): Regular polytopes, 3rd edn. Dover, NY

21. Coxeter, H.S.M., Moser, W.O.J. (1980): Generators and relations for discrete groups, 4th edn. Springer, NY

22. Forney, G.D. Jr. (1988): Coset codes – part I: introduction and geometrical classification. IEEE Trans. Information Theory 34:1123–1151

23. Forney, G.D. Jr. (1988): Coset codes – part II: binary lattices and related codes. IEEE Trans. Information Theory 34:1151–1187

24. Forney, G.D. Jr., Gallager, R.G., Lang, G.R., Longstaff, F.M., Qureshi, S.U. (1984): Efficient modulation for band-limited channels. IEEE J. Selected Areas Commun. 2:632–647

25. Fourer, R., Gay, D.M., Kernighan, B.W. (1990): A modeling language for mathematical programming, Management Science 36:519–554

26. Gosset, T. (1900): On the regular and semi-regular figures in space of n dimensions. Messenger of Math. 29:43–48

27. Grosswald, E. (1985): Representations of integers as sums of squares. Springer, NY
28. Grove, L.C., Benson, C.T. (1985): Finite reflection groups, 2nd edn. Springer
29. Grünbaum, B. (1967): Convex polytopes. Wiley, NY
30. Hazewinkel, M., Hesselink, W., Siersma, D., Veldkamp, F.D. (1977): The ubiquity of Coxeter-Dynkin diagrams (an introduction to the $A-D-E$ problem). Nieuw Arch. Wisk. 25:257–307
31. Henderson, A. (1972): The twenty-seven lines upon the cubic surface. Cambridge Univ. Press, 1911, Hafner, NY
32. Hesse, L.O. (1855): Über die Doppeltangenten der Curven vierter Ordnung. J. Reine Angew. Math. 49:279–332 (Gesam. Werke, pp. 345–404)
33. Humphreys, J.E. (1972): Introduction to Lie algebras and representation theory. Springer, NY
34. Kantor, W.M. (1981): Generation of linear groups. In: Davis, C., Grünbaum, B., Sherk, F.A. (ed.) The geometric vein. Springer, NY, pp. 497–509
35. Murtagh, B.A., Saunders, M.A. (1987): MINOS 5.1 user's guide, Technical Report SOL 83-20R, Dept. Operations Research. Stanford Univ., Stanford CA
36. Rogers, C.A. (1964): Packing and covering. Cambridge Univ. Press
37. Ryskov, S.S., Baranovskii, E.P. (1976): C-types of n-dimensional lattices and 5-dimensional primitive parallelohedra (with application to the theory of coverings). Trudy. Mat. Inst. Steklova 137: No. 4
38. Schläfli, L. (1858): An attempt to determine the twenty-seven lines upon a surface of the third order, and to divide such surfaces into species in reference to the reality of the lines upon that surface. Quart. J. Math. 2:110–120
39. Serre, J-P. (1988): Cours d'arithmétique, 3rd edn. Presses Universitaires de France
40. Sullivan, J.M. (1990): A crystalline approximation theorem for hypersurfaces, Ph. D. Dissertation, Princeton Univ., 1990
41. Vetchinkin, N.M. (1980): Uniqueness of the classes of positive quadratic forms on which the values of Hermite constants are attained for $6 \leq n \leq 8$. Trudy Mat. Inst. Steklova 152:34–86
42. Worley, R.T. (1987): The Voronoi region of E_6^*. J. Australian Math. Soc. Ser. A 48:268–278
43. Worley, R.T. (1988): The Voronoi region of E_7^*. SIAM J. Discrete Math. 1:134–141
44. Wythoff, W.A. (1918): A relation between the polytopes of the C_{600}-family. Proc. Royal Acad. Sci. (Amsterdam) 20:966–970

Added in proof. The Voronoi cells for E_6^* and E_7^* have been independently derived by E. Pervin, "The Voronoi cells of the E_6^* and E_7^* lattices", preprint, 1990.

Beno Eckmann

Von der Studierstube in die Öffentlichkeit

Wissenschaftliches Publizieren aus der Sicht
des Mathematikers als Forscher, Autor und Herausgeber

1. PROLOG

Es gehört zur »déformation professionelle« des Mathematikers, daß
er jeweils das Feld seiner Betrachtungen möglichst genau absteckt
mit Axiomen, Definitionen und vielem andern. Er weiß, daß er
dabei natürlich nur bestimmte Teile der Realitäten erfassen kann.
Was sich ergibt, nennen viele »abstrakt«; man kann es auch »kon-
kret« nennen, da wenigstens klargestellt wird, wovon man spricht.
So soll also der Bereich wissenschaftlicher Publikation, von dem
im folgenden die Rede ist, genau begrenzt sein.

Es handelt sich um die Publikation, die sich an Fachkollegen
wendet; an die auf verwandten Gebieten tätigen Forscher, an solche,
die Ergebnisse anwenden möchten, an Studenten des Fachs – kurz
an einen im Prinzip auf das Verstehen vorbereiteten Kreis. Es sind
Publikationen in Fachzeitschriften, in Monographien und in
Büchern. Der Weg, der von dort in eine breitere Öffentlichkeit,
ja in die Gesellschaft führt, ist sehr lang, beschwerlich, unsicher.
Man darf aber nicht vergessen, daß vom Standpunkt der Information aus gerade die hier gemeinte Publikation im Fachkreis den
entscheidenden Schritt bedeutet: Es ist der Schritt vom Individuum
zur weltweiten wissenschaftlichen Gemeinschaft. Was der einzelne
gedacht, entdeckt, erfunden, bewiesen hat, wird in dieser Gemeinschaft ein Teil allgemeinen Wissens. Es muß sich dort bewähren
und zu weiterer Blüte gelangen und kann schließlich (vielleicht)
in der Form konkreter Anwendungen oder allgemein zugänglicher
Erkenntnisse der Gesellschaft bewußt und für sie relevant werden;
ob das eintrifft oder nicht, und wann, ist im Einzelfall kaum je
vorauszusehen.

Von dieser entscheidenden Schnittstelle zwischen Forschung und
Öffentlichkeit ist also hier die Rede. An der dort geleisteten Arbeit
sind sozusagen alle forschenden Mathematiker beteiligt, sind sie
doch in der Regel gleichzeitig Autoren, Referenten und Herausgeber.

Beteiligt ist natürlich auch der Wissenschaftsverleger; ohne ihn, sein Verständnis, seinen intensiven persönlichen Einsatz und seine Risikobereitschaft wäre dieser Übergang gar nicht denkbar. So ist es denn wohl am Platz, in einem Heinz Götze gewidmeten Band einige der Gedanken festzuhalten, die manchen Autor – Referenten – Herausgeber bewegen dürften; wie so vieles andere, bleiben sie ja zumeist ungesagt.

2. RELEVANZ

Damit dieser entscheidende Schritt von der Studierstube in die wissenschaftliche Öffentlichkeit gelingt, muß jede Arbeit nach strengen Kriterien eine ganze Reihe von Hürden überwinden: zunächst muß sich ein Ergebnis im kleinen Team, im Institut durchsetzen; dann muß es vor größeren wissenschaftlichen Gremien an Symposien und Kongressen bestehen können; und dann kommt die schwierigste Hürde, nämlich vor den Herausgebern einer Publikationsreihe oder einer Zeitschrift Gnade zu finden.

Welches sind diese Kriterien? Auf den ersten Blick sind sie höchst einfach. Ich zitiere wörtlich aus den Richtlinien (editorial policy) der Proceedings American Mathematical Society: »A paper must be correct, new, non-trivial and significant. Further it must be well written and of interest to a substantial number of mathematicians.« Es handelt sich also im wesentlichen um drei Aspekte (a) richtig, (b) neu, (c) interessant-signifikant.

Bevor ich näher darauf eingehe, sei ein Aspekt hervorgehoben, der nicht erwähnt wird und offenbar keine Rolle spielen darf: die Relevanz. Von der gesellschaftlichen Bedeutung in technischer, ökonomischer, politischer, ökologischer Hinsicht ist nicht die Rede. Das mag für Außenstehende etwas erstaunlich, wenn nicht sogar enttäuschend sein. Mancher vermutet gerade an dieser Stelle eine gewisse Kontrolle, ein Abwägen relevanter Prioritäten, einen Arm der Wissenschaftspolitik – um nur einige der gängigen Schlagworte zu nennen. Es ist eine Erfahrung, die sich durch die ganze Wissenschaftsgeschichte hindurchzieht, daß so etwas in diesem Stadium gar nicht möglich ist, jedenfalls in der Mathematik. Die Konsequenzen mathematischer Erkenntnisse können ganz einfach unvorhersehbar sein. Und es ist klar, daß dieser Charakter der Unvorhersehbarkeit jedes auf Relevanz zielende Kriterium verbietet.

Einige Beispiele, dem Mathematiker zwar wohlbekannt, einem weiteren Kreis aber weniger geläufig, sollen dies illustrieren. Sie handeln von unerwarteten Anwendungen zu späteren Zeiten; sie zeigen auch, daß so manches in unserer Welt auf mathematische Methoden und Erkenntnisse zurückgeht, ohne daß dies dem breiten Publikum bewußt wäre – gar oft nicht einmal dem Anwender selbst, der Naturwissenschafter, Ingenieur, Philosoph oder vieles andere sein kann.

Galois und seine Zeitgenossen in der ersten Hälfte des 19. Jahrhunderts konnten bestimmt nicht ahnen, daß ihre Ideen zur Lösung eines klassischen Problems (des Auflösens von Gleichungen mit Hilfe von Wurzelzeichen) dereinst in Form der Theorie der Gruppen und ihrer Darstellungen wesentlich dazu beitragen würden, das Rätsel der Atomspektren und des periodischen Systems der Elemente zu enthüllen. Oder: Wußten die Mathematiker, die sich mit der Wellengleichung befaßten, daß viel später einmal aufgrund ihrer Resultate und der Maxwellschen Gleichungen Hertz die elektromagnetischen Wellen entdecken und damit das Zeitalter der Radiotechnik einleiten sollte? Oder: Die Blütezeit der algebraischen Topologie, die in den dreißiger Jahren ihren Anfang nahm, entstand aus dem Bemühen, geometrisch-topologische Eigenschaften mit algebraischen Hilfsmitteln zu erfassen – wer hätte vorausgesagt, daß hier sehr bedeutungsvolle Werkzeuge der modernen theoretischen Physik bereitgestellt würden (Homologie, Faserungen, charakteristische Klassen); daß durch die homologische Algebra Methoden entstanden, die in alle Anwendungen eindringen würden; daß die damals entstandene Theorie der Kategorien für formale Sprachen, Computer und Automaten wesentlich sein sollte? Wir können auch an die Existenz- und Eindeutigkeitssätze für Lösungen von Differentialgleichungen denken: Sie sind das Substrat für die Aussage der klassischen Mechanik, daß Kraftfeld und Anfangsdaten (Ort und Geschwindigkeit) die Bewegung eines Massenpunktes vollständig bestimmen – einer der Ausgangspunkte für eine deterministische und mechanistische Weltanschauung.

Selbst in mathematischen Gebieten und Untersuchungen, die direkt anwendungsorientiert sind, ist das Wort »relevant« mit großer Vorsicht zu benützen. Was man für wichtig und relevant hält, kann es vielleicht wirklich sein – aber nur zu oft in einer Richtung, an die weder Autor noch Herausgeber dachten. Es wäre sehr bedenklich, wegen gewisser im Augenblick wichtig scheinender

Anwendungen einzelnen Gebieten, etwa Partiellen Differentialgleichungen oder Numerischer Methoden, vor andern den Vorzug zu
geben.

Sicher wird, sogar in der »reinsten« Mathematik, Relevanz und
Weltanschauung zur Motivation des Forschers wesentlich beitragen
und in der späteren Entwicklung der Ideen eine Rolle spielen; aber
sie darf niemals als Kriterium dienen, und das ist der Anlaß für
diese etwas länger geratene Vorbemerkung!

3. Richtlinien der Herausgeber

Doch zurück, nach diesem Exkurs, zu den Kriterien, auf die der
Entscheid der Referenten und Herausgeber sich stützen soll. Sie
sind in den zitierten Richtlinien durch die Stichworte »richtig«,
»neu«, »interessant-signifikant« charakterisiert. Wie das System
funktioniert, liegt auf der Hand: Unter Beizug vieler Spezialisten
sucht der Herausgeber nach bestem Wissen und Gewissen diese
Kriterien anzuwenden und zu einer Entscheidung zu kommen.
Manchmal ist er zu nachsichtig, manchmal zu streng; manchmal
steht er unter Zeitdruck, und es kommen Pannen vor. Im großen
und ganzen ist das Vorgehen jedoch erfolgreich und anerkannt.
Es hat in hohem Maße den Charakter einer Zusammenarbeit der
ganzen wissenschaftlichen Gemeinschaft.

So weit, so gut. Aber so einfach wie sie auf den ersten Blick
aussehen, sind die Kriterien nicht. Und wenn ich jetzt auch gewisse
negative oder zumindest problematische Seiten betrachte, so deshalb, weil sich notwendigerweise einige Gedanken anknüpfen, die
man nicht übergehen sollte. Sie streifen vieles, das heute als »Philosophie der Mathematik« zu einem Diskussionsthema geworden ist,
welches manche Mathematiker, und nicht nur diese, zu faszinieren
scheint, und aus welchem wir den Aspekt der Relevanz ausgeklam-
mert haben.

Richtig, neu, interessant. Beginnen wir mit »richtig«. Der Laie
meint ja allgemein, daß es in der Mathematik ohne weiteres klar
und feststellbar ist, ob ein Ergebnis richtig oder falsch ist. Das ist
leider nicht immer der Fall. Ein tiefes Ergebnis benötigt unter
Umständen eine derartige Fülle von Einzelheiten aus ganz verschie-
denartigen Überlegungen, daß eine genaue Kontrolle jedes Schrittes
(und der dabei benützten früheren Resultate!) sehr zeitraubend ist.

Hier muß auf die Erfahrung des Referenten abgestellt werden, auf bekannte analoge Ergebnisse, auf Plausibilitätsbetrachtungen, auf Originalität, ja auch die Schönheit der Idee. Es gibt Fälle, in denen der Autor eine Einsicht intuitiv richtig erfaßt hat, obwohl sein Beweis nicht stichhaltig ist – oder es kann auch sein, daß er nicht alle seine Überlegungen genau verrät. Oder umgekehrt: Ein Ergebnis erweist sich offensichtlich als falsch, Methode und Überlegungen geben aber Anlaß zu bedeutenden andern Entwicklungen – dies kommt vor, wenn es auch selten der Fall sein dürfte!

Wie sagt Erich Kästner:

> Irrtümer haben ihren Wert
> jedoch nur hie und da,
> nicht jeder der nach Indien fährt
> entdeckt Amerika.

Wie aber den guten und schöpferischen Irrtum vom schlechten unterscheiden? Es kommen Betrachtungen heuristischer, philosophischer, ästhetischer Art ins Spiel, die durchaus nicht den rationalen Charakter haben, den man der Mathematik zuschreibt.

Eine neue Art von Problematik liegt in den »Computer-assisted«-Beweisen, wie sie gerade in letzter Zeit in subtilen Problemen der Analysis aktuell geworden sind; z.B. im Beweis der Feigenbaum-Vermutung betreffend den Übergang eines dynamischen Systems in chaotisches Verhalten. Die mathematischen Überlegungen, die das Problem in eine Computer-Verifikation transformieren (Intervall-Rechnen u.a.) sind wie übliche Beweise kontrollierbar. Wie aber steht es mit dem Computer-Programm selbst, mit der Zuverlässigkeit der Hardware, mit allfälligen Störungen im Ablauf? Aber ist es nicht ähnlich bei einem reinen »brain-assisted«-Beweis, wenn dieser über hundert Seiten beansprucht und sich erst noch auf andere, ebenso lange und noch kompliziertere Beweise stützt?

Noch schwieriger kann die Situation bei Forschungsberichten aus der Numerischen Analysis sein. Es kann sich ja dort manchmal um Algorithmen zur Approximation von Lösungen handeln, bei denen – wenigstens fürs erste – auf einen strengen Beweis im üblichen Sinne gar nicht Anspruch erhoben wird; vielmehr soll der Erfolg im Sinne guter Näherung, wie sie für praktische Anwendungen wichtig ist, demonstriert werden.

4. Neu versus interessant

»Neu«. Daß es sich nicht um eine Kopie, möglicherweise mit äußerlichen Änderungen, handeln darf, ist klar; darum geht es nicht. Aber etwas Neues knüpft immer an Altes, Bekanntes an. Wie neu aber dann ein Gedanke wirklich ist, läßt sich nur selten eindeutig feststellen. Er kann an ganz anderem Ort und in anderer Einkleidung schon vorhanden sein, wie das bei der unerhörten Verästelung der mathematischen Gebiete vorkommen kann. Aber gerade solche Zusammenhänge dürften in vielen Fällen den Reiz oder die tiefere Bedeutung ausmachen. Und ist ein Gedanke völlig neu, so ist es um so schwieriger festzustellen, ob er wirklich richtig und interessant ist. Allzu neue Gedanken werden oft nicht verstanden; vor allem treffen sie nicht immer auf allzu großes Interesse.

Um wieder Beispiele hierzu aus meinem engen Interessenkreis zu zitieren, die wenig bekannt sind: Als Heinz Hopf 1930 die wesentlichen Abbildungen der dreidimensionalen Sphäre auf die Kugelfläche entdeckte, war – wie er selbst erzählte – das Interesse der Kollegen sehr gering. Und doch handelte es sich um den entscheidenden Ausgangspunkt für die gesamte Homotopietheorie, die Theorie der Homotopiegruppen und der Faserungen, und vieles andere. Und noch früher: Die beiden Arbeiten zur »kombinatorischen« (eigentlich »algebraischen«) Topologie von Hermann Weyl 1924/1925 – die erste handelt im Speziellen von Graphen und elektrischen Netzwerken, die zweite von höheren Dimensionen – schienen dem Verfasser vom allgemeinen Interesse soweit weg zu liegen, daß er sie gemäß seiner eigenen Aussage in wenig bekannten Zeitschriften veröffentlichte.

So ist man auch bei »interessant« noch weiter von einem objektiven Kriterium entfernt. Gewiß ist das Interesse, das eine große Zahl von Mathematikern bezeugt, ein guter Hinweis. Wenn Gebiete, Methoden und Theorien häufig zitiert werden, so stehen sie sicher im Mittelpunkt des Interesses. Aber wie überall gibt es auch hier so etwas wie eine Mode, die kommt und vergeht. Es gibt Schulen, deren Mitglieder sich gegenseitig hochspielen (man spricht manchmal von einer »Société d'admiration mutuelle«) und die früher oder später von andern abgelöst werden.

Es kann also sehr wohl etwas Interessantes nicht neu sein und etwas Neues im Moment nicht interessant scheinen, etwas Falsches bedeutungsvoll und etwas Richtiges unbedeutend sein. Und in dieser Weise geht man mit den Kriterien im Kreise herum, ohne wirklich zu einer begründeten Entscheidung zu gelangen. Natürlich, das sei wiederholt, gibt es genügend Fälle, in denen ihre Anwendung durchaus einfach und schlüssig ist. Aber es muß doch nachdenklich stimmen, in wie hohem Maße es bei einem so wichtigen Schritt des Informationsvorganges, beim Übergang vom Individuum in die Öffentlichkeit, auf Dinge ankommt, die mit der Rationalität der Wissenschaft wenig verträglich zu sein scheinen.

5. QUALITÄT

Entscheidend ist schließlich nach wie vor die Qualität. Damit sind wir bei einem heiklen Wort angelangt. Qualität ist zu verstehen in einem Sinne, der weitgehend auf Gefühl, Erfahrung und Autorität beruht – Autorität, ein zweites heikles Wort. Autorität beruht natürlich – und dies ganz besonders in der Mathematik – weder auf Alter, noch auf Titel, noch auf Stellung, vielmehr auf dem Instinkt für die Tiefe der Ideen. Man lasse sich durch den zumeist kalten und sachlichen Charakter heutiger mathematischer Publikation nicht täuschen, der all das verschweigt, was zum Ergebnis und zur endgültigen Fassung führte. In die Arbeit des Forschers ist seine ganze Persönlichkeit in all ihrer Irrationalität einbezogen. Intuitives Erkennen, Ringen um Klarheit und Einheit, weltanschauliche Leitgedanken gehören ebenso dazu wie das Wissen um Vergangenes, um Lehrer und Schulen; wie die geschichtliche Entwicklung; wie das Erbe derer, die sich tastend um eine Einsicht bemühten, ohne sie wirklich ganz fassen zu können. Am Anfang scheinen Fragen und Antworten klar und einfach zu sein; je weiter man sich engagiert, um so komplexer wird das Bild. Immer wieder steht man an jener Wegkreuzung zwischen Bindung und Freiheit, welche ja nicht nur das Wesen der Mathematik, sondern überhaupt der menschlichen Natur ausmacht; es ist das, was Jeanne Hersch die »condition humaine« nennt. So sind wir versucht, an das eindrückliche Zitat (T.S. Eliot) zu denken, welches Hermann Weyl seiner »Philosophy of Mathematics and Natural Science« als Motto vorangestellt hat:

> Home is where one starts from. As one grows older
> the world becomes stranger, the pattern more complicated
> of dead and living.

6. STIL UND SPRACHE

In den zitierten Richtlinien der Proceedings A.M.S. kommt ein weiterer Punkt hinzu: es ist die Frage des Stils und der Sprache. Ich glaube, daß wissenschaftliche Einsichten in jedem Falle »sprachlos« sind, so daß es einer besondern Bemühung bedarf, sie in Worte zu fassen. Zunächst: die Mathematik schafft ja im Zuge der Untersuchungen gleichzeitig die Formelsprache und Ausdrucksweise, in der man festzuhalten versucht, was man meint. Diese ist aber von Gebiet zu Gebiet zum Teil recht verschieden und variabel, so daß immer wieder Übersetzungen nötig werden.

Dieser Formalismus allein ist natürlich mit der genannten Richtlinie nicht gemeint. Es handelt sich vielmehr um die verbindenden Texte, mit denen man sich an einen Leser wendet, den man interessieren und überzeugen will. Logischer Ablauf der Gedanken gehört auf alle Fälle dazu. In lapidarer Art hat George Polya zwei Stilregeln formuliert: Erstens, man muß etwas zu sagen haben; und zweitens, wenn man zwei Dinge zu sagen hat, so sage man zuerst das eine und dann das andere.

Ich möchte noch weitergehen und in der sprachlichen Gestaltung mehr sehen als all dies. Die Bemühung, in der Wissenschaft und anderswo, einzelne Tatsachen aus dem verschwommenen und turbulenten Strom unseres Lebens herauszuschälen, und die Bemühung, die adäquate Sprache zu finden, um die Tatsachen einander mitzuteilen – dies sind schöpferische Akte des Menschen, die Hand in Hand gehen müssen. Es ist nicht vielen gegeben, es darin zur Meisterschaft zu bringen; ihre Bücher und Arbeiten zu lesen ist ein Genuß, der Inhalt ebenso wie Ausdruckskraft umfaßt. Es ist sicher nicht Sache eines Herausgebers oder Referenten, in seiner eigenen Unzulänglichkeit so etwas vom Autor zu verlangen. Gerade bei besonders neuen und interessanten Gedanken ist die ideale Form im ersten Anlauf noch nicht zu erwarten; sie muß späteren verbesserten Versionen vorbehalten bleiben, und gerade deshalb ist die Möglichkeit »vorläufiger« Publikation so wichtig, wie sie sich in den letzten Jahrzehnten eingebürgert hat (»Lecture Notes« aller Arten). Es ist bemerkenswert festzustellen, wie die spätere definitive Fassung oft präziser, eleganter – und viel kürzer ist als die vorläufige!

Angemessene Kürze ist sicher der Weitschweifigkeit vorzuziehen, selbst wenn sie zeitraubend ist und einiges Nachdenken erfordert.

Um noch einmal Kästner zu zitieren:

> Wer was zu sagen hat,
> hat keine Eile,
> er läßt sich Zeit
> und sagt's in einer Zeile.

Ich fürchte, daß ich mich hier nicht daran gehalten habe, wobei mir die etwas diffuse Materie als Entschuldigung dienen soll. Wie denn überhaupt alle Kriterien und Richtlinien ihre Tücken haben. Wenn der Herausgeber sein anderes Ich hervorholt und wieder zum Autor wird, dann merkt er bald, wie schwer es ihm fällt, sich an seine eigenen Regeln zu halten. So wird er auch dem Autor manche Abweichungen und Kompromisse zubilligen.

7. Epilog

Überhaupt muß man nach all dem Gesagten mehr und mehr einsehen, daß die wissenschaftliche Publikation, das Buch, der Zeitschriftenartikel, einen eigenen Charakter hat, der sich wesentlich von dem unterscheidet, was den Forscher bewegte und was in seinem Geist vorging; und auch von dem, was dann der Leser herausliest. Die Dreiheit Idee–Buch–Leser hat ihre augenfällige Parallele in der Musik. Auch dort steht das Werk, das Opus, in der Mitte zwischen kreativem Einfall und Interpretation: Idee und Einfall des Komponisten kommt und vergeht; das Werk bleibt; der ausführende Musiker macht daraus etwas Neues, das von seiner Art das Werk zu lesen und zu interpretieren abhängt. Um zur mathematischen Publikation zurückzukehren: Sie ist eine bleibende Zwischenstufe, die je nach dem Umfeld des Lesers zu neuem Leben, zu neuen Intuitionen oder Anwendungen (in jedem Sinne des Wortes) Anlaß gibt. Diese Zwischenstufe mit ihrem eigenen Charakter kann im Grunde genommen durch die beschriebenen Kriterien nur schwer beurteilt werden. Ist es nicht nur allzu bekannt, daß Bücher, die den Richtlinien wenig entsprachen, die weitere Entwicklung und das Denken ganzer Generationen besonders stark beeinflußt haben? Wenn ein solches als Exempel genannt werden darf: W.V.D. Hodge »The Theory and Applications of Harmonic Integrals«. Trotz mancher unbestreitbaren Unzulänglichkeit hat Hermann Weyl es als das wichtigste mathematische Buch der Zeit bezeichnet.

Das Irrationale der Situation, die ich hier zu beschreiben versucht habe, ist nicht leicht zu durchschauen. Und doch scheint es in der Natur der Sache zu liegen. In der wissenschaftlichen Publikation steckt eben mehr als nur das Vermitteln von einzelnen Fakten und Ideen; ich sehe darin auch die Persönlichkeit des Autors, sein berufliches und menschliches Umfeld, seine freundschaftlichen und wissenschaftlichen Kontakte – und auch die Bemühungen des Verlages und der Herstellung und Gestaltung. Wenn Artikel oder Buch dem Leser etwas von dem weitergeben können, was von allen Beteiligten, und nicht zuletzt vom Verleger, an menschlicher und intellektueller Bemühung hineingelegt wurde, dann kann er, der Leser, zum Interpreten werden und in faszinierender Weise die Gedanken des Forschers immer wieder zu neuem Leben erwecken.

Ludvig D. Faddeev

A Mathematician's View
of the Development of Physics *

Mathematics in its clean form is the product of the free human mind.

Physics is a natural science with just a single goal – uncovering the structure of matter.

In their quest physicists naturally use mathematical tools to correlate data, to express the laws found by means of formulas and to make relevant calculations. To a greater or smaller extent this is done in all sciences. And there is no *a priori* reason for such a distinguished role of mathematics in physics which we are witnessing nowadays – namely that of an imminent language of physical theory.

I shall not elaborate on the examples to prove this role. Everybody in his profession can choose his favorite. It is enough to recall that such purely mathematical structures as Riemann geometry or Lie groups theory are indispensable in modern theory of gravity, formulated by Einstein, or in the description of kinematical and dynamical symmetries of any physical system.

This role of mathematics as the language of physics is taken by physicists with mixed feelings of admiration and irritation. Take for example the title of the famous essay by Wigner – "On the unreasonable effectiveness of mathematics in natural sciences". The complex formed by these feelings is sometimes resolved in malicious jokes on mathematics which some great men allowed themselves to tell. I shall not comment more on this.

Instead, I shall take seriously the stated role of mathematics as a fact and try to present in this spirit the analysis of modern trends in physics. To do this I shall need some formalized framework and I proceed now to its description.

In the description of the physical system we use two main notions: those of observables and those of states. The set of observ-

* The original source of this article is: Proceedings of the 25th Anniversary Conference – Frontiers in Physics, High Technology & Mathematics, eds. HA Cerdeira & SO Lundqvist (1990) pp 238–246

ables $\mathfrak{A}$ comprises all physical entities A, B, C, ... constituting the system. The set of states Ω, with elements ω, μ, ..., describes the possible results of the measurements of observables. More formally, each state ω gives to each observable A its probability distribution – a nonnegative, monotone increasing function $\omega_A(\lambda)$ of a real variable λ, $-\infty < \lambda < \infty$, normalized by the conditions $\omega_A(-\infty) = 0$, $\omega_A(\infty) = 1$. In particular, the mean value of an observable A in a state ω is given by

$$\langle \omega | A \rangle = \int\limits_{-\infty}^{\infty} \lambda \, d\omega_A(\lambda).$$

The completeness of such a description is expressed in the requirement that states separate observables. Namely, if two observables A and B have the same mean value in all states, then they coincide. This is a formal expression of the main epistemological principle of the ability of cognition of the universe.

Mathematically this principle introduces some structure in the set of observables:

1. $\mathfrak{A}$ is a real linear space.

Indeed, observables $A + B$ and kA for real k are defined as having the mean values

$$\langle \omega | A + B \rangle = \langle \omega | A \rangle + \langle \omega | B \rangle$$

and

$$\langle \omega | kA \rangle = k \langle \omega | A \rangle$$

in all states ω.

2. For each real-valued function $\varphi(\lambda)$ of the real-variable λ and each observable A we can construct the observable $\varphi(A)$ by means of the formula

$$\langle \omega | \varphi(A) \rangle = \int\limits_{-\infty}^{\infty} \varphi(\lambda) \, d\omega_A(\lambda),$$

valid for any state ω. Alternatively, we can say that the probability distribution of $\varphi(A)$ is given by

$$\omega_{\varphi(A)}(\lambda) = \omega_A(\varphi^{-1}(\lambda)).$$

To introduce the dynamics of the system we are to describe the notion of motions or one-parameter automorphisms $A \to A(s)$ in the set of observables. This is done by means of a binary operation (bracket) $\{A, B\}$ which allows one to associate a particular motion $A \to A(s)$ with every observable B by means of the differential equation

$$\frac{dA(s)}{ds} = \{B, A(s)\}, \quad A(0) = A.$$

It is natural to require that the generating observable B does not change, so that $\{B, B\} = 0$. The compatibility with the linear structure implies that $\{,\}$ must be a Lie bracket, namely it must be linear

$$\{kA + lB, C\} = k\{A, C\} + l\{B, C\}$$

and satisfy the Jacobi identity

$$\{A, \{B, C\}\} + \{B, \{C, A\}\} + \{C, \{A, B\}\} = 0.$$

Moreover, the notion of function is to commute with motion

$$\varphi(A(s)) = \varphi(A)(s).$$

This is essentially all that we need to provide a general framework for the description of a physical system. I admit that the dynamical principle is less intuitive than the kinematical one. However, I do not see any other way of formalizing the experience we have until now.

The existing physical theories give us concrete realizations of this general scheme. Take classical mechanics. The basic notion there is that of the phase space Γ consisting of the generalized coordinates q and momenta p. Observables are real-valued functions $f(p, q)$ on Γ. The linear structure is evident, $\varphi(f)$ is understood as the superposition of the functions f and φ. The remarkable mathematical theorem of Markov then defines the set of states Ω in a unique manner as that of normalized measures ω on Γ. The distribution function $\omega_f(\lambda)$ is given by

$$\omega_f(\lambda) = \int\limits_{f(p,q) \le \lambda} d\omega = \int\limits_{\Gamma} \theta(f - \lambda)\, d\omega,$$

where $\theta(\lambda)$ is the Heaviside function.

The dynamical Lie operation is given by the Poisson bracket which looks in the canonical variables p, q as follows

$$\{f, g\} = \sum \left(\frac{\partial f}{\partial p} \frac{\partial g}{\partial q} - \frac{\partial f}{\partial q} \frac{\partial g}{\partial p} \right).$$

A particular motion corresponding to evolution or time development is given by the Hamilton equation

$$\frac{df}{dt} = \{H, f\},$$

where the observable H is called the energy.

Quantum mechanics is just another realization of our general scheme. In the usual description of quantum mechanics the role of observables is played by selfadjoint operators $A, B, \ldots$ in some (auxillary) Hilbert space $\mathfrak{H}$. The states are given by positive operators M with trace equal to 1. The distribution of A in the state M is given by

$$\omega_A(\lambda) = \mathrm{tr}(MP_A(\lambda)),$$

where $P_A(\lambda)$ is the spectral function of A (which can be formally written as $P_A(\lambda) = \Theta(\lambda - A)$). The definition of the function $\varphi(A)$ of A is given by

$$\varphi(A) = \int\limits_{-\infty}^{\infty} \varphi(\lambda)\, dP_A(\lambda),$$

and the number of degrees of freedom is equal to the number of functionally independent commuting observables. The Lie bracket is given by

$$\{A, B\} = \frac{i}{\hbar}(AB - BA),$$

where $i = \sqrt{-1}$ and $\hbar$ is a fixed parameter of dimension

$$[\hbar] = [p][q],$$

the famous Planck constant.

Let us mention that in both realizations the notion of function could be introduced purely algebraically by means of the associative product existing in the set of observables. It is not clear if there exists a more general realization where such a product does not appear at all.

Another mathematical observation, already vaguely mentioned, is that the structure of the set of observables is sufficient to define the set of states as the dual object – the convex set of positive functionals.

I am now ready to proceed to speculations on the development of physics. However, it is worth making some general comments on what was already said.

1. The scheme was formulated already after the advent of quantum mechanics. The main role here belongs to Dirac who in particular introduced the term "observables". In the particular way of arranging the formulas and notions I am influenced by my mathematical colleagues, I. Segal and G. Mackey.

2. The fact that classical mechanics bears all the essential features of the scheme makes it plausible that it could already be formulated in the previous century for example by Hamilton or Gibbs. Also, the mathematics available at that time allowed for the search for other realizations, leading to quantum mechanics. However, history did not take that path.

3. It is often said that quantum mechanics is "indeterministic" because the notion of probability is used in its formulation. This is a misleading statement. We have seen that distribution functions in the role of states appeared already in classical mechanics. The specific feature of the classical case is that all observables are exact in pure states, whereas in the quantum case they are exact only in eigenstates, which is, however, enough to describe them completely. So it is just an undeserved luxury what we have in classical mechanics and it is only justified that it is eliminated during the passage to quantum mechanics.

After these comments I return to my main goal and use the general scheme to analyze the relation between classical and quantum mechanics. It will be convenient to describe both theories by means of similar objects. It is enough to stick with the observables, because we know that states are defined by observables. So we shall use the possibility to describe the quantum mechanical observables by means of functions on the phase space. In this description the function $f(p, q)$ is the symbol of the operator A_f. In the simplest case of linear phase space with one degree of freedom we can express A_f as an integral operator in $L_2(\mathbb{R})$ in terms of its kernel $A_f(x, y)$ by the Weyl formula

$$A_f(x, y) = \frac{1}{2\pi\hbar} \int_{-\infty}^{\infty} f\left(p, \frac{x+y}{2}\right) e^{ip(x, y)/\hbar} dp$$

(note the explicit presence of $\hbar$). It is clear that the main structures of quantum observables $\{f, g\}$ and $\varphi(f)$ are different from those of the classical ones. However, the former converge to the latter in the limit $\hbar \to 0$. More exactly, we have the expansion

$$\{f, g\}_\hbar = \{f, g\}_0 + \hbar \{f, g\}^{(1)} + \hbar^2 \{f, g\}^{(2)} + \ldots$$

$$\varphi(f)_\hbar = \varphi(f)_0 + \hbar \varphi(f)^{(1)} + \hbar^2 \varphi(f)^{(2)} + \ldots,$$

where we specified the quantum operations by subscript $\hbar$ and the classical ones by 0. Both these expansions are corollaries of the formula for the product of the symbols f and g induced by the operator product of A_f and A_g. If we use the Weyl correspondence between symbols and operators we have for the product $f * g$ the following expansion in powers of $\hbar$

$$f * g = fg + \frac{\hbar}{2i} \{f, g\}_0 + \ldots$$

which in turn leads to the corresponding expansions for the Lie bracket

$$\{f, g\}_\hbar = \frac{i}{\hbar}(f * g - g * f) + \ldots$$

and the definition of the function $\varphi(f)$ of a given observable f.

Of course, in nature the Planck constant $\hbar$ has a given value, $\hbar \cong 10^{-27}$ g·cm^2 s^{-1}; the existence of a family of quantum mechanics, labeled by the parameter $\hbar$ is a mathematical play of mind. Mathematicians use the term "deformation" of structure in such cases. Using this term we can say that quantum mechanics is a deformation of the classical one with $\hbar$ playing the role of deformation parameter.

This statement is the shortest and most adequate formulation of the correspondence principle. For a professional mathematician there is nothing to add or to delete in this formulation. Many words used in popular literature to explain the correspondence principle is nothing but "belletristics".

Now we are coming to the most important place in our exposition. The matter is that in the mathematical theory of deformations of structures there exists the notion of stability. We say that the given algebraic structure is stable if all deformations of it lying near to it are equivalent to it. Now, the following fact is true: the structure of quantum mechanics underlined in the general scheme above is stable. On the contrary, classical mechanics is not stable – it has in its vicinity a nonequivalent deformation – quantum mechanics. The degeneracy of classical mechanics is connected with its overdeterministic nature realized in the exactness of pure states for all observables. The passage to quantum mechanics removes this degeneracy and leads to a stable "generic" theory which is nondeformable in the framework of our general scheme.

Thus we are led to an important conclusion: whereas the change of classical mechanics into the quantum one is fully justified, we have no reasons to predict any change of the latter in the future.

The application of the mathematical theory of deformation of structures to analysis of the evolution of the physical picture of nature is not exhausted by this radical statement. We can describe in a similar fashion the passage from the nonrelativistic dynamics to the relativistic one. In fact, it is even easier.

From the mathematical point of view, this passage was connected with the change of the dynamical group – or the group of symmetries of the space-time – from the Galilei transformations to those of Lorentz-Poincaré.

Both these groups contain 10 parameters. Let us compare the changes of coordinates $\mathbf{x}$ and time t, corresponding to the change of the reference frame with the relative velocity $\mathbf{v}$. The Galileo transformation is given by

$$\mathbf{x} \to \mathbf{x} + \mathbf{v}t; \quad t \to t$$

whereas in the Lorentz-Poincaré case we have

$$\mathbf{x}^{\|} \to \frac{\mathbf{x}^{\|} + \mathbf{v}t}{\sqrt{1 - v^2/c^2}}, \quad \mathbf{x}^{\perp} \to \mathbf{x}^{\perp}$$

$$t \to \frac{t + (\mathbf{v} \cdot \mathbf{x})/c^2}{\sqrt{1 - v^2/c^2}}$$

$$\mathbf{x}^{\|} = (\mathbf{x} \cdot \mathbf{v})\, \mathbf{v}/v^2; \quad \mathbf{x}^{\perp} = \mathbf{x} - \mathbf{x}^{\|}.$$

We see that the Lorentz transformation contains a fixed parameter c the velocity of light with the value $c = 3 \cdot 10^{10}$ cm/s. It it clear that Lorentz transformations go into the Galilei in the limit $c \to \infty$. (To do this we are to imagine a mathematical fiction – the existence of the worlds with different values of c in the same fashion as we have done above in the case of Planck constant $\hbar$.) In the meantime both transformations constitute the same mathematical structure – a Lie group. Thus we see another manifestation of the deformation in the development of physics: the group of relativistic dynamics is a deformation of the group of nonrelativistic motion. The role of the deformation parameter is played by $1/c^2$. Now an important statement follows: the passage to relativism is into a stable structure. In other words, the Lorentz group is stable and does not allow any nontrivial deformations.

Let us now turn off the current of mathematical consciousness and look at the results of our analysis. We see that the two main revolutions in physics (and in all modern natural sciences) from the mathematical point of view are deformations from unstable structures into stable ones. The fashionable words on the change of paradigms are not highly relevant from this point of view. But nothing is lost in the realization of this fact. I think that our statement is a short and most adequate description of the evolution of our views on the theory of the structure of matter.

Now a natural question must be asked: does the analysis of the past of physics allow us to say something about its future? A historically minded person would answer no. However, those who believe in the existence of the mathematical structure underlying the world around us could try to make some predictions. So I cannot help but use the opportunity to do so; I realize how speculative such predictions can be.

I have already stressed that the parameters of deformation $\hbar$ and $1/c^2$ have physical dimensions. In terms of basic dimensions – mass M, length L and time T we have $[\hbar] = M[L]^2[T]^{-1}$; $[1/c^2] = [L]^{-2}[T]^2$. It is clear, that only one dimensional parameter is lacking and one could think, that one more deformation of our description of matter is to be sought for with such a parameter playing the role of that of deformation.

In fact, we can admit that such a deformation is already realized. Indeed, the third main achievement of physical theory in our century – the Einstein theory of gravity – can be considered as such a deformation in the stable direction. This theory is based on the use of the curvilinear pseudo-Riemann space as a space-time manifold instead of the linear Minkowski space. It is clear, that in the set of Riemann spaces the Minkowski space is a kind of degeneracy, whereas a generic Riemannian space is stable so that in its vicinity all spaces are curvilinear. The measure of the deformation is the gravitation constant γ, entering the Hilbert-Einstein equations of gravity, which was known in physics from the time of Newton. The gravitational constant γ has dimension functionally independent from those of $\hbar$ and c and together with them constitutes the basis for all dimensional parameters.

It is clear what I'm driving at: the unification of relativism, quantum principles and gravity must give us the ultimate physical theory, which is stable and not submittable to changes without really drastically breaking down the established framework, which has a very general character.

I do not see any wrong in the prediction that one more natural science may find a final formulation. Chemistry has already found its fundamental formulation on the basis of the quantum mechanics of many electron systems. So it is not surprising that physics is going to share the same fate.

Unfortunately, the natural synthesis of relativism, quantum principles and gravity is not achieved yet in modern theoretical physics. We have a reasonably complete quantum field theory in Minkowski space ($\hbar$ and c) or classical theory of gravity (c and γ). However, there exists no theory which incorporates all three parameters $\hbar$, c and γ in a natural manner. The main efforts of a numerous army of theoretical and mathematical physicists are directed to the realization of this unification. Not so many participants in this quest would agree that their labour will lead to the end of fundamental physics which will be formulated by means of an adequate mathematical language. But for some of them, including this author, this idea is self-evident, inevitable and, what is most important, a guiding one.

The Methods of the Theory of Functions of Several Complex Variables

Meinem verehrten Herrn Verleger Heinz Götze
gewidmet

It is more difficult to construct holomorphic and meromorphic functions of several complex variables than those of one variable. Up to today they occur only as modular functions or Feynmanintegrals or functions with group symmetries if they depend properly on more than one variable. In general, the detailed analysis of such functions is very difficult. It is no wonder that till now essentially only general facts were obtained. *Complex analysis* (of several complex variables) is rather a special kind of geometry than an analysis of properties of functions.

1

Actually, research on complex analysis started not before the turn of the century. Hartogs discovered a phenomenon which was unknown in the classical theory of functions. As all people know the symbol $\mathbb{C}^n$ denotes the space of n-tuples of complex numbers. The domain $H \subset \mathbb{C}^n$ the Hartogs figure, is defined as follows:

$$H = \{z = (z_1, z_2): |z_1| < \tfrac{1}{2}, \ |z_2| < 1 \ \text{or} \ |z_1| < 1, \ \tfrac{1}{2} < |z_2| < 1\}.$$

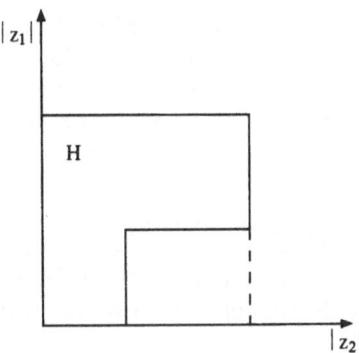

Then every function f holomorphic in H can be continued analytically to a holomorphic function in the unit polydisc $P = \{z: |z_1| < 1, |z_2| < 1\}$. The proof was done simply by use of the Cauchy integral in one complex variable. So no holomorphic function f can be singular at the boundary points ∂H which are in the interior interior of P. We call a *connected open subset G of $\mathbb{C}^n$ a domain of holomorphy* if there is a holomorphic function f in G which is singular at all of the boundary points (from all possible sides). Therefore, our H is not such a domain of holomorphy. On the other hand people know that every domain is a domain of holomorphy in the one dimensional z-plane.

Next E.E. Levi proposed the following question: Assume that $B \subset \mathbb{C}^n$ is an open set and that $M \subset B$ is a (real) twice continuously differentiable smooth hypersurface, thus a surface of codimension 1. Assume $O \in M$ and that M decomposes B into a positive side B^+ and a negative side B^-. When does there exist an open neighborhood $U(O) \subset B$ and a holomorphic function in the intersection $U \cap B^-$, which is singular at O? One calls M then a *natural boundary at O*.

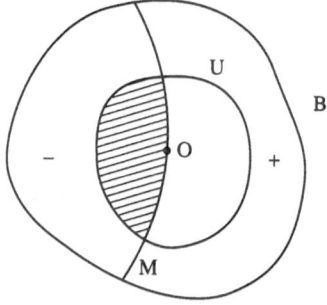

In general, there is no such function. The surface M has to possess a special property of convexity, it has to be *pseudoconvex*. This property is much weaker than the ordinary euclidian convexity, namely, it is invariant under holomorphic coordinate transformations. It is possible to characterize it by an expression from differential geometry. Namely, if we choose U sufficiently small, there is a twice continuously differentiable real function ψ in U, such that the total differential $d\psi$ is $\neq 0$ everywhere and that $\psi < 0$ on the negative side of M and $\psi > 0$ on the positive side of M and $\psi = 0$ on M itself. We obtain the *Levi differential expression*:

$$L(\psi) = \sum \psi_{z_\nu z_\mu}(z) \cdot d z_\nu \, d \bar{z}_\mu.$$

It is a hermitian form for every fixed z. If this form is positive semidefinite at $z = O$ for all vectors dz with

$$\sum \psi_{z_\mu}(O) \, dz_\mu = 0,$$

then M is called pseudoconvex at O from the negative side, if moreover the form is positive definite for the special vectors the surface M is said to be *strongly pseudoconvex* at O. We say that M itself is *(strongly) pseudoconvex* if M is (strongly) pseudoconvex at all of its points.

A deep theorem tells us that M is a natural boundary at O if M is pseudoconvex in a full neighborhood of O. Its proof under these general conditions was not obtained before the famous K. Oka did it in the year 1953. But if M is strongly pseudoconvex in O our result is nearly trivial. In this case there is a smooth (complex) 1-codimensional analytic surface $A = \{f(z) = 0\} \subset V(O)$ through O which does not enter into the negative side of M. We call such a surface a *supporting surface*. Here, V is an open neighborhood of O and f a holomorphic function in V with $df \neq 0$, everywhere. The function which is singular in O is simply the meromorphic function $1/f$.

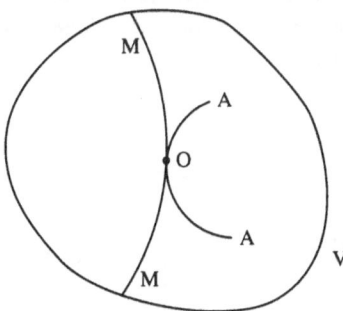

2

We see that in the case $n > 1$ the polar set of our meromorphic function is not a discrete set any longer but an extended surface. Therefore, it became necessary to describe the surfaces $\{f = 0\}$, more precisely. Of course, we have to assume that f does not vanish identically in any neighborhood of O. It proved to be convenient to transform the function f by a unitary rotation into general position. This makes f z_1-*general*, that means that the 1 variable function $f(z_1, 0, \ldots, 0)$ does not vanish identically in any neighborhood of O. After the rotation the variable z_1 will be distinguished. We put $z_1 = w$. We can prove then that in a neighborhood of O an equation of the following kind is valid:

$$f(z) \equiv e(z) \cdot \omega \quad \text{with} \quad \omega = w^b + a_1 w^{b-1} + \ldots + a_b.$$

In this e is a holomorphic function in a neighborhood of O which does not vanish there and

$$a_1(z'), \ldots, a_b(z')$$

are holomorphic functions near to O' with $z' = (z_2, \ldots, z_n)$ and O' is the $(n-1)$-tuple consisting of zeros. The number b is the order of zero of the function $f(w, 0, \ldots, 0)$ at O. After having made the rotation, the decomposition of f is uniquely determined. The zero set $\{f = 0\}$ is identical with $\{\omega = 0\}$ in a neighborhood of O. But the function ω is a normalized polynomial of degree b. By counting points with multiplicity we find that there are just b points in $\{\omega = 0\}$ over every point z'. By discriminant theory it can be proved that there is a branching locus D, which is a local analytic set of dimension $n-2$ in the variables z'. If z' is not in D all the b points over z' are different.

All this means that the *zero set of f* in a neighborhood of O is a *branched finite covering over a neighborhood of O'*. Its form can be studied in this way.

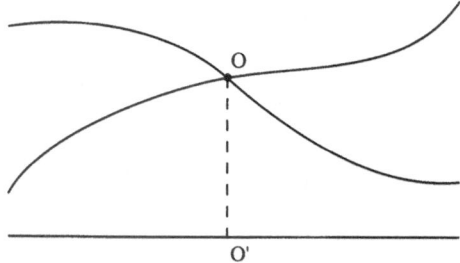

Therefore, the theorem so obtained is also called the *preparation theorem*, the *Weierstrass preparation theorem*. It can be proved most simply by using the methods of power series. In any case the result shows that for $n > 1$ the zero set of any non constant holomorphic function is an extended surface of complex codimension 1. However, in general this surface may not be smooth.

In the theory of functions of 1 complex variable we have the theorem of Mittag-Leffler. Assume that principal parts of certain meromorphic functions are given at the points of a discrete subset of an open domain $B \subset \mathbb{C}$. Then, we can construct a meromorphic function f in B, which has the given principal parts at those points. In the case of several complex variables we get new complications. As with the zeros of the denominators, the polar sets are now extended and cannot contain isolated points. Therefore, it is advisable to proceed as follows. We take an open covering $\mathfrak{U}$ of B and assume that a meromorphic function f_U is given on every element $U \in \mathfrak{U}$ in U. If $U, V \in \mathfrak{U}$ are 2 elements, we assume that the difference $f_V - f_U$ is holomorphic in the intersection $U \cap V$. We may say that f_U and f_V have the same principal parts in $U \cap V$ and call such a distribution a *Cousin-I distribution* in B. This gives a well defined distribution of principal parts everywhere in B. The question is if there is a meromorphic function in B with the given principal parts. It was discovered very early that this is not the case. H. Cartan was even able to prove that in the case of 2 complex dimensions B is a domain of holomorphy if every Cousin-I distribution has such a solution f. However, it turns out that this proposition is wrong in higher complex dimensions. But in the thirties people conjectured that every Cousin-I distribution in any domain of holomorphy of $\mathbb{C}^n$ has a solution. Finally, Oka found new methods and proved that this conjecture is true.

In the case of 1 variable a holomorphic function can be constructed in any domain B having a given discrete set of zeros (with multiplicities!).

In several complex variables the zero sets are extended. In order to get a good distribution of zeros we use an open covering $\mathfrak{U}$ of B, again. We take in every element $U \in \mathfrak{U}$ a holomorphic function f_U, which is nowhere $\equiv 0$. We assume that in the intersection of any 2 elements the quotient of the 2 given functions always is holomorphic and $\neq 0$, everywhere. Then we call such a system a *Cousin-II distribution* in B. Again, the question is: Is there a holomorphic function in B with the given zeros? Unfortunately, this is even not the case in every domain of holomorphy. After many years of research by others Oka was able to prove this: If there is a continuous complex function in B with the given zeros then there is a holomorphic solution, too. That meant that in domains of holomorphy the solvability depends on topological conditions, only. The theory for this was developed by K. Stein.

By his strong methods Oka derived many more similar results. His papers were very difficult to read. They also contained minor errors. It seemed to be necessary to combine these results in a clearly arranged theory. This was done by Cartan and J-P. Serre. They introduced the *analytic sheaves*, a notion which is very near to the Cousin distributions. By an analytic sheaf $\mathscr{S}$ (more precisely a presheaf) over B we understand a system which attaches to every open subset $U \subset B$ a module $\mathscr{S}(U)$ over the ring $\mathcal{O}(U)$ of holomorphic functions in U. For general sheaves also other algebraic structures are possible. The algebraic structure can be put in the name of the sheaf. Therefore, we call an analytic sheaf often a *sheaf of $\mathcal{O}$-modules*. – For all sheaves we have the restriction homomorphisms: If $V \subset U$ is an open subset we have a well defined homomorphism $r_V^U : \mathscr{S}(U) \to \mathscr{S}(V)$. The set of these homomorphisms has the obvious properties. – If $B' \subset B$ is an open subset there is a restriction $\mathscr{S}|B'$.

In the case of a Cousin-I distribution all $\mathscr{S}(U)$ consist of the module of meromorphic functions over U. The r_V^U are just the usual restrictions from U to V. Even more essential is the analytic *sheaf* $\mathcal{O} = \{\mathcal{O}(U)\}$ *of local holomorphic functions*. We need the cohomology theory with coefficients in $\mathcal{O}$ and more general in $\mathscr{S}$. For this only analytic sheaves $\mathscr{S}$ are suitable for which locally there is a good "connection" between its objects: they have to be *coherent*. The proof of coherence is very important, therefore. Theorems of this kind can only be derived in long papers, in general. Up to now, mainly the following results were obtained:

1. The sheaf $\mathcal{O}$ is coherent (Oka).
2. If $A \subset B$ denotes an analytic subset (for definition see a later paragraph) and I_A is the analytic sheaf of local holomorphic functions which vanish on A then I_A is coherent (Oka, Cartan).
3. The sheaf of normalization on a complex space X (for definition see a later paragraph) is coherent (Oka).
4. Assume that $\psi : X \to Y$ is a proper holomorphic map of complex spaces. Then for $\mu = 0, 1, \ldots$ the direct image sheaves $R^\mu \psi_* \mathscr{S}$ in Y of coherent analytic sheaves $\mathscr{S}$ in X are again coherent (Grauert).

As said already, there is a cohomology theory with coefficients in a sheaf $\mathscr{S}$. We can do this for any sheaf $\mathscr{S}$ of modules over a ring R. An analytic sheaf also is a sheaf of modules over the ring $\mathcal{O}(B)$ of holomorphic functions in B. We have (in general non trivial) cohomology groups for dimensions $\mu = 0, 1, 2, \ldots, n$. They are denoted by $H^\mu(B, \mathcal{S})$ and are modules over R. In the case of an analytic sheaf these are complex vector spaces since $\mathbb{C} \subset R = \mathcal{O}(B)$ is a submodule.

Now the solvability of a Cousin-I distribution follows from the vanishing of the first cohomology group of B with coefficients on $\mathcal{O}$. This is true for domains of holomorphy. There are more results of this kind. They were combined in 2 theorems. For stating them we always assume that B is a domain of holomorphy and that $\mathcal{S}$ is a coherent analytic sheaf on B:

THÉORÈME A. The set $\mathcal{S}(B)$ generates locally the sheaf $\mathcal{S}$ in a neighborhood of any point of B by the module structure of $\mathcal{S}(U)$ over $\mathcal{O}(U)$.

THÉORÈME B. All cohomology groups $H^\mu(B, \mathcal{S})$ for dimensions $\mu \geq 1$ vanish.

3

In the function theory of 1 complex variable we have the so called *abstract Riemann surfaces*. There have to be analoguous objects in several complex variables! These are the *complex spaces*. For their definition our sheaf theory is excellently suitable. We start with the definition of an analytic set $A \subset B$:

$A \subset B$ is a closed subset in B and for every point $z \in B$ there is an open neighborhood $U(z) \subset B$ with holomorphic functions $f_1, \ldots, f_l$, such that $U \cap A = \{f_1 = \ldots = f_l = 0\}$.

We can determine the form of A by applying the preparation theorem of Weierstrass iteratively.

By an ideal sheaf I for an analytic set $A \subset B$ we understand a coherent subsheaf of *the sheaf* I_A of local holomorphic functions which vanish on A such that the restriction of I to $B - A$ is the structure sheaf $\mathcal{O}$. If $f_1, \ldots, f_l$ are holomorphic functions in B whose common zero set is A then these functions generate such an ideal sheaf by the structure of modules in $\mathcal{O}$. Like $\mathcal{O}$ the quotient sheaf $\mathcal{O}/I$ is a sheaf of $\mathbb{C}$-algebras. It is coherent and on $B - A$ it is the zero sheaf. We therefore restrict it to A: if $U' \subset A$ is open we take an open subset $U \subset B$ with $U \cap A = U'$ and put $H(U') = \mathcal{O}(U)/I(U)$

and obtain a unique module and by these modules a sheaf of $\mathbb{C}$ algebras H on A. We take this for *structure sheaf* of A. The topological (Hausdorff-)space A together with its structure sheaf is the local prototype of our complex spaces.

Now we have arrived at the notion of complex space. A complex space X is simply a Hausdorff space together with a structure sheaf H, i.e. a sheaf of $\mathbb{C}$-algebras on it, such that locally X is isomorphic to a prototype. The sections $f \in H(U)$ are called the *holomorphic functions* on U. In general, the f cannot be considered as complex valued functions on U, but there is a natural homomorphism τ: $H(U) \to C(U)$, where $C(U)$ denotes the $\mathbb{C}$-algebra of continuous complex functions on U. To every holomorphic function on U a continuous complex function $[f] = \tau(f)$ is attached.

But, in general, τ is not injective. If τ is always injective, we call our complex space X *reduced*. If X is reduced and $H(U)$ always integrally closed in its quotient ring (we use for multiplicative system the set of nowhere identically vanishing holomorphic functions $f \in H(U) \subset C(U)$) then X is said to be a *normal complex space*. – In the case of 1 complex dimension the normal complex spaces just are the abstract Riemann surfaces.

K. Stein transfered the notion of domain of holomorphy to (at first, only a special class of) complex spaces. Today a complex space X is called a *Stein space*, if it has the 2 properties:

1. For every point $x \in X$ there are finitely many holomorphic functions $f_1, \ldots, f_l$ on X, such that x is an isolated point of the set $\{[f_1] = \ldots = [f_l] = 0\}$.

2. For every infinite discrete set $D \subset X$ there is a holomorphic function f in X such that $[f]|D$ is not bounded.

The condition (1) is satisfied for every domain $B \subset \mathbb{C}^n$, the condition (2) means that X is *holomorphically convex*. By a theorem of Cartan and Thullen this characterizes domains of holomorphy. By (1) alone we see that such a complex space is never compact.

For Stein spaces the théorèmes A and B are valid. We can apply both theorems to open coverings of arbitrary complex spaces by Stein spaces (so called *Stein coverings*). By this we obtain important results for more general complex spaces. It follows that on a compact complex space X all cohomology groups with coefficients in a coherent analytic sheaf are finite dimensional complex vector spaces (théorème de finitude). In the case of 1 complex dimension this implies that X is an algebraic curve, where the (generic) meromorphic functions are just the rational functions. This curve can be embedded into a complex projective space P_N and hence it can be treated by algebraic methods.

All these means consist properly speaking in a sophisticated composition of integrations of 1 complex variable and an application of the Weierstrass preparation theorem (which also uses 1 variable). So once upon a time Ph. Griffith called this methodical compound: *one variable a time*.

4

We call a point x of a normal complex space X a *smooth point* of X if there is an open neighborhood $U(x) \subset X$ which is isomorphic to a domain $G \subset \mathbb{C}^n$. The points of G then are holomorphic coordinates in U. All other points are called *singular* points of X. Their totality is an analytic subset $S(X)$ of X which everywhere has codimension 2, at least. An isolated point of $S(X)$ is said to be an *isolated singularity* of X. M. Artin proved that all isolated singularities are algebraic: If $x \in X$ is an isolated singularity then to x there is an affine normal algebraic space Y over the field $\mathbb{C}$. The space Y gives a unique normal complex space Y' such that every regular function on Y becomes a holomorphic function on Y'. Moreover, there is a point $O \in Y'$ and an open neighborhood $V(O)$ which is isomorphic to an open neighborhood $U(x) \subset X$. So we can look upon the isolated singularity x as being an isolated singularity of Y and we can treat this one by purely algebraic methods. In this way Renée Elkik succeeded in constructing the semi universal flat deformation of the germ of X in x. Since her result is algebraic it has certain global properties, different from the analoguous analytic statements.

As mentioned already 1-dimensional compact complex spaces always are algebraic curves. So the algebraic results concerning these curves can be used in complex analysis.

Assume that X is a compact connected normal complex space of dimension n. Then the field K of all meromorphic functions on X is an algebraic function field. That is a result which in this generality was first proved by R. Remmert. Very often it happens that there are only constant meromorphic functions on X. In no case do more than n algebraically and hence analytically independent meromorphic functions exist on X. So we are in a special case if there are n algebraically independent meromorphic functions on X. We call X then a *Moishezon space*. A complex torus is a Moishezon space if and only if the period relations are satisfied. This again is the case exactly if X is a projective algebraic space.

In general, a Moishezon space is not a projective algebraic. However, M. Artin proved that it carries a well defined étale algebraic structure. Among other properties: for every point $x \in X$ there is an unbranched, but in general non proper covering Y such that x lies under Y and Y is an affine algebraic space. Such spaces which have an étale algebraic structure are the most general possible algebraic spaces. It is always possible to treat them by algebraic methods.

We call a normal complex space a *complex manifold* if it consists entirely of smooth points. During the last years the famous *Kodaira vanishing theorem* for holomorphic line bundles F on such compact complex manifolds X could even be proved by algebraic methods (by Deligne and Illusie: if F exists then X is projective algebraic). But it was first proved by Kodaira using elliptic systems of linear partial differential equations.

5

Nowadays, it is impossible to take real analysis out of complex analysis. First steps in applying real methods were taken by S. Bochner during the beginning of the fifties. Using his ideas Kodaira developed the theory of compact *Kähler manifolds*. These are complex manifolds with a special kind of Hermitian metric:

$$ds^2 = \sum_{i,k=1,\ldots,n} g_{i,k}(x)\, dz_i\, d\bar{z}_k.$$

In this formula $z_1, \ldots, z_n$ are local holomorphic coordinates in X. Metrics of our special kind were introduced by E. Kähler already during the thirties. They have in addition the following property. We associate to every Hermitian metric a real 2-dimensional exterior form of type $(1, 1)$:

$$\omega = i \cdot \sum g_{i,k}\, dz_i \wedge d\bar{z}_k.$$

(A form is of type (p, q) with $p, q = 0, \ldots, n$ if each of its summands contains p differentials dz and q differentials $d\bar{z}$. If ψ is a r-dimensional exterior form on X then ψ decomposes in a unique sum of forms of type (p, q) with $p + q = r$.) Now, our metric is said to be Kählerian if ω is closed, that means that the total derivative $d\omega$ is 0. On compact Kähler manifolds Kodaira developed the theory of *harmonic forms* of type (p, q). We denote the space of these harmonic forms by $H^{p,q}(X)$. It has finite dimension. Moreover, it follows from the theory of de Rham that the (topological) r-th cohomology group with coefficients in $\mathbb{C}$ of X is isomorphic with the following direct sum:

$$\sum_{p+q=r} H^{p,q}(X).$$

This decomposition is called the *Hodge decomposition*. Furthermore, Kodaira observed that $H^{p,q}(X)$ is isomorphic with $H^{q,p}(X)$. The decomposition is independent of the choice of the Kähler metric. Then, some properties of the cohomology groups of Kähler manifolds followed. Every projective algebraic manifold is Kählerian, every complex torus is Kählerian. But there are compact non Kählerian manifolds.

Kodaira could generalize the theorem on period relations for compact complex tori. By this he found a new proof for the period relations. He used *holomorphic line bundles*. A holomorphic line bundle F on a complex space is a holomorphic fibration over X whose fibres are 1-dimensional complex vector spaces such that locally F always is a cartesian product of a neighborhood $U \subset X$ with the complex vector space $\mathbb{C}$. The 0-points of the fibres form a holomorphic section in F. This is called the zero section 0.

We see easily that the local holomorphic sections in F give a coherent analytic sheaf F on X. Assume now that X is compact. We then call the line bundle F *negative* if 0 can be *blown down* to a point. That means: we can take 0 out of F and replace it in a direct way by a single point such that we get a new complex space. Kodaira proved by this methods:

If the compact complex manifold X has a negative holomorphic line bundle F then X is a projective algebraic manifold. The vector spaces $H^\mu(X, F)$ for $\mu \neq n$ have dimension 0.

We see again that the statement of the vanishing theorem is of algebraic nature. But it is false in algebraic geometry of characteristic $p \neq 0$. As stated it was proved in characteristic 0 by algebraic methods. The people who did it used Frobenius transforms in characteristic p but with large p.

A Kählerian metric on a normal complex space X is called a *Hodge metric* if the attached exterior form ω has integrals over the 2-dimensional closed cycles of X, which are integers. It can be shown that some negative line bundles F always belong to such a metric. Therefore, from the existence of a Hodge metric on a compact complex manifold X it follows that X is projective algebraic. We can see very easily that this is nothing else but a generalization of the theorem on period relations for complex tori. By the way, this theorem also was proved by purely sheaf theoretic methods even more generally for normal complex spaces (Grauert).

The groups $H^{p,q}(X)$ and $H^\mu(X, F)$ are objects which reflect essential properties of the complex structure of X. Kodaira first constructed harmonic forms and then succeeded, using the compactness of X, in proving that these forms have certain holomorphic properties. But harmonic forms mostly can be constructed simply by use of an optimization. Later on others generalized this method to more general cases. Assume for instance, that X and Y are compact complex Hermitian manifolds. Then a differentiable map $\alpha: X \to Y$ is called a *harmonic map* if it minimizes the energy integral. In many important cases the existence of such an α could be proved and sometimes it was also possible to prove that α is holomorphic (see Eels, Siu, Diederich). Siu and Yau proved the theorem that every compact complex manifold X which has a negative cotangent bundle is the complex projective space P_n.

Harmonic forms are also important for "bounded" complex manifolds. Such complex manifolds have a C^∞-smooth boundary and the union of both is compact (C^∞ means ∞ continuously differentiable). If X is such a complex manifold whose boundary is strongly pseudoconvex, we can define on X the (good) harmonic forms of type (p, q). These are differentiable up to the boundary and satisfy certain boundary conditions. They again represent the cohomology (J. Kohn, H. Rossi).

6

In this paragraph we mostly assume that $X = G$ is a bounded domain in $\mathbb{C}^n$ with C^∞-smooth boundary. Assume that G and B are 2 of these domains and that $P \in \partial G$ and $Q \in \partial B$ are boundary points. The following question came up: When are there neighborhoods $U(P)$ and $V(Q)$ and a biholomorphic map $\psi: U \xrightarrow{\sim} V$ such that ∂G corresponds to ∂B? That means, when are ∂G and ∂B locally isomorphic at P and Q? For instance, when is ∂G (near P) biholomorphically equivalent with the boundary of a hyperball? In 1 complex dimension this always is the case. In higher dimensions nobody did expect it. Finally, Chern and Moser published a long paper. They built up the theory of invariants for the boundaries.

By a generic C^∞-smooth embedding in $\mathbb{C}^n$ every differentiable surface F gets a special structure. We call it CR-structure, nowadays. It consists of a field of complex tangents. Later on abstract differentiable manifolds with CR-structure were treated, with respect to their Chern-Moser invariants and some other properties. A very vast amount of literature was created in this way.

Of course, domains $G \subset \mathbb{C}^n$ with a smooth strongly pseudoconvex boundary were especially important. If P is a boundary point then there is a neighborhood $U(P)$ with a smooth quadratic complex 1-codimensional surface A in it, which passes through P and does not enter in G but contacts ∂G to precisely first order. This surface is a supporting surface in P to ∂G. There was a conjecture that in the general (not necessarily strict) pseudoconvex case a (not necessarily quadratic) supporting surface (which may contact of higher order) always exists. But in the seventies Kohn and L. Nirenberg found a counterexample.

In bounded domains G we have the *Bergman kernel function*. It is especially important for studying the boundary behaviour of certain holomorphic functions in G. In our case ∂G is smooth and strongly pseudoconvex. Therefore, the kernel function should possess a well determined boundary behaviour. This was first considered by Diederich and then finally by Fefferman. The result of Fefferman gave a theorem on the boundary behaviour of biholomorphic maps. The following statement was already known: Assume that G and B are bounded complex 1-dimensional domains with real analytically smooth boundary and that α is a biholomorphic map $B \to G$. Then α can be analytically continued to a biholomorphic map α' of open neighborhoods of the closures of B and G. In the case of several complex variables people had conjectured the analogous theorem. Then, first by the theorem of Fefferman in the strongly pseudoconvex case the continuation of α to a C^∞-map of the closures of G and B was obtained. After this

Hans Levi and Pinçuk found a statement which was in analogy to the Schwarz reflexion principle. Using this it was rather easy to prove the analytic continuation.

If ∂G is only pseudoconvex or even more general, completely new phenomenae occur (as in many other cases). The analysis which had to be used got much more difficult. The problem of analytic continuation was even treated in the case of not injective proper holomorphic maps by Diederich, Fornaess, Baouendi and Rothschild.

If $G \subset \mathbb{C}$ is a simply connected bounded domain with smooth real analytic boundary (for simplicity) then G is a *Runge domain*. If f is a holomorphic function in G (or in a neighborhood of the closure of G) then by the Runge theorem f can be approximated by a sequence of polynomials, which converges in G (resp. in a neighborhood of the closure of G) locally uniformly to f. In several complex variables we have to suppose that G is pseudoconvex. We would like the conjecture then that Runge's approximation theorem should be valid if G is of the topological type of the hyperball. But there are counterexamples (Fornaess, he relied on results by J. Wermer). There is even a real analytic family of such strongly pseudoconvex domains $G(t)$, $t \in R$, such that $G(t)$ is Runge for small t, but no longer for big t. There is a parameter t' such that $G = G(t')$ still is Runge, but the closure of G is not.

According to the old "Ergebnisbericht" written by H. Behnke and P. Thullen, every domain $G \subset \mathbb{C}^n$ has a hull of holomorphy $H(G)$. Every holomorphic function in G can be analytically extended to a holomorphic function in $H(G)$. The question of whether every holomorphic function on a small neighborhood of the closure of G is extendable to an even larger domain remained open. The smallest open domain into which this extension is possible is uniquely determined and called the *Nebenhülle* $N(G)$ of G. So the question was: do we always have $H(G) = N(G)$? Now, Diederich and Fornaess found domains with a C^∞-smooth even pseudoconvex boundary where this is not true. Thus we have $G = H(G) \neq N(G)$. If ∂G is strongly pseudoconvex we always can prove $H(G) = N(G)$. This is also true if ∂G is only pseudoconvex, but real analytic. Diederich and Fornaess proved the last statement around 1977.

If $G \subset \mathbb{C}^n$ is a pseudoconvex domain, then G can be exhausted by a sequence of strongly pseudoconvex domains $G(\mu) \Subset G$ with $G(\mu) \Subset G(\mu + 1)$ and $\mu = 1, 2, 3, \ldots$. For this we have to use the Euclidian metric of $\mathbb{C}^n$, essentially. The proofs of the most important theorems of sheaf theory utilise this exhaustion. Therefore, a generalization to complex spaces would be desirable. But here Diederich and Fornaess very quickly constructed a bounded manifold with a smooth real analytic pseudoconvex boundary, which cannot be exhausted in this way (around 1980). On the other hand a complex manifold which can be exhausted by a sequence of relatively compact strongly pseudoconvex open Stein subsets is not necessarily Stein (Fornaess).

Since a long time people have known that there is a well defined natural notion of distance on most of the Riemann surfaces. We say that these Riemann surfaces have their *hyperbolic geometry*. Around the end of the fifties E. Calabi conjectured that something similar also is true in higher dimensions. Only some years ago Yau proved that this really is the case (existence of a Kähler-Einstein metric). There are many applications of this result. So many other open problems were solved by Yau.

In complex analysis it is impossible to work without Hirzebruch's generalization of the classical theorem of Riemann-Roch to higher dimensional projective algebraic manifolds. The old theorem tells us something on the number of meromorphic functions on compact Riemann surfaces. In 2 dimensions the extension contributed to the classification of algebraic surfaces. Recently, it was used for research of 3-dimensional algebraic manifolds. Then, during the same years the theorem of Riemann-Roch was proved in the context of global differential geometry and most generally in that of pseudodifferential operators (Atiyah and Singer). In this form it turned to be useful even for the theory of elementary particles.

All this shows that the theory of functions of several complex variables led to many new surprising phenomenae. It is self-contained and independent of classical function theory.

Peter Hilton

The Mathematical Component
of a Good Education

Dedicated to Dr. Heinz Götze,
a truly educated man,
in respect and affection

1. INTRODUCTION

The main thesis of this article is that mathematics is, like music,
worth doing for its own sake. This is, of course, not to deny the
great usefulness of mathematics in pure and applied science; indeed,
it was never more useful than it is today, when so many areas
of human enquiry hitherto immune to mathematical contamination
now find mathematics an essential tool in achieving progress and
an essential language for the expression of their pertinent concepts
and results.

However, the usefulness of mathematics, which is a theme receiving perfectly adequate attention in influential circles today, is not
the theme of this article, except insofar as it tends to conceal and
disguise the cultural aspect of mathematics. The role of music suffers
no such distortion, for it is clearly an art whose exercise enriches
composer, performer and audience; and music does not need to
be justified by its contribution to some other aspect of human
existence. Nobody asks, after listening to a Beethoven symphony,
'What is the use of that?'. The usefulness of mathematics, while
perhaps contributing to the relative affluence of its skilled practi-
tioners, does have very deleterious effects on the nature of mathematical education, which it would be well to describe quite explicit-
ly. However, before detailing these distortions, it must be stated
quite unequivocally that there is no gain for the utility of mathematics in committing them – on the contrary, for an appreciation of
mathematics and an understanding of its inherent dynamic are necessary in order to be able to apply it effectively.

The first serious error is the confusion of education with training.
This error, of course, goes far beyond mathematics education[1] –

---

[1] It is a wry commentary on the value-system in the United States that
one speaks there of 'teacher training' and 'driver education'!

our bureaucrats and politicians now use the two terms quite synonymously – but it is particularly meretricious when applied to mathematics. For students, and their parents, believe that mathematics education should consist exclusively of the acquisition of a set of skills which prove useful in their later careers; so the skills must be learnt, that is, committed to memory, and no real understanding need occur. Of course, we cannot, in fact, predict what skills the student will need. What we can predict is that those skills will change and that the student will need to understand and not merely to remember. Adaptability to change is itself a hallmark of successful education, and it is change, not technology, which most aptly characterizes life today and in the foreseeable future. A genuine education enables one to acquire, for oneself, the skills one happens, at a given stage of one's life, to need. A training, on its own, contributes almost nothing to education and produces distressingly ephemeral advantages. Unfortunately some of the most influential formers of opinion in the English-speaking world – notably, Ronald Reagan and Margaret Thatcher – are vociferous advocates of the view that it is the function of our educational institutions to train, and that the success of their mission can therefore be measured by instruments appropriate to the market-place, employing the criteria of cost-benefit analysis.

The usefulness of mathematics leads to other, related abuses. Since mathematics is useful its acquisition must be tested. Since, in the perverse view we are deprecating, it is a skill, it is tested as a skill. Since it is useful, it must be taught to all. Thus the testing problem becomes enormous, and grading by machine becomes commonplace. The result is that the standard tests have almost nothing to do with the acquisition of mathematical understanding and put a premium on brute knowledge and memory, speed and slickness. They provide no opportunity for the student to explain his or her answer and treat all 'wrong' answers as equally wrong. Thus their effect is to distort the teaching and learning processes and the curriculum, in the direction of unmitigated skill-acquisition. They are, in short, inimical to mathematics itself. Let me be quite explicit and unequivocal – a testing procedure which gives the student no opportunity to explain his, or her, answer should have no place in an enlightened curriculum.

Further, the study of mathematics starts with the teaching of arithmetic, a horrible, wretched subject, far removed from real mathematics, but perceived to be useful. So vast numbers of intelligent people become 'mathematics avoiders' although they have never met mathematics. Their desire to avoid the tedium of elementary arithmetic, with its boring, unappetising algorithms and pointless drill-calculations, is perfectly natural and healthy (despite the practice of referring to mathematics avoidance in the language of pathology – 'mathophobia', 'math clinics', 'math anxiety'). Arithmetic is the cholesterol of elementary education, clogging the arteries of learning; if there ever was any justification for inflicting this unpalatable diet on our innocent children, there is none today, with the increasingly ubiquitous availability of low-priced hand-calculators. Yet, as we say, the practice continues, so that very few 'educated' people even understand what mathematics is, let alone have an appreciation of its potential role in enriching their lives and their culture[2].

Thus, to some, it must seem absurd to liken mathematics to music as an art to be savoured and enjoyed even in one's leisure time. Yet that is how it should appear and could appear if it were playing its proper role in our (otherwise) civilized society. Just as an appreciation of music is a hallmark of the educated person, so should be an appreciation of mathematics.

2. An Educated Person

There is, we claim, a valid and valuable concept of an educated person. The ancient Greeks had this concept, and it included for them an appreciation of mathematics, especially geometry; on the other hand, the Romans, conspicuously, did not. As Philip Howard writes, reporting on the 1989 meeting of the British Classical Association: 'The Romans were bad at science. They were practical men who followed intellectual pursuits only if they were useful and profitable, or, in the uncharming vogue phrase, "bankable skills". It is an attitude that is still with us'. Indeed it is! How comfortable

---

[2] 'And what will you make, when you are queen?'
It was silly, this, really – I mean, if any of one's friends could hear ...
'No more maths'
'Ah. That is difficult for the banks and the shops and the men of business. Never mind, we arrange'.
                                        Penelope Lively, The French Exchange.

Mrs. Thatcher would have been in Roman society – if only it had accorded to women access to political dominance.

The broader concept of education was certainly again current in the 17th and 18th centuries in Britain and animated those who founded the Royal Society of London; other nations, too, in Europe and elsewhere in the world, have had their Enlightenments, their Renaissances. However, the concept began to undergo a curious transition in Victorian England. Certainly, it continued to connote the desire and the ability to go on learning, by reading and other forms of study; and it implied a familiarity with, and appreciation for, poetry, literature, music, the arts and architecture. It suggested a philosophical, reflective turn of mind. However, when the transition was complete, it carried two rather unfortunate connotations as well. The term tended to be applied to members of the leisured class (and, naturally but sadly, predominantly to the masculine sex); and there was no implication of a knowledge or appreciation of science.

This last feature largely persists to this day in Britain. Exasperation with its manifestations led C.P. Snow to deliver his celebrated Rede Lecture, 'The Two Cultures', in which he deplored the prevalence, in positions of prominence and influence, of people having no knowledge or understanding of the Second Law of Thermodynamics. Of course, Snow was not the first to remark on this phenomenon, but his own popularity as a writer and reputation as a thinker and man of affairs undoubtedly broadened the discussion, if it did not always succeed in deepening it. It is important to recall that Snow's viewpoint that a person was only to be considered educated if he or she was versed in the arts *and* the sciences was by no means universally accepted at the time his lecture was delivered and published. However, we believe that today the problem of acceptance is very different in certain contemporary societies where cultural philistinism is rampant. Now it is often necessary to argue that a technologically advanced society needs people with an understanding of history, an appreciation of language (theirs and other people's) and an awareness of the 'higher purposes' to which their increased affluence and computerized efficiency give them access. The proneness, to which we have already drawn attention, to confuse education with training has led, at least in the English-speaking world, to a marked down-grading of the study of the arts and the humanities, and to the emergence of the dangerous illusion that a modern industrial society should encourage applied science at the expense of pure science. Such an attitude, had it been widespread 20 years ago, would, for example, have seriously impeded the development of 'medium temperature' superconductors. It is surely clear, moreover, that an educated person should have some understanding of both pure and applied aspects of science. If, for example, he or she is is to appreciate the actual and potential roles of the computer in our and future societies, then the educated person must appreciate and comprehend significant parts of science, technology, logic and mathematics.

3. WHAT IS MATHEMATICS?

Let me at this stage abandon, at least for the time being, my role of Cassandra and discuss briefly the essential qualifications of the educated person. Such a person should, of course, have all the traditional qualifications. In addition such a person must bestride 'The Two Cultures' understanding that both are vital to the individual and to society. With a better and deeper understanding of other civilizations, separated by time or space, or both, from our own, we are the more likely to take a more long-term view of our fundamental purposes on this earth, thus avoiding the glaring errors that arise today from short-term greed allied to technological skill and ignorant pride. However, it is my special case that mathematics is common to the 'two cultures', and the educated person should appreciate it.

It is not reasonable to expect lay persons to understand the details of sophisticated mathematical reasoning. Nevertheless, enough has surely been said to imply that our educated person must appreciate the role of mathematics in science and technology. Richard Feynman, echoing the thought of Galileo, has said: 'Nature talks to us in the language of mathematics', and it behoves educated people to understand just what this profound aphorism implies. Certainly such an understanding cannot be achieved without a far better insight into what mathematics actually is than is commonly found even among university-trained people today. Yet even such an insight, however essential, would not, in my opinion, be adequate; for mathematics grows and develops in many ways unrelated to science, and thus plays a crucial role in the history

of human thought. So I argue that the educated person must understand what mathematics is – but not in the sense of a dictionary
definition. Such a person must have an appreciation of mathematical reasoning and of the role of mathematics in the evolution and
development of human society. Such an appreciation requires one
to understand something of what mathematicians *do* – this would
provide a much better working description of what mathematics
is, in practice, than any dictionary could be expected to provide[3].

Unfortunately, as we will argue below, very few people have
this kind of appreciation of the true nature of mathematics. The
most common fallacy, even among otherwise well-informed people,
is, as we have said, to confuse mathematics with elementary arithmetic, and to suppose that progress in mathematics consists of performing ever more complicated calculations with speed, dexterity
and accuracy. Thus, for example, Dustin Hoffman received an Oscar
in 1989 for his portrayal of the autistic brother in the film 'The
Rain Man'. This person is an 'idiot savant', capable of performing
rapid, totally unmotivated mental calculations such as 341×127,
$\sqrt{19}$ to 10 decimal places. However, he is described by various crit-
ics, in their reviews of the film, as a genius. It is surely unnecessary
for me to belabour the point further that such an extraordinary
ability, far from being evidence of genius, is usually an indication
of stupidity – as in this case. There have been rare exceptions,
such as Gauss, the British civil engineer George Parker Bidder,
and the Scottish statistician A.C. Aitken. However, it is interesting
and significant to note that Gauss' powers of mental calculation
declined as his genius grew, thus testifying to the antithesis between
calculation and mathematical insight which we are claiming to exist.

In real life a characteristic example of the idiot savant was the
Derbyshire agricultural labourer Jedediah Buxton, who was able
to demonstrate that the Fermat number $2^{2^5} + 1$ is not prime by
actually factorizing it when it was given to him in decimal notation.
He performed this feat in his head while carrying out his everyday
duties. Buxton was brought to London to be examined by a group
of Fellows of the Royal Society. He was taken to the theatre to
see Garrick perform, to see how he would react to the experience.

---

[3] Bertrand Russell's famous dictum that 'mathematics is the subject in
which you don't know what you're talking about, and don't care whether
what you say is true' is merely a philosophical joke, though a good one!

He reacted by compulsively counting the number of steps Garrick took during the performance! Thus indeed did Buxton symbolically demonstrate that arithmetical skill, of however high an order, is no part of our culture.

This justified conviction, on the part of many sensitive and 'educated' people, that arithmetic cannot be regarded as a part of the individual's cultural equipment, together with the erroneous belief that arithmetic is the essence of mathematics, has led to the widely-held view that mathematics itself is not to be regarded as a component of a liberal education. Thus many aesthetes are to be found positively glorying in their ignorance of, and ineptitude in, mathematics. Such people may proudly announce that they do not understand railway timetables, and are merely vexed by their difficulty in computing the tip in a restaurant. There are not to be found educated people who glory in their inability to use their language[4] or to read properly; anybody with such a difficulty would doubtless seek to conceal it.

Genuine mathematics, then, its methods and its concepts, by contrast with soulless calculation, constitute one of the finest expressions of the human spirit. The great areas of mathematics – algebra, real analysis, complex analysis, number theory, combinatorics, probability theory, statistics, topology, geometry, and so on – have undoubtedly arisen from our experience of the world around us, in order to systematize that experience, to give it order and coherence, and thereby to enable us to predict and perhaps control future events. However, within each of these areas, and between these areas, progress is very often made with no reference to the real world, but in response to what might be called the mathematician's apprehension of the natural dynamic of mathematics itself. Let us develop this theme further.

Mathematics, while essential to science, as Feynman has so vividly testified, has its own internal dynamic, powerful and subtle. Often, and today most especially, mathematics moves forward not under the stimulus of science but under the stimulus of its own recent advances. Applied mathematicians will often find a piece

---

[4] Regrettably, statistical evidence is accumulating to indicate that students in the United States, offered training rather than education, and fascinated by the potential of modern technology, are increasingly unable to use the English language properly to convey their ideas.

of mathematics, *developed for its own sake*, the precise tool they need for the expression and elucidation of their scientific problem.

All this is commonplace to the mathematician – but it is not for the research mathematicians that I write. Thus I will allow myself to offer a schema for the expert approach to a scientific problem. We may represent it diagrammatically as follows:

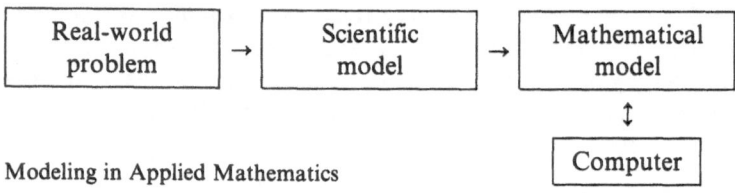

Modeling in Applied Mathematics

Thus, one first constructs a scientific model of the real-world situation one is studying. If the problem requires a physical model, then, typically, the scientific model will be concerned with physical entities (not necessarily observed or observable, perhaps merely postulated) and physical laws (for example, laws of conservation of energy and momentum). One then constructs a mathematical model to reason about the scientific model (for example, a differential equation such as the Navier-Stokes equation or the KdV equation, or an infinite-dimensional Lie algebra in superstring theory). One does calculations, perhaps based on experiment, in special cases and feeds the information back into the mathematical model to assist one in formulating plausible hypotheses and conjecturing general solutions.

Of course, this is a gross oversimplification of the process, which is nothing like as linear as our diagram above suggests. One may well refer back to the real world at some stage and decide that it is necessary to modify the scientific model or the mathematical model. Also one may make a mathematical model of the mathematical model, that is, one may embed the mathematical model in a broader class of mathematical problems for which there already exists a rather substantial theory; for example, the qualitative theory of differential equations of certain types may lead one into the generalization which consists of the study of vector fields on manifolds.

One feature of our modeling schema to which I wish to draw attention is that, with fairly minor modifications, it may be applied to 'pure' mathematics also. In that case the problem comes, of course, from within mathematics. To solve it, however, one may well need, or wish, to generalize the problem in order to be able to apply an existing theory or in order to have a theoretical framework within which to develop a solution. One may also conduct 'experiments', in the sense of considering special cases obtained by specifying certain variables or by simplifying the model (without distorting it); in the first case, it may well be necessary to do some well-motivated calculations.

Thus it emerges that there is no great difference between the procedures of pure and applied mathematics – there is really only one mathematics. Of course there is the difference that the source of the problem comes in one case from mathematics itself and in the other from the real world; but even here this difference is confined to the *original* source of the problem – the applied mathematician grappling with a differential equation is, at that point, behaving in a manner indistinguishable from that of a pure mathematician.

Indeed, to strike a controversial note, it could be argued that the pure mathematician has opportunities for application which transcend those of the applied mathematician. One can apply mathematics to solve problems in physics – but it is difficult (though not, perhaps, absolutely impossible) to conceive of applying physics to solve problems in mathematics. However, within mathematics, it is perfectly clear, indeed commonplace, that one may, for example, apply algebra to solve a problem in geometry, or apply geometry to solve a problem in algebra.

The foregoing discussion is designed to show, in outline, what mathematics is. My own position, as a mathematician, is to be suitably humble about my own contributions to mathematics, but not modest at all about my claims for mathematics itself. This was the position adopted by my teacher and friend, the great British topologist Henry Whitehead – though he had far less justification for his humility! Whitehead argued that there are relatively few pursuits in life which are inherently worth while – he instanced the making of music and the design of elegant and useful furniture – and that doing, or at least appreciating, mathematics is one of them. It is surely reasonable to equate Whitehead's concept of intrinsically valuable pursuits with our own concept of the desiderata of the educated person. There is, in fact, no doubt in my mind that mathematical appreciation is not only a component part of the education of civilized people, but a pillar of that education. I long for the day when, indeed, mathematics will be appreciated and enjoyed by educated laymen as an art and also respected as the mainstay of science. It has been so in the past, but it is not so now. Is it too optimistic to hope it might be so again?

Fame, Sweet and Bitter

雲　の　ごとく　高く
雲　の　ごとく　輝きて
雲　の　ごとく　とらわれず
小川　未明

Mimei Ogawa, 1882–1961,
novelist

As high as clouds
as bright as clouds
as free as clouds

One late afternoon in the spring of 1989, I paid a social call on Dr. Koh Hirasawa (平沢 興) at his Kyoto office where I enjoyed listening to his reminiscences of his years in Niigata as a young anatomist sixty some years ago. He then recalled those three exciting years of research (1927–30) he spent abroad, especially those studies in Germany on pyramidal and extrapyramidal nervous systems, and his later years at Kyoto University where he had held a professorship for seventeen years including six years as a president until his retirement in 1963. The meeting was totally relaxed and very pleasant.

He had a large and beautiful office in a Kyoto-based publishing company, of which he was a "distinguished" consultant and an "honorary" advisor on text books and educational materials that the company produced. His office, the best in the corporate premises, commanded a superb view of the sky aglow with sunset clouds, thanks to wide panoramic windows on two sides. That day, the clouds reminded me of a Spanish poet and philosopher who, according to my friend in mathematics Jose-Manuel Aroca, once said "Clouds are forever." When I casually mentioned this, Dr. Hirasawa seemed deeply impressed and became more interested. He told me of his long-held belief that "Clouds are the universe." Instantly then I felt convinced of a common understanding between us that the whole universe might be one huge hierarchical structure consisting of many different forms of clouds, like nerve systems, networks of blood vessels, interaction of bodies, communication of minds, atmospheric layers, energy levels, galaxies, etc. But after I left his office, I began doubting if I understood what he had really meant and then wondering what images he had been picturing in his mind. I resolved to ask him next time what had been meant by his saying "clouds are the universe." Regretably the meeting was our last since the fine man passed away on June 14 of the same year.

In a sense I had known him for many years. The first time I recognized him was in the spring of 1957 when he became the president of Kyoto University of which I was a graduate student working towards a master's degree in mathematics. Needless to say, the recognition was strictly one way. As students, we heard much of Dr. Hirasawa and often talked about him. His popular book *Progress of medical science* (医学のあゆみ) was so down to earth and accessible to non-medical students that it was widely read even by high school students of the 1950s. When he was elected President of Kyoto University, we were told that he was an excellent anatomist and renown worldwide for his research on pyramidal and extrapyramidal tract systems in connection with the voluntary and involuntary movements of the body.

I left Kyoto University and Japan early in the autumn of 1957 to extend my graduate study under the guidance of Professor Oscar Zariski at Harvard University. After my Ph.D. in mathematics, I chose to take a teaching and research career in America and kept my base in New England for many years. My thoughts were kept confined to mathematics and mathematicians and entirely off from Dr. Hirasawa for nearly two decades. After my election to the Japan Academy in 1976, I began sharing the same table as the medical scientist at the Academy and it was at this time that we became acquainted as colleagues. Naturally I got to know more about his personality which I regarded as being "high, bright and free." For instance, one day he approached me and said "Thank you, Mr. Hironaka." Caught by surprise, I asked "For what, sensei?" He told me of the joy he had felt by understanding a piece of pure mathematics during my short presentation on the work of Masaki Kashiwara. The academician, senior to me by 30 years, appeared completely free of the sense of seniority. My admiration towards him grew faster as I learned more of his activities outside academic

life, particularly his contribution to the promotion of scientific collaboration between Japan and the United States and the interest
he took in the education of Japanese children. Actually, I looked
upon him as a role model for a person dedicated to a wider perspec-
tive of education and I gained inspiration from this image which
encouraged me to embark on a new venture, or rather an adventure.

He said he had been wishing that I would take over his position
at the Kyoto Publishers. He attained his highly privileged position
in the company, since the publishers, editors and readers all loved
his personality. I admiried and envied him, but I declined his offer.
I knew well of my own temperamental karma that would never
make me looked up to as being "so high and so free" within an
established corporation. At the same time I had pride and confidence, partly due to my inexperience and ignorance of the world
outside of academia. I wanted to create and lead my own educational programs, and I began moving fast before thinking things
through, especially after I became "famous" in Japan.

In 1975, I was awarded the *Order of Culture* by Emperor Showa
(昭和) of Japan, which was one of the most envied honors for a
Japanese of any age. The prize was not just a reputed medal to
decorate with but it included a life time pension, annually paid
from the year of the award and about at the level of a retired
professor from a national university. I was considered unusually
young for an Order of Culture in comparison with the normal
awardees in their sixties and seventies. Moreover, special attention
was paid to my case because I was the first awardee of the honor
from among those who were born after the enthronement of Emperor Showa himself. This made quite a lot of publicity through
television, newspapers and popular journals, and I suddenly became
well-known in Japan far beyond a mathematician's normal expecta-
tions.

Fame is sweet when fresh. A new-born reputation is greeted
with lavish praise by the mass media and people do not bother
about whether the popularity is well merited or not. Congratulatory
empathy by close friends dominated any apathy or antipathy of
the envious few. Once born, a fame grows and travels far among
curious people in spite of those serious scholars who look down
on all fame as being spurious. For one thing, an honored fame
provides a quick and easy access to many other famous people.
Speaking of myself, I liked new experiences and enjoyed meeting

VIPs of all kinds, not only those in academia but also many outside my professional circle, such as politicians, musicians, architects, novelists, industrialists, etc. At some point, I began to feel that my good fortune and fame should be put to work not only for myself but also for other people, especially younger mathematicians and junior scientists who would be needier than myself.

I had two thoughts in mind. One thought was to help young people who were interested in exploring foreign countries and becoming acquainted with people of different cultural backgrounds. The other was to provide places and occasions in which young people could have the opportunity to work together in an educationally stimulating environment. Thinking was easy but deeds were not. I needed funds to transform my thoughts into actions and so I looked for ways to trade my fame for hard currency. Unscholarly? Distasteful? Stupid even? Perhaps so but I did not give a damn.

During the academic year 1977–1978, I spent my sabatical year in Japan and discovered three young men who were willing to work with me to raise money, The first of the three collaboraters, Mr. Shohei Nagatsuji (長辻 象平), was a science journalist working for Sankei Newspapers and had been helping me write a series of short mathematical articles for high school students. Such was his scholarly mind, it would have been better, I thought at times, if he had chosen a professorial job in the Ivory Tower. Mr. Minoru Tanaka (田中 穰), my second collaborater, was a producer and organizer of public events and fairs who naturally had many connections with people in the business of publicity and broadcasting. He was a born optimist. The third, Mr. Tatsuya Kobayashi (小林 達也), was a self-made businessman and the owner of several small but profitable corporations. He was rich but financially cautious. When Mr. Tanaka turned too optimistic, Mr. Kobayashi taught us caution and prudence. When Mr. Nagatsuji fell into pessimism after unsuccessful negociations, Mr. Tanaka's optimistic new ideas brightened our mind. Mr. Nagatsuji was totally dedicated to my fund-raising campaign. In many cases of soliciting donations he had to work alone in gathering information, making appointments and going out to negociate on the details.

The three agreed to help me inaugurate the two educational programmes that I had been considering for some time. The first project which I called JMS or the Educational Project for Japanese

Mathematical Scientists (数理科学者育成事業) was intended to support Japanese mathematical scientists under the age of 30 who wished to pursue their studies abroad or to participate in international conferences. The second was what we called Suuri-no-Tsubasa (数理の翼) which literally means "wings of mathematical sciences." This was an annual programme of residential seminars for bright high school students with a strong interest in mathematics and sciences.

At first we naively thought that programmes like JMS should receive some funding from the Japanese government or from the Ministry of Education. After all, JMS was an educational program for Japanese nationals, and in view of Japan's current economic wealth, the government could easily support its fellow students in their studies abroad. The conditions were different when I left Japan for the first time. After the defeat of Japan in the World War II, Japan was poor and her currency was low at the foreign exchange. Even then many Japanese mathematicians and scientists could go overseas and advance their studies, all thanks to financial aid by the host countries, such as fellowships by Galioa, Fulbright, private foundations in the US, French governmental bourses, etc. I was one of those aspiring Japanese youths for whom foreign scholarships were the only means to go and study abroad, and I was proud of my selection to a Fulbright Fellowship. But that was in 1957. Since then, Japan has changed and moved up in the ranks of economic powers. I began to feel that scholars of a wealthy nation such as Japan after 1970s should not be proud of snatching away foreign grant-in-aid that could be otherwise allocated to candidates from much poorer countries. At any rate, we thought that the Ministry of Education could be convinced to support our JMS project. However, talking with some officials at the Ministry, we realised our ignorance of ministrial systems that could not be easily moved to act on a new proposal or even to take one into serious consideration.

Three major objections were raised against our proposal:

1. The education in our JMS proposal was strongly tied with the policies of foreign institutions upon which the Ministry had no power of supervision.
2. The proposal was not made through the proper channel.
3. The proposal was supported by only a few proponents (myself and my collaboraters).

I did not have much faith in the so-called "proper channels" nor in the "broad-ranged consensus" which prevails in the academic world. More seriously, I had no intention of modifying our programme in order to a commodate the first objection.

We soon realised that our funding would have to rely exclusively on donations from private sources, at least for the first few years. We therefore began discussions with business people and financiers to assess the extent to which they would support our JMS proposal. I could contact Mr. Takeda (武田 豊), a Vice-President and later President of the Japan Steel Corporation (新日本製鉄株式会社), who was very sympathetic to our problems and who has continued to be helpful for a long time since. I met several senior executives of the Matsushita Corporation (松下電器産業株式会社), in particular its founder Mr. Konosuke Matsushita (松下幸之助) and his longtime adviser Mr. Kitae Ogawa (小川 鍛). Mr. Ogawa proved extremely helpful in the beginning and later at many crucial turningpoints. It was Mr. Ogawa who first introduced me to several important representatives of the Keidanren (経済団体連合会). Talking with them was very educational to us especially with regard to their explanations of the structure of the federation formed by large Japanese businesses. We received much important advice from Mr. Hanamura (花村仁八郎, 経団連副会長) who was known as the most informed and influential man in all kind of fund-raising through Keidanren. It soon transpired that it would be extremely difficult to raise any substantial amount of funding without firstly obtaining tax exempt status from the Japanese government.

We also discovered that the creation of a charitable organisation with tax exemption status in Japan was far more difficult than in the United States. An offical at the Ministry of Education informed us that:

1. The Ministry of Education would not "approve" a new organization unless, to begin with, the organisation has a fixed fund of at least ¥ 100,000,000 which is certified by banks. The "approval" means that the Ministry agrees to be the governmental supervisor which any charaitable organization must have by law. Incidentally, the initial fund is not exempt from tax.

2. Tax exemption status can be granted by the Ministry of Finance only after the Ministry of Education has approved the organi-

zation and then proposed its tax exemption to the Ministry of Finance.

3. The Ministry of Education would not propose its tax exemption to the Ministry of Finance unless, after the approval, the organization has survived a trial period of at least a year and has proven its ability to function properly by using only the interest and dividend of its fixed fund.

At first it looked like an impossible task and it left us in a dilemma. On the one hand we needed a tax exemption status to attract sizeable donations and on the other we had to have a large sum of money to obtain this tax exemption.

This left us with only one option: Go straight to the top! We began making appointments with some heads of industry and political leaders. If this proved difficult, we looked for special connections. I first wanted to meet Mr. Ohira (大平 正芳), the Prime Minister of the time. Our meeting was arranged by his personal assistant Mr. Morita (森田 一) and took place on September 4, 1979. I had earlier been introduced to Mr. Morita by Mr. Nakahara (中原 伸之), then Senior Executive (常務取締役) and now President of Tonen Ltd. (東亜燃料株式会社). I had known Mr. Nakahara since 1958 when we were both graduate students at Harvard University. The Prime Minister was very relaxed and willingly listened to my views and wishes. It was my first meeting face to face with any prime minister and yet I did not feel any uneasiness. Although Mr. Ohira could not promise anything, I gained an extremely reassuring impression of him because of his sincere and friendly attitude, all of which was contrary to my earlier images of politicians in high office.

On March 27, 1980, after having made further contact with several business executives and having received some pledges of donations, I had the pleasure of having a second meeting with the Prime Minister. This time, I was accompanied by two congressmen: Mr. Nikaido (二階堂 進, 衆議院議員) who was an important figure in the leading political party LDP (Liberal Democratic Party, 自民党) and Mr. Takeiri (竹入 義勝) who was the leader (委員長) of an opposition party Komeito (公明党). Incidentally, I had met and become acquainted with Mr. Takeiri as early as August 1979 and he later introduced me to Mr. Nikaido. They appeared to be good friends although they belonged to parties which are politically opposed to each other. In that last meeting, Mr. Ohira telephoned Mr.

Takeshita (竹下 登) and arranged an appointment for me with the man who himself became Prime Minister several years later. At the time Mr. Takeshita was the Minister of Finance in Mr. Ohira's cabinet. To my sadness, this meeting with Mr. Ohira was to be the last because of his sudden death soon after.

Two days later, Mr. Takeiri and I paid a visit to Mr. Takeshita at his office. Mr. Takeshita was very considerate and the meeting was pleasant. I had just one aim: to obtain government authorization granting tax exemption status to our JMS project. The difficulties explained by Mr. Takeshita and by his associates were more or less the same as those that had been explained to us again and again by the officials in charge at the Ministry of Education. It became obvious to me that no minister, even a prime minister, has the power to sweep away those bureaucratic obstacles that were mined in the mountains of laws, precedents, customs, specifications, orders, agreements, understandings and the like.

Nonetheless, the meeting with the Minister of Finance was informative and indeed educational to me as I could learn the politicians' view of the nature of our problem which was different from the bureaucrats' view. More importantly, the meeting was productive. He called an official at the Ministry of Education and came up with a brand new idea; a possible solution to our problem. The idea was to investigate the possibility of adding our JMS programme to an old, established organization which already had been granted tax exemption. I was told that there were some inactive charitable organizations whose assets had been far too reduced by inflation and mismanagement over many years. Such organizations may welcome our program if it meant an injection of new funds. However, the old directors and officials may then wish to remain, knowing that new donors were available. Since they have their own wishes, customs, rules and precedents, there could be confrontations between the old timers and us, the new comers. The one that welcomes us readily might be the one that troubles us most later. In any case, it was very difficult to select one with confidence. One ideal host organization we thought of was the Gakushin (学術振興会) or the JSPS, the Japan Association for Promotion of Science. This was almost totally financed by the Japanese government and was closely affiliated with the Ministry of Education. It was very sound, active, highly respectable and well known even to many foreign scientists who had been invited to Japan as Gakushin scholars. We would be happy with the JSPS but we would not be welcome unless we obtained strong backing from the Ministry of Education. As is customary with any wistful proposition, there was some level of difficulty associated with the JSPS rules.

Nevertheless, we wanted to push ahead with the JSPS idea with all the means at our disposal. My collaborators, Mr. Nagatsuji (長辻 象平) and Mr. Tanaka (田中 穰) along with our first administrative assistant Ms. Miyazaki (宮崎 道子), made numerous contacts with officials in charge at the Ministry of Education and in the JSPS. Mr. Nagatsuji decided to work full time with me and he often had to work alone in establishing contact and developing connections with ministorial officials on the one hand and potential donors on the other. His negociation with the government officials appeared to be often perplexing and occasionally painful. Ms. Miyazaki, excellent and dedicated, was an indispensable assistant to Mr. Nagatsuji and myself. Fortunately, the Ministry of Education had its sympathizers too, for instance the unforgetable Mr. Osaki (大崎 仁) whose clear understanding and sincere support were outstanding. On September 1, 1980, after the details and specifications were worked out between Mr. Nagatsuji and the governmental officials, I had a formal meeting with the Minister of Education, Mr. Tatsuo Tanaka (田中 龍夫). Our JMS programme was at last established within the framework of the JSPS with a term of duration set at 10 years from 1980 to 1990 (see Appendix 1).

Our fund-raising campaign for the JMS programme began officially and in earnest from that point. We were very pleased and much encouraged by an immediate and generous donation from Sumitomo Metal, Ltd. (住友金属株式会社), that was made possible by the strong support of its Chairman Mr. Hyuga (日向 方斎) who was also heading Kankeiren (関西経済連合会会長), the federation of businesses in the Kansai area (headquarter in Osaka). Thanks to this donation, the JMS scholarship programme began at once with four selected JMS fellows.

Two of the four went to Harvard University: Mr. T. Shiota (塩田 隆比呂) from Nagoya University, and Mr. A. Yukie (雪江 明彦) from Tokyo University. The other two went to Princeton University: Mr. Y. Suzuki (鈴木 康正) from Tokyo University and Mr. S. Bando (板東 重稔) from Tohoku University.

In October, 1980, we received two major donations: one from Matsushita Corporation (松下電器産業株式会社) and another from Mr. T. Taniguchi (谷口 豊三郎). "Taniguchi" is a very familiar name among mathematicians of many nationalities due to a series of highly respected international conferences in Katata (堅田) near Kyoto, held annually for more than two decades and all financed by the Taniguchi Foundation. Incidentally, Mr. Taniguchi's support for mathematicians stretches back more than half a century and is still continuing today. It began with his personal subsidy to mathematical studies of Mr. Yasuo Akizuki (秋月 康夫). At that time, Mr. Taniguchi was a youthful executive of Toyo Textile Industry, Ltd. (東洋紡績株式会社) while Prof. Akizuki, once his classmate, was a mathematician at Kyoto University. Mr. Taniguchi once said that he did not understand any advanced mathematical works but he had been fascinated by mathematicians' passionate devotion. He once spoke of his belief that if meetings were to be truly creative they would have to be informal and intimate. The residential conferences at Katata have been unique in that spirit of Mr. Taniguchi. Mr. Taniguchi's generosity extended to mathematicians in many other occasions. Very recently, for instance, his support was important in connection with the 1990 International Congress of Mathematicians held in Kyoto. Speaking personally, I am deeply indebted to Mr. Taniguchi not only monetarily but also intellectually. Having been on the board of directors of the Taniguchi Foundation and having had many opportunities to hear Mr. Taniguchi's opinion, I learned a lot in terms of how to manage a charitable organization and about what a private foundation should do or should not do. Mr. Taniguchi was one of those I looked up as being "high, bright and free."

To raise funds for the JMS Project, we were willing to do anything that a mathematician and his limited number of collaborators could do. For instance, we helped various commercial advertisements for those companies which offered extraordinary donations to the JMS Project. The total donations to our JMS programme within JSPS amounted to ¥ 181,111,000. All the donations were used as scholarships to more than hundred young Japanese mathematicians in support of their studies overseas. Not a single yen out of those donations was used for the administration of the JMS programme, whether in the office of JSPS or at our place for managerial works. To cover inevitable administrative cost, Mr. Kobayashi helped us to incorporate a private firm called Hironaka Kyoiku Kenkyusho, Ltd. (株式会社広中教育研究所). This provided us with an office and two administrative assistants solely for the purpose of managing the JMS Programme and the second project which I am going to explain next.

The second project was Suuri-no-Tsubasa (数理の翼), the annual residential seminars for selected high school students. We anticipated that it would be even more difficult to obtain a swift governmental authorization of this project, and we decided to go ahead completely by our own means and independently on our own formulation. Here the major financial backing was offered by the private company (株式会社) H.K.K. (広中教育研究所) which was incorporated on August 31, 1979, after a year's planning and preparation. Its corporate activity was headed by Mr. Kobayashi (小林 達也) who was by far the largest stock holder. Mr. Tanaka (田中 穣) joined as a secondary stock holder of the corporation whose business in those days was to create and sell educational materials and services. Mr. Nagatsuji (長辻 象平) left the Sankei Newspaper company in order to work full-time with me and he was put on the payroll of the H.K.K. Thanks to this hidden support, all of the donations to the JMS fund could be used for scholarships and none were lost on administrative expenses. The Suuri-no-tsubasa seminars were in fact financed by the H.K.K. and by local donors from wherever the seminars were held.

The first meeting was held in a seminar house in the Nasu Heights (那須高原), Aug. 8–12, 1980, and was attended by 47 selected and invited participants consisting of 30 high school students, 13 college students and 4 graduate students. These four were those awarded the first JMS scholarships. It was planned that the main body of participants would consist of high school students, aged between 14 and 19. We wrote to about 300 high schools all over Japan, requesting that one or two students showing strong interest in mathematics and sciences be recommended. High school students could not apply on their own behalf. The invited 30 students were selected by us from only those recommended and the selection was done by us using special evaluation forms filled in by their teachers. The university participants, however, were selected on the strength of their own applications which were in response to our leaflets. We distributed leaflets to more than 50 national and private universities in Japan. Participation was completely free of charge but strictly limited to those we selected and invited. Full board and lodging were provided for all participants as well as their full travelling expenses.

One of the main objectives of these residential seminars was to give young, bright high school students the experience of living and studying together in a new environment. Coming from different provinces, their mutual stimulation could be fresher. Having shared room and board, their friendship might last longer. We also included university students who we hoped would provide junior participants with an insight into their future life and studies.

Since 1980 the seminar has been held annually during the summer vacation and the eleventh was completed in August 1990. We have allowed the number of participants to grow slightly but there have been yearly fluctuations. We have selected a new seminar site each year, and the numbers have often depended on the size of the available budget and accommodation. We tried not to change the basic style of the seminar during the first decade, but yet a slow and steady evolution has been taking place. Immediately after the fourth meeting in 1983, volunteers from the list of past participants formed an alumni association which I named "Yugain Club" (湧源クラブ). The word "yugain" (湧源) literally means the source of spring. The club members translated it as "un puits de science" and chose this as the title of their semi-annual bulletin. Looking at the membership list of the Yugain Club of September 1989, I found that the number of members had grown to 421 and that 73 of them were already working professionally. Among these professionals, there were 3 assistant professors and 4 lecturers in universities. I noted 3 practicing medical doctors and also an impressive list of corporate names that the others were working with. Student members of the Yugain Club totalled 197 in colleges and 66 in graduate schools. Of these university students, about 42% were majoring in mathematics and basic sciences, while 24% were in engineering and 20% in medical science. Others were divided among different disciplines such as law, literature, economics and sociology. The rest were still high school students. From the 5th year, club member began assisting us in planning and organising the seminars. In recent years some senior members of the club have begun to appear on the list of speakers at the seminars.

At the eleventh meeting, to which 57 new high school students and 10 new college students were invited, four lecturers and six staff members were volunteers from the Yugain Club. A few more from the Club came to assist the guests from the Harvard University, Professor Sheldon Glashow, physicist, and Dr. Ruth Laurence, mathematician, both of whom were invited by the Japan Association for Mathematical Sciences (数理科学振興会, see below) to give lectures.

The Suuri-no-tsubasa seminars have been normally held in Japan, though the location has changed from year to year. There have been two exceptions that took place overseas. The fifth seminar was held in Seoul, Korea, where 21 Korean students joined us with the 40 students from Japan. The tenth anniversary seminar in the summer of 1989 was held in Australia and 10 Australian college students were invited to join 128 Japanese participants. Most of these participants from Japan were members of the Yugain Club and had attended in order to celebrate its anniversary.

Early in 1984, our organization that had originated in Tokyo was expanded to include a Kyoto office which was to be managed by my new associate Mr. Shiozaki (塩崎 武男). He had been extremely helpful in my fund-raising campaign in the kansai (関西) area, especially in Kyoto and Osaka. We were also lucky enough to find two more excellent and dedicated administrative assistants, Ms. Minami (南 知子) for the Kyoto office and Ms. Yoshida (吉田 津多枝) for the Tokyo office in addition to Ms. Miyazaki (宮崎 道子) who had been working with me in Tokyo since 1979. Mr. Udo (宇土 条治) also joined forces with us after his retirement from the directorship of the Kyoto Imperial House. He was the best choice for this position of heavy responsibility, supervising our entire administrative and accounting operations. On December 4, 1984, the Ministry of Education officially recognized our group as an incorporated charitable organization (財団法人), which I named JAMS or Japan Association for Mathematical Sciences (数理科学振興会). We were lucky and very pleased that Mr. Koukichi Nadao (灘尾 弘吉) agreed to serve as the Chairman of the Board of Directors. He had been appointed twice as Minister of Education (文部大臣) and once as Speaker of the House of Representatives (衆議院議長). He was then a retired politician and highly respected by incumbent political leaders, bureaucrats, businessmen and scholars. On April 25 of the following year, less than half a year after the incorporation, we were granted tax exemption status by the Ministry of Education. Thanks to Mr. Nadao, our communication with government ministries became smoother and our fundraising campaign worked more effectively. Moreover the semi-annual meetings of the Board of Directors and Trustees were run very efficiently. He was completely selfless with us, serving the foundation for six years with absolutely no momentary gain or tangible reward.

Thanks to the tax-exempt status, JAMS was able to expand beyond its pre-1985 activities. In 1985 we began awarding grants and fellowships to mathematicians who were not necessarily under 30, and in 1988 we inaugurated a new annual residential seminar called the Japan-US JAMS Seminars to which we invited about 30 college and graduate students, mostly American, to Japan and about an equal number of Japanese students. Again all travel expense, board and lodgings were provided by donations to JAMS. In 1990, the Japan-US JAMS seminar was attended by 26 participants from the United States, 5 from the Soviet Union, and 1 from the People's Republic of China. Unfortunately, two students from Beijing who applied and were accepted were unable to participate because they could not obtain passports.

This spring Mr. Nadao, after his 90th birthday, finally stepped down from the JAMS Chairmanship and we could not find a suitable person to replace him. Until we find a right person, I acceded to the position in spite of the terrible difficulty that my professorship at Harvard University would keep me away from Japan. The JAMS foundation is still very small with a fixed fund of only ¥ 460,000,000, which must be kept in safe and sound investments. Only the income from these investments is available for the internal and external operations of the foundation, unless of course new funds were constantly poured into it. JAMS has never had any special tie with any big corporation nor with any rich patron. This is good for its administrative independence but not so for its financial freedom. The foundation was essentially created by one crazy mathematician, originally from a poor family, and by a few non-mathematicians who happened to be crazy enough to work so hard and so long for the cause proposed by the mathematician. The sad fact is that the income from the JAMS's fixed fund is barely enough for its internal expenses or the cost of maintaining offices and secretaries. That of course leaves no money to support mathematicians. This is ridiculous because all our dedication and sacrifice to create the foundation had been motivated and upheld by a single purpose that was to support mathematicians and to promote mathematical sciences.

In spite of all that, JAMS has been spending a substantial amount of money externally, i.e., on scholarships, grants, seminars, and so on. How can this be? The answer is very simple. JAMS gets new donations every year and for that JAMS' tax exemption status is very helpful. Is the tax exemption status safe and forever? Definitely not forever, but we hope it is safe for a while so long as our foundation is deemed to maintain good standing by the Ministry of Education. In other words, JAMS must maintain its current level of grants and scholarships. For this, we must obtain new donations every year. In Japan, tax exemption is granted for only two years at a time. It must be renewed every two years, and the renewal could involve extensive effort and tough negociations between our staff and the officials in charge at the Ministry of Education. Last year for instance, JAMS temporarily lost this status due to the "rigor and care" in the bureaucratic processing of our renewal application. Let us assume that our tax exemption status would be safe. Why should anybody then give new donations every year? The answer is not simple. My collaborators, none of whom are mathematicians, have been working very hard year after year to obtain new donations for JAMS. Why should they do this? I really don't know. For one thing, my fame in Japan has been very helpful and I am glad that I have been able to transform at least some of this fame into financial support for other mathematicians. If this is the case, why should I care if my fame is vacuous or fake? JAMS has been struggling to squeeze out a small amount of grants to pure mathematicians. This granting has also been necessary for JAMS to survive as a charitable organization. At the same time, it was not so easy to obtain large donations simply to support pure mathematicians. It has been much easier to find sponsors for more conspicuous events like the Suuri-no-Tsubasa seminars and the Japan-US JAMS seminars, because these could catch a better attention and a broader recognition by the public media.

A crazy mathematician and crazy non-mathematicians obviously do not mind being considered or even treated like beggars. I just think of Gessen The Beggar (乞食 月僊), see Appendix 2) and I do not feel too bad. But if the mathematician loses faith in the value of what he has been doing, then he will meet with serious trouble. He could turn into a "patient diagnosed of terminal illness" similar to the way in which the famous psychologist Elizabeth Kübler-Ross described in her book "On Death and Dying" (Travistock Publications, 1970). The stages that a patient sentenced death normally undergoes are: Stage I *Denial*, Stage II *Anger*, Stage III *Bargaining*, Stage IV *Depression*, and the final Stage V *Acceptance*. First he would strongly deny the meaninglessness of his past endeavor, though he would feel somewhat guilty of not having done better. An internal conflict will definitely be felt. Then perhaps some kind of loneliness, being sick of emptiness in fame. The whole thing might just have been "much ado about nothing," or perhaps a comedy of false fame. Deep fatigue after long struggle. Then comes the stage of anger. Complaints and even accusations begin to surface. "Why me?" Anger is first felt against some of his close friends, then against himself and finally against everybody indiscriminately. After a while he becomes calmer and more logical, and then he begins to bargain. For instance, he might say, "let us abandon one of the expensive projects" just to save the scholarship programme. "No, let us wind down the scholarship programme to offset the increasing cost of the rent and salaries." "Sacrificing the tax-exemption status?" "What else?" Well, give up one of the offices even though it would cause great inconvenience to the future fund raising campaign." Nothing works if he tries to please everybody inside the foundation and outside. He then may fall into the stage of depression.

Finally and hopefully he may come to some sort of Buddhist resignation, simply accepting everything as it is and allowing events ot their natual course. He stops questioning whether right or wrong. After all, the programmes may continue as they are, until the foundation get resolved by itself. He quietly waits for the time to fade away. Until such a state of mind settles in, an aged fame can be felt bitter from time to time.

I remember that on October 28, 1983, I met with Mr. Stanford R. Ovshinsky in Tokyo. He was an internationally renowned scientist and entrepreneur who had made numerous important contributions to amorphous materials. That meeting was the very first time we met but I was strongly impressed with his fine personality. I was so pleased that our conversation continued for hours through a diner. Many of his words were impressive. The one thing he said and which I never forget was: "Some persons were chosen never to say: Hell with it." Whenever I think or say "Hell with it," I remember Stan Ovshinsky.

My adventure with JAMS might have been a lost decade of my life, but thanks to heaven I discovered many beautiful people, high, bright and yet free, outside my small professional circle.

Heisuke Hironaka
January 6, 1991

APPENDIX 1: A List of Important Names for the Creation of JMS.

The following is a very brief list of people I personally met and with them I discussed our JMS programme:

Year	Name	Position
1979 June 31	武田 豊 Yutaka Takeda	新日本製鉄株式会社副社長 the V-P of the Japan Steel Corporation
Aug 17	松下 幸之助 Kounosuke Matsushita	松下電器産業株式会社創業者 the founder of Matsushita Corporation
	小川 鍛 Ogawa	松下電器産業株式会社常任顧問 the standing adviser of Matsushita Corporation
Aug 19	北条 浩 Hiroshi Hojo	創価学会会長 the President of Soka-Gakkai
Aug 30	花村 仁八郎 Jinpatiro Hanamura	経済団体連合会副会長 the V-P of Keidanren
	小川 鍛 Ogawa	松下電器産業株式会社常任顧問 the standing adviser of Matsushita Corporation

Year	Name	Position
Aug 31	中原 伸之 Nobuyuki Nakahara	東亜燃料株式会社常務取締役 a managing director of Toa Nenryo
	北条 浩 Hiroshi Hojo	創価学会会長 the President of Soka-Gakkai
	竹入 義勝 Yoshikatsu Takeiri	公明党委員長 the chairman of the Komeito party
	（株）広中教育研究所設立	資本金 ¥ 50,000,000 理事長（取締役社長） 小林 達也
	the establishment of Hironaka research	institute of education with a capital of fifty milion yen the chairman of the board of directors Tasuya Kobayashi
Sept 4	大平 正芳 Masayoshi Ohira	内閣総理大臣 the Prime Minister
	森田 … Hajime Morita	内閣総理大臣秘書官 the private assistant to the Prime Minister
Oct 30	岸 信介 Nobusuke Kish	元内閣総理大臣 the former prime minister
	小川 鍛 Ogawa	松下電器産業株式会社常任顧問 the standing adviser of Matsushita Corporation
Nov 22	日向 方斎 Hosai Hyuga	関西経済連合会会長 the President of Kansai Keizai Rengokai (Kansai Economic Federatopm)
	古川 普 Susumu Furukawa	関西経済連合会専務理事 the managing director of Kansai Keizai Rengokai

Year	Name	Position
Nov 23	弥永 昌吉 Shokiti Ienaga 吉田 耕作 Kosaku Yoshida	数学者 Mathematician 数学者 Mathematician
Nov 24	伊藤 清 Kiyoshi Ito	数学者 Mathematician
Dec 20	永野 重雄 Shigeo Nagano	東京商工会議所会頭 the President of the Chamber Commerce and Industry Tokyo
Dec 21	佐治 敬三 Keizo Saji 日向 方斎 Hosai Hyuga 谷口 豊三郎 Toyozaburo Taniguchi	Suntory 株式会社社長 the President of Suntory Co. 関西経済連合会会長 the President of Kansai Keizai Rengokai 東洋紡績株式会社相談役 the Adviser of Toyo-Boseki Co.
Dec 23	秋月 康夫 Yasuo Akizuki	数学者 Mathematician
1980 March 27	大平 正芳 Masayoshi Ohira 二階堂 進 Susumu Nikaidoa 竹入 義勝 Yoshikatsu Takeiri	内閣総理大臣 the Prime Minister 衆議院議員 member of the House of Representatives 公明党委員長 the chairman of the Komeito party
March 29	竹下 登 Noboru Takeshita 竹入 義勝 Yoshikatsu Takeiri	大蔵大臣 the Minister of Finance 公明党委員長 the chairman of the Komeito party

Year	Name	Position

(The JSPS method was suggested)

April 8 （長辻 象平 面会）　篠沢 公平 文部省学術国際局局長
An interview with Shohei Nagatsuji
(Kohei Shinozawa, the director
of the international and
science bureau of the
Ministry of Education)

April 10 （長辻 象平 相談）　数学者 小平 邦彦, 吉田 耕作,
伊藤 清, 弥永 昌吉, 秋月 康夫
An interview with Shohei Nagatsuji
(Kunihiko Kodaira, Kosaku
Yoshida, Shokichi Ienaga,
Yasuo Akitsuki:
mathematicians)

May 23 町村 鉄雄　　　住友銀行専務取締役
Tetsuo Machimura the managing director
of Sumitomo Bank

Aug 28 本田 宗一郎　　本田技研株式会社
Soihiro Honda Honda Giken Co.

Sept 1 田中 龍夫　　　文部大臣
Tatsuo Tanaka   the Ministry of education
JSPS（学術振興会)内 新規事業項目
JMS（数理科学者育成事業)発足

APPENDIX 2: The Legend of Gessen the Beggar

There once was a monk by the name of Gessen (月僊), born in
1741 and died in 1809, whose unusual talent for paintings was
widely known. Many people, however, referred to him as Gessen
the Beggar (乞食 月僊) because when he was requested to paint,
he never failed to ask how much the customer was willing to pay.
Moreover, he preferred to receive payments in advance. He never
gave in to bargaining on completing the work, not even to a good
friend who may have previously been generous to him. Comissioned
by wealthy daimyos (大名, feudal lords), he produced paintings for
a fee of 50 ryo (両, one ryo contained 13 grams of gold which
was 65% of the total weight, the rest being silver). Ordinary people
who required inexpensive pieces of work would receive hastely completed paintings for a fee of one or two shu (朱, one 16th of one
ryo). Naturally he painted a lot, perhaps the largest number of
paintings by one artist in the history of Japanese paintings. It has
been said that he produced several tens of thousands of paintings.

One famous story about Gessen the Beggar involves a geisha
Shozan (松山). One day she commissioned a scroll painting by Gessen to decorate her tokonoma (床の間), a sort of alcove or niche
in a living room, following a suggestion by her favourite consort,
Kuwanaya (桑名屋), a rich rice dealer of the time. Gessen asked
how much she was willing to pay and Shozan replied she was
willing to pay any price. A few days later, Gessen took his work
to the geisha house where Shozan was having a dinner party with
Kuwanaya and his associates. The painter displayed his painstaking
and beautifully completed work, and Shozan liked it. But Shozan
was outraged when Gessen requested 10 ryo for the work, much
more than she wished to pay, but Gessen of course refused to bargain. After she paid with Kuwanaya's help, she declared: "Now
that Gessen's painting is in my possession, am I not able to do
anything I like with it?" She undressed before the guests, hung
up her koshimaki (腰巻, or undergarment) in the tokonoma and
then wrapped herself up in the scroll. She said, "I have gained
a pretty tunic instead of an excellent scroll haven't I?" The monk
artist observed Shozan's whole act quietly without losing his
temper. Then, with the 10 ryo in his pocket, he thanked everyone
and left the geisha house.

Daigado (大雅堂) or Ikeno-daiga (池大雅, born in 1723 and died in 1776), an esteemed artist of the day, appreciated Gessen's talent. Gessen meanwhile had looked up to Daigado as one of his mentors. One day Daigado asked Gessen why he had deliberately damaged his name by behaving so avariciously. The monk for the first time confessed that he had just three wishes in his life for which he needed all the money he could make and he did not mind what he was called in the process. The first of these wishes was to accomplish his temple master's lifetime ambition. Gessen's master had died without having realized this wish (寺再建) to reconstruct the hondo (本堂) of his Buddhist temple, Ise Jakushouji (伊勢の寂照寺), which had been badly damaged by a fire. The second of Gessen's wishes was the so-called michibushin (道普請) or reconstruction of the sando (参道), a roadway for worshippers which lead to the most revered of all Shinto (神道) shrines, Ise Jingu (伊勢神宮). Part of the road led through a forest which was long neglected and even dangerous. He wanted to renovate the sando for Shinto worshippers and also for Buddhist believers. Finally, he told of the last wish which was to aid the poor (貧民救済) of the then Tenmei era (天明). He found so many homeless beggars, hungry and sick, owing to a national famine of the Tenmei era (天明飢饉) that lasted for four years from 1783. People suffered also from a great fire in the Ise district in 1804. He wanted to help them by giving them shelters and as much food as he could. Daigado was said to be deeply impressed by the sincerity of Gessen's intentions. The reconstruction of the Jakushouji's hondo began in 1791 and was completed in 1797. The renovation of the Ise Sando began in around 1781 and was extended to building a new bridge over the river Miyakawa (宮川) all of which was financed by Gessen. After having helped many homeless poor, he established an aid fund of 1500 ryo. Just before his death at the age of 69, he distributed all his fortune among his disciples and employees. It was said that even the rice maid inherited as much as 10 ryo.

1. 再建 (saiken)＝rebuild
2. 普請 (fushin)＝civil enginering, in particular building or renovating michi (道＝roads) for public use
3. 救済 (kyusai)＝aid to the misfortuned

Friedrich Hirzebruch

Centennial of the German Mathematical Society (Bremen, September 16th–22nd 1990)

VORBEMERKUNG

Ich freue mich sehr, daß meine Rede zum 100jährigen Bestehen der Deutschen Mathematiker-Vereinigung in diesem Heinz Götze gewidmeten Band erscheinen und ich damit ein Zeichen des Dankes und der Verbundenheit mit ihm setzen kann*.

Heinz Götze hat die Entwicklung der Mathematik seit Jahrzehnten mit Interesse und Anteilnahme verfolgt. Unsere Wissenschaft steht ihm nah. Man braucht nur an sein Buch über Castel del Monte und sein Studium symmetrischer mathematischer Objekte, wie der platonischen Körper, zu denken. Durch Heinz Götze wurde die mathematische Tradition im Springer-Verlag weitergeführt. Viele Mathematik-Programme hat er inspiriert. Neue Ideen für Publikationen greift er immer gern auf. Seine Arbeit findet im internationalen Rahmen statt. Mathematiker in der ganzen Welt sind ihm zu Dank verpflichtet. Mich verbindet jahrzehntelange Zusammenarbeit mit dem Springer-Verlag und mit Heinz Götze.

OPENING ADDRESS

Ladies and gentlemen, dear colleagues: I want to extend a warm welcome to all of you who came here from far away and in such large numbers. I also want to thank all the previous speakers cordially for all their understanding words on the situation in mathematics. I want to thank the local organizers of the meeting for all the efforts they put into the preparation of this meeting. I want to thank the University and City of Bremen for their hospitality,

---

* Als Vorsitzender der Deutschen Mathematiker-Vereinigung hielt ich am 17. September 1990 während der festlichen Eröffnung der Jubiläumstagung eine Ansprache, die von Herrn Professor P. Hilton und Frau Dr. L.C. Kappe ins Englische übertragen wurde. Beiden möchte ich meinen herzlichen Dank aussprechen.

and I want to give my cordial thanks to the Federal Minister for Research and Technology (BMFT), who sent Mr. Knoerich as his deputy, for his birthday present. We are happy to hear that we will now receive additional support directly from the BMFT; it is now up to us to prepare and submit grant proposals and we can be sure that in many cases we will receive a positive reply. I want to thank Volker Banfield, the pianist, for his passionate and magnificent introduction to the works of György Ligeti.

Many scientific organizations and foundations are represented here. I surely cannot recognize them all individually, so, in the name of them all, I want to welcome here Professor Wilke, the vice president of the Max-Planck-Society, and Mr. Möller, the secretary general of the Volkswagen Foundation, as well as Dr. Pfeiffer, the secretary general of the Alexander-von-Humboldt Foundation.

In a few days, the 116th annual meeting of the Society of German Natural Scientists and Physicians will get under way in Berlin. In its program notes, this society reminds us that it is the "mother" of many other specialized societies, in which debate is now carried on. Yes, indeed! Between September 15 and 20, 1890, the society met here in Bremen. Out of its Section I (Mathematics and Astronomy) came the DMV, the German Mathematical Union. A photo shows the 31 members of Section I, among them the first president of the DMV, Georg Cantor, the founder of set theory.

At the first meeting of the DMV in Halle in 1891 (at the same time as the 64th meeting of the Society of German Natural Scientists and Physicians) the by-laws were adopted. The purpose of the union is:

"... to expand and develop our science in all directions through a cooperative effort, to put its different parts and various organizations into a vivid union and interaction, to enhance its position in the cultural life of the nation, to give its representatives and disciples the opportunity of a free exchange of their experiences, wishes and ideas..."

Under the by-laws, it is the obligation of the governing board

"... to prepare the annual meeting by setting up a detailed program, into which, in as much as possible, presentations on the development of some of the areas should be incorporated..."

I hope that the governing board of the DMV has fulfilled its obligations in the year 1990 through the selection of the nineteen main lectures. At the 63rd meeting of the Natural Scientists here in Bremen a hundred years ago, when the DMV was founded, the list of speakers included the following: Georg Cantor, Paul Gordan, David Hilbert, Felix Klein, Hermann Minkowski, Eduard Study and Heinrich Weber. Maybe our 19 main speakers look at these names as a challenge.

In 1867, Alfred Clebsch, mathematical physicist and algebraic geometer, was the first to suggest the founding of the DMV, on the occasion of a lecture on binary forms which he gave at the meeting of the Natural Scientists in Frankfurt. The mathematicians recognized that it had become necessary to organize regular meetings in their discipline and to found new journals for the publication of their research. The outcome of a two-day hike along the Bergstrasse by twenty mathematicians, originally thought to give the necessary impetus to found the DMV, was, in fact, the founding of the "Mathematische Annalen" (Alfred Clebsch and Carl Neumann were the first editors), which published its 288th volume in 1990. The founding of the DMV was repeatedly postponed. Alfred Clebsch died in 1872 of diphtheria at the age of 39. His student Felix Klein, 23 years old at the time, held on to the idea. Much later, the decisive initiative originated with Georg Cantor.

Since its founding the DMV has published the "Jahresbericht," devoted chiefly to survey articles about areas in pure and applied mathematics. A look at the volumes of the first 10 years shows how the founders handled this task: Reports on soil pressure, Fachwerk, photogrammetry, mechanics, probability, kinematics and oscillating functions clearly show the role of applications; invariant theory, algebraic functions, number theory and Cantor's set theory are topics belonging to pure mathematics. In one of the issues, in an article of 370 pages, Hilbert gave a new development of the theory of algebraic number fields in his "Zahlbericht." His report is the basis for a cornucopia of research up to the present day. In the volumes of 1989 and 1990 the topics range over the following: frontiers between geometry and physics, free boundary value problems, elliptic curves, minimal surfaces, statistics, control theory, homotopy theory and number theory, theory of complexity, spherepacking, the finite-element method in the mechanics of solids. We see that there is still today a healthy mixture between pure and applied mathematics.

Despite the support of applied mathematics within the DMV, the Society for Applied Mathematics and Mechanics (GAMM) originated in 1922 with the following mission:

"... to cultivate and support scientific research in all those areas of mechanics, applied mathematics and physics which belong to the foundations of engineering sciences..."

The DMV considers itself as the mother of GAMM, as the society sees itself as the daughter of the society of German Natural Scientists and Physicians. Felix Klein was an honorary member of GAMM. Once he said:

"I strongly hope that the words of Leonardo da Vinci prove themselves again – that mechanics is the paradise of mathematicians."

From the founding years of the DMV up to Nazi times, mathematics in Germany was leading internationally. Some remarks on the work and life of Cantor, Klein and Hilbert will follow here to indicate the high standards of those times.

GEORG CANTOR (1845–1918)
President of the DMV 1890–1893

In the year 1878, his paper, "A contribution to the theory of manifolds" appeared, (he calls a set a "manifold") in which he introduced the cardinality (the power or the number of elements of a set), and with it he discovered or invented the transfinite numbers, which can be compared and with which one can do actual calculations. In this paper he conjectures that an infinite set of points on the real line is either countable or has the same cardinality as all points of the line. This is the famous continuum hypothesis. At the end of the paper he says:

"The more detailed investigation of this question will be postponed to a later publication."

At the meeting in Halle in 1891 he gave his famous lecture, in which he proved that the power 2^M for every cardinal M is bigger than M by using the diagonal method now bearing his name. Thus for every cardinal there exists a bigger one. His lecture ends as follows:

"The infinite cardinals represent a unique and at the same time necessary generalization of the finite 'cardinal numbers.' They are nothing else than the actually-infinite numbers and they are as real as the finite ones, only the rules governing them, i.e. their 'number theory,' are different from the finite case. The further exploration of this area is a task for the future."

Cantor had to endure a lot of hostility from his colleagues; in particular, they called his result, stating that the points on the line and in the plane had the same cardinality, an absurdity. As early as 1895 Cantor had discovered some of the antinomies of set theory, contradictions which arise by using excessively strange sets, and he was in communication with Hilbert on this. The axiomatisation of set theory by Zermelo and Fraenkel was an attempt to overcome the crisis in the foundations. In 1940, Kurt Gödel proved that the continuum hypothesis was consistent with the axioms of Zermelo-Fraenkel set theory (including the axiom of choice), and in 1964 Paul J. Cohen proved that the continuum hypothesis is independent of these axioms. For this achievement he was awarded the Fields Medal at the International Mathematical Congress in Moscow in 1966. Set theory has become a deep and important research area in the foundations of mathematics. The talk of Ronald Jensen about large cardinals next Friday will surely confirm this.

FELIX KLEIN (1849–1925)
President of the DMV 1897, 1903, 1908
Honorary President 1919

On the occasion of the golden anniversary of his doctorate his colleagues dedicated to him the following laudatio:

„Hochverehrter Herr Geheimrat[1]!

On December 12, 1918, fifty years will have passed, since you as a 19-year-old were awarded the degree of Doctor of Philosophy by the University of Bonn... Shortly after your dissertation appeared in publication... you, together with Lie, started your first group-theoretical investigations. In the seventies this was followed by your deep research in the area of Non-Euclidean Geometry,

---

[1] Most honorable privy councillor!

and at the start of your professorship in Erlangen you unveiled your pioneering 'Erlanger Programm' to the public. What became the hallmark of a shining series of successful investigations in later years, was already visible then: the penetration and mutual revitalisation of different mathematical disciplines, your ingenuity in seeing the innermost connections. A culminating point of this, your very own way of thought in approaching mathematical investigations, is your magnificent paper on the transformations of seventh order elliptic functions, which always will be revered as a gem of mathematical research. Your way of thought has proven itself to be tremendously successful in many other ways, in particular if one thinks of the rich body of papers concerned with applications of mathematics to science..."

By the way, some of Klein's papers have a lot in common with the contents of the talk by Egbert Brieskorn on Saturday.

Klein was the leading politician on behalf of science among the mathematicians of his time: Plans for merging universities and technical universities, cooperation with industry and the founding of the research institute for aerodynamics in Göttingen under the direction of Prandtl (wind tunnel for aerodynamical research, since 1918 a part of the Kaiser-Wilhelm-Society). Furthermore, Klein collaborated in editing the Encyclopedia of Mathematical Sciences, he was the chairman of the International Commission on Mathematics Education, he was instrumental in reforming the teaching of science and mathematics in schools as well as universities. All these are activities which are as important today as they were then. The "penetration and mutual revitalization of different mathematical disciplines," mentioned in the quote above, as well as the "applications to the natural sciences" (today to be extended to technology and computer science) have never been of more importance than they are today. The split of mathematics into pure and applied mathematics is arbitrary and detrimental. In the past decades this has led to many developments in the wrong direction. Felix Klein would be happy today to see that this split has, to a great extent, been overcome recently, and that mathematics presents itself again as a unity. Klein's interest in the wind tunnel would be replaced today by an interest in mathematical theories leading to exact modeling techniques, so that, for example, the optimal form of an airplane wing could be determined with the help of a super computer.

One final remark: Klein obtained his doctorate at the age of nineteen – a challenge to us to lower the average age when people receive their doctorate nowadays, which is currently 29 years.

DAVID HILBERT (1862–1943)
President of the DMV 1900

During the year of his presidency of the DMV he presented his 23 problems to the International Mathematical Congress in Paris. Hilbert says:

"What new methods and new facts in the wide and rich field of mathematical thought will the new century disclose?"

Hilbert's problems, which represent a glimpse into the future, start with Cantor's continuum hypothesis (Problem 1) and end with extensions of methods in the calculus of variations (Problem 23). Some of the problems have been solved. In Problem 8 (the distribution of primes) the central question still remains unanswered as to whether the Riemann hypothesis is true or not. Hilbert died in 1943 in the middle of the war and the years of the Nazi terror. Hermann Weyl, president of the DMV in 1932 and the successor in Hilbert's chair in Göttingen, who emigrated with his Jewish wife to Princeton in 1933, writes as follows in an obituary, published in the middle of the war by the Royal Society and by the American Philosophical Society:

"At the begining of this year died in Göttingen, Germany, David Hilbert, upon whom the world looked during the last decades as the greatest of the living mathematicians... Hilbert and Minkowski were the real heroes of the great and brilliant period which mathematics experienced during the first decade of this century in Göttingen... Among the authors of the great number of valuable dissertations... written under Hilbert's guidance we find many Anglo-Saxon names, names of men who subsequently have played a considerable role in the development of American Mathematics..."

In 1921, Richard Courant became the successor of Klein in his chair in Göttingen. Courant was instrumental in obtaining a grant from the Rockefeller Foundation for a mathematical institute in Göttingen, which was considered at that time as the mathematical center of the world. This institute, dedicated in 1929, would have been a dream come true for Felix Klein.

Hermann Weyl continues:

"But soon the Nazi storm broke and those who had laid the foundations and who taught there besides Hilbert were scattered over the earth and the years after 1933 became for Hilbert years of ever deepening tragic loneliness..."

For Hilbert and his successor Weyl, and for Klein and his successor Courant, close relationships between mathematics and physics in research and teaching were always a matter of fact. After decades of drifting apart, the last ten years have seen a dramatic development: mathematics and physics are growing towards each other again, and certain branches of mathematics (e.g. algebraic topology and algebraic geometry), which were considered as lacking in any applications, now play a great role in physics. On the other hand, physics has stimulated mathematics to develop some fascinating theories leading to the solution of some classical problems within mathematics itself. A symbol for this growing together of mathematics and physics is the award of the Fields Medal to Edward Witten, a physicist at Princeton, at the International Mathematical Congress a few weeks ago in Kyoto. This is a reaffirmation of what Felix Klein said at the International Mathematical Congress at Heidelberg in 1904:

"For science to thrive, without any doubt, the unimpeded development of all of its parts is an absolute necessity. In this process, applied mathematics plays a double role: to convey new ideas to the core parts of mathematics from the outside, and on the other hand, to deliver the harvest of pure mathematics to the areas of application."

The Commemorative Volume

On the occasion of this meeting, the DMV will present a commemorative volume "A century of mathematics 1890–1990." The DMV is very grateful to all authors who had the courage to contribute an article in one of the areas, and we are very much aware of the fact that in contrast to the all-encompassing claim of the title, this volume can only bring a highly individualized selection of topics. Despite all its shortcomings, this has become an excellent book. The introduction consists of an article entitled "Professional organizations – the institutes – the nation – spotlights on the relationship between mathematics, society and politics in Germany since 1890 with special consideration of the times of National

Socialism." The authors Norbert Schappacher and Martin Kneser refer to other articles on the history of the DMV, in particular the one by Helmuth Gericke on the occasion of the 75th anniversary of the DMV, which was also very helpful for me in preparing this address.

The Time of the Third Reich

Schappacher and Kneser wrote their article giving special consideration to the time of National Socialism. I want to say a few words here about those horrifying times of the so-called Third Reich. The law on the restoration of the career civil service of April 1933 and the ensuing tightening of the regulations led to the dismissal of Jewish academics and all civil servants who were considered politically unreliable in the opinion of the ruling party. Disguised by legal formalities, that was the beginning of the Nazi atrocities, which ended in genocide. Many scientists left Germany (Hermann Weyl said: "[they] were scattered over the earth"), many lost their lives. We can recognize this terrible past at our own universities and should be deeply aware of it. In the case of Bonn I think of Felix Hausdorff, who in 1942 together with his wife took his own life to escape deportation to a concentration camp, and of Otto Toeplitz, who emigrated to Jerusalem in 1939 and died soon afterwards. Max Pinl is the author of a report, entitled "Academics in dark times" which appeared in the "Jahresberichte" of the DMV between 1969 and 1974. The year 1934 saw fierce debates within the DMV. I feel that a letter by a prominent mathematician to one of the editors of the "Jahresberichte" is especially shameful. In this letter the writer claims that the DMV is not in a position to accomplish its goals, as set out in its by-laws, i.e. "to further the position (of mathematics) in all aspects of the cultural life of the nation," since forty percent of its members do not belong to the German race. Therefore, he deems a change of the by-laws as necessary, which clearly should spell out the position of the DMV within the national socialist state. – However, it proved possible to defeat the move to incorporate the 'Führerprinzip'[2] into the by-laws of the DMV; and those forces in the leadership of the DMV (after 1937 Wilhelm Süss), which saw mathematics as the first priority, prevailed. After 1938, according to a decree of the education minis-

---

[2] Leadership principle, rule by one leader (Führer)

ter, only German subjects, the so-called 'Reichsbürger,' could be members, and all Jewish members were expelled. The DMV and its members did not behave like heroes during the Nazi regime. But who of us can claim that they would have acted differently?

The Postwar Years

After World War I, Mathematics was still on a high level, but after the Second World War, it lost its internationally leading role. The DMV was refounded with Erich Kamke as its president; Kamke also became vice president of the International Mathematical Union in 1950. In September of 1946 the first meeting after the war took place in Tübingen. But first, let me make some remarks of a highly personal flavor about the time after 1945, which covers almost half the life span of the DMV.

After labor service, military service and some time as a prisoner of war, I returned home on July 1, 1945. I was forced by the Department of Labor to work in one of the barracks of the Allied Forces, for example scrubbing the toilets there. An officer, who spoke German fluently, talked to me and found out that I wanted to study mathematics. On the spot, he drove me home in his Jeep: I should do mathematics. Thus this work detail was over in less than a day. In the winter of 1945 I started studying with Heinrich Behnke and Heinrich Scholz at the University of Münster. I know that I was unbelievably lucky. Many of my fellow students were ten or more years older than I when they started their studies. Already in 1947 there were foreign visitors giving colloquium lectures, and in 1948 I went for the first time to Switzerland, doing harvest labor and visiting Heinz Hopf. Since 1949 I have studied with Hopf and Eckmann in Zürich, and, for instance, I met Hermann Weyl there. When I came to Princeton in 1952, I met Einstein, Gödel, John von Neumann, Hermann Weyl, Emil Artin, Bargmann, Bochner, Feller, Wigner, Eilenberg, Rademacher, Samelson, Calabi, Bott, and in next-door New York Richard Courant and his institute. Everywhere I got a warm welcome and was treated as a colleague. Could it be any clearer to me what Germany and Europe had lost?

The golden fifties I spent in Princeton and in Münster, where the school of complex analysis had made big strides forward under the leadership of Heinrich Behnke (President of the DMV 1964 and 1965) and Karl Stein (President of the DMV 1966); and since

1956 I have been professor in Bonn, where during my negotiations in Düsseldorf, the government officials matter-of-factly replied to most of my wishes by saying: "Yes, of course, Herr Professor."

Support of Mathematics outside the University in the Federal Republic of Germany

During the fifties I was able to participate in many meetings at the Mathematical Institute in Oberwolfach. Naturally there were always many international guests at these conferences. The Oberwolfach Institute had been founded by Wilhelm Süss in 1944 and he was its director until shortly before his death. With the help of two officers of the Allied Forces, G.E.H. Reuter and John Todd, both mathematicians, the institute could be saved after the war, and already in 1946 international guests could be welcomed there (Ehresmann, Heinz Hopf, Henri Cartan). Since 1959 the finances of the institute have been on a sound foundation; it is indebted to a donation of the Volkswagen Foundation for its facilities. Since 1963 the Oberwolfach Institute has been under the leadership of Martin Barner (President of the DMV 1968–1975) and it has become a treasure for which we are envied by the rest of the mathematical world.

Since the end of the sixties, the DFG, the German Science Foundation, has supported mathematics in the Federal Republic by establishing various special research areas (Sonderforschungsbereiche) at the universities of Bonn, Heidelberg, Göttingen and Bielefeld; the Max-Planck-Institute for Mathematics in Bonn has its origins in such a special research area. The special research area "Stochastic Mathematical Models" in Heidelberg works in close cooperation with the newly founded "Center for Scientific Computations." The special research areas and the Max-Planck-Institute always have researchers from all over the world as visitors. Looking at other forms of support by the DFG, the mathematicians appear to be less favored. In 1989, the DFG only awarded 73 grants to mathematics, out of a total of 4,627. In the same year the DFG spent about 1 Billion DM for research support, out of which only 17 million (1.7%) went towards mathematics. The Max-Planck Society only spends 0.4% of its total budget on its Mathematical Institute.

The Humboldt Foundation is always happy to receive applications from mathematicians. During the years 1980–1989 five percent of their grants went to foreign mathematicians, from many countries. Likewise, during the same period, about five percent of the dissertation fellowships of the Studienstiftung[3] went towards mathematics.

The Stifterverband[4] supports the newly founded Institute for Discrete Mathematics at the University of Bonn. With a start-up grant from the Volkswagen Foundation the newly founded Institute for Experimental Mathematics at the University of Essen opened its doors in February 1990; there mathematicians, computer experts and communication engineers work in close cooperation. The computer has very much enhanced the creative power of mathematics (e.g. by testing conjectures in number theory and geometry – I refer to the exhibits at this meeting). On the other hand, mathematics is indispensible for the progress of computer science. We hope the two new institutes will bear this out!

The Situation Today

The foundations and research organizations mentioned above, and some others, are very open towards the needs of mathematics. It is *our* obligation to write more grant applications and to justify the necessity of increased support for the mathematical sciences. It is very difficult to plan mathematical research. Nevertheless we have to try to increase our planning for projects and forward our proposals to the support organizations.

Our accomplishments can live up to international standards. A highlight was the award of the Fields Medal to Gerd Faltings four years ago. Faltings was a winner in the Federal Mathematics Competition for secondary schools and was supported by the DFG as a student.

The future of the new generation of scientists is an area of great concern, but things should ease up soon because of the upcoming wave of retirements. With the help of the Heisenberg Fellowships of the DFG it was possible to keep good mathematicians at the university. Of the 694 Heisenberg Fellowships (since 1978) 65 went

---

[3] German National Scholarship Foundation
[4] Donor's Association for the Promotion of Science and Humanities in Germany

towards mathematics. We also should mention here that the DMV seminars for junior mathematicians have had considerable success. These seminars were started by Gerd Fischer (President of the DMV 1980 and 1981) with financial help from the Volkswagen Foundation.

Concerning the professional situation of mathematicians outside the university, I only want to mention that mathematicians with a diploma or with a Ph.D. have excellent chances in industry, and soon there will again be a shortage of mathematics teachers at the secondary school level. Two of my recent Ph.D. students are working successfully at computer chip construction and computer-aided design for car manufacturing.

Mathematics in the DDR, *the Union of* DMV *and* MGDDR

About 75 mathematicians from the DDR are participating in this anniversary meeting. I very warmly welcome them here. The changes in Eastern, Southeastern and Central Europe, in particular the peaceful revolution in the DDR, the abolition of the communist dictatorship and the upcoming reunification present a challenge and special opportunities to us. Last Wednesday, during its annual congress in Dresden, the membership of the MGDDR, the Mathematical Society of the DDR, voted unanimously that MGDDR and DMV should be merged. The union of the two societies will be an item on the agenda at the upcoming general meeting of the DMV this Thursday. Since February of 1990 three consultations between the Mathematical Society of the DDR and the DMV have taken place. The president of the MGDDR has been, since February, a permanent guest of the governing board of the DMV. It is a very moving moment for me to announce here the forthcoming union of the two societies.

I was, once before, president of the DMV, that is, I was elected on September 20, 1961, at the last joint DMV-meeting. This meeting took place in Halle in the DDR, shortly after the Berlin Wall went up, and exactly seventy years after the very first DMV-meeting, which coincidently also was held in Halle. My current tenure as president of the DMV started shortly after the breach in the wall.

In between there have been almost 30 years of enforced separation of the DDR from the West. Nevertheless many West German mathematicians have had contacts with mathematicians from the

DDR, and participated in conferences and colloquia in the DDR and tried in return to invite colleagues from the DDR to come here, though with only limited success. Also I have frequently visited the DDR, e.g. in 1958 I gave three lectures on the theorem of Riemann-Roch-Grothendieck at the Humboldt University, and, since 1964, I have participated in the meetings of the Leopoldina in Halle on a fairly regular basis. We know, through these never totally interrupted scientific contacts, that mathematics is in good shape in the DDR and has some excellent achievements to its credit under very difficult conditions. There were institutes in which human decency and scientific achievements were in higher esteem than accommodation and servility to the party.

But back to the meeting in Halle of 1961: How should I, as the newly elected president, summon a meeting of the board? There were three kinds of Germans: West Germans, East Germans and West Berliners. The East Germans had to stay in East Germany, West Germans could go to East Berlin, but West Berliners could not. After briefly toying with the idea of a meeting in Prague or Warsaw, it was decided to have two meetings with the same agenda, one in East Berlin (December 16, 1961) and one in West Berlin (December 17, 1961). Every member of the board could participate in at least one of the meetings. Thus, shortly before Christmas 1961, I went through the wall for the first time. Apart from the board members from West Germany, Hans Reichardt (East Berlin) and Ott-Heinrich Keller (Halle), my predecessor in the office of president of the DMV, also participated in the meeting. But the split could not be avoided any longer. It was instigated exclusively by the government of the DDR, in particular by the Academy of Science, which wanted a separate representation, appointed by itself, in the International Mathematical Union. During that time, mathematicians in the DDR often tried to communicate to me the view that the joint representation of German mathematicians by the DMV in the International Mathematical Union should be preserved. In June of 1962 the Mathematical Society of the DDR was founded and it developed rapidly. This organization has currently 1,300 members, many secondary school teachers belonging to it. In contrast, the DMV has only 2,100 members. The mathematical congresses of the DDR took place about every four years with about a thousand participants every time. I participated in the congress in Rostock four years ago, as well as in the congress in Dresden last week.

The union of the two German societies should be viewed in the context of the ever closer cooperation in Europe. Next month, in Warsaw the European Mathematical Society will be founded[5]. All mathematical societies can become members. Of course, we are dealing here with all of Europe, not only Western Europe. At the double meeting of the board of the DMV in Berlin, as the anticipated split was already weighing heavily on us, item 6 on the agenda (somewhat ironical) read "Coordination of mathematical studies at European Universities." Here, a circle of mathematicians around Henri Cartan planned to introduce a European Student's Record, of which later many copies were printed, but very few were actually used. This booklet specifies the minimum content of important courses. Professor X in country A certifies that the student has taken such a course and has mastered the minimal requirements. Thus, Professor Y in country B knows in which course this student should now be placed. Here is this thirty-year-old European Student's Record. Perhaps the European Mathematical Society can reintroduce it in a somewhat modified form.

The Future of the DMV *and the Grötschel Initiative*

Where do we go from here? It is clear that we have to promote individual cooperation between the mathematicians here and those in the DDR. If it is their desire, we should help them to carry out necessary reforms at the institutions in the DDR, and in the immediate future give them our support and advice in taking an inventory as planned by the Wissenschaftsrat (Science Advisory Board).

We should try to start an exchange program between here and the DDR. Some of us should teach for a semester in the new Länder in the eastern part of Germany and, in return, some of our colleagues there should take over our teaching duties here.

The new society created by the union of DMV and MGDDR will presumably be called "Deutsche Mathematische Vereinigung," adding a new meaning to the word "Vereinigung" (union). But what will be the tasks of the new DMV? Already, before the

---

[5] The EMS was founded on October 28, 1990. See the Article of Michael Atiyah in this volume

'Wende,' the turning point in the affairs of the DDR, Martin Grötschel, as a member of the governing board of the DMV, formulated an initiative concerning the goals and tasks of the DMV and how the organization should present itself. In January 1990, the "Mitteilungen"[6] reported about this initiative, and at the business session of this meeting it will be presented to the membership of the current DMV; but at the end this should become an initiative of the new combined DMV. In essence, by promoting mathematics in all directions, the DMV should address all mathematicians, including those in industry and those teaching in secondary schools. In this way it should evolve into an organization representing all mathematicians. In no way does the DMV want to be in competition with other organizations, whose goals concern the special interests of their members. Also students of mathematics at the university level should be recruited as members. The publications of the DMV, i.e., the "Jahresbericht" and "Mitteilungen," should be made more attractive and interesting for all members. An essential part of the Grötschel Initiative will be a documentation about the current situation of mathematics in Germany, the significance of mathematics as part of our civilization reaching back six thousand years, and mathematics as the foundation for the natural sciences, technology and computer science. As a model for such an undertaking we are looking at the two reports "Renewing U.S. Mathematics" (David Report I, 1984, and David Report II, 1990). From David Report I, let me quote here one sentence:

"When we entered the era of high technology, we entered the era of mathematical technology."

David Report II gives an account of exemplary mathematical research activities, and it describes applications to physics, chemistry, biology, technology, computer science, medicine as well as economics. Our report should suggest promising research projects which would benefit from further mathematical penetration.

The Mathematical Society of the DDR has lots to offer to the new DMV. It is already open to teachers and mathematicians in industry. Their publication, the "Mitteilungshefte," will serve as stimulation for the new "Mitteilungen" oft the DMV. The Mathematical Society of the DDR has special sections and interest groups, in particular for some of the areas on the fringes of mathematics,

---

[6] Communications

which seem to be very valuable in communicating mathematics to the general public. The society has organized, on a regular basis, small conferences about new developments in certain areas, which were also well received internationally. These conferences should be continued in order to give some relief to the Oberwolfach Institute, which is completely booked up, and as a supplement to DMV seminars.

Mathematics is Art

I want to quote here verbatim an introductory remark from the collection of examples in the David Report:

"The unification and cross-fertilization of areas within core mathematics, increased... applications (which often uncover unusual and unexpected uses of mathematics), and the growing role of computers are all themes that are illustrated in these descriptions."

It should be mentioned that in the laudatio of the year 1918 for Felix Klein you find the words "penetration and mutual revitalization" instead of "unification and cross-fertilization." The David report also talks about "core mathematics," and "core" means "the innermost," "heart," "kernel." Felix Klein talks about "central parts." This refers to the mathematics which, irrespective of applications, discovers and creates works of art, the mathematics which its "representatives and disciples" love and develop because of its beauty, and which is pursued by mathematicians in the same spirit as that in which artists paint and musicians compose. Mathematics is an art as well.

Despite our love and commitment to mathematics, we did not succeed in convincing the Federal Post Minister to issue a commemorative stamp on the occasion of the 100th anniversary of the DMV. We have to step up our efforts to persuade the public that mathematics is something very much alive. We surely can appreciate the commemorative stamps "75 years of the German Homemakers' Federation" and "500 years of Riesling." It is uplifting to see stamps with a mathematical theme which other countries issue much more frequently, as well as seeing Euler on the Swiss 10-Frank note, and Gauss on the newly planned 10-Mark note. However, as already said, since mathematics is also art, we consider ourselves honored by the commemorative stamp "100 years of the Artist Colony Worpswede." On Wednesday we will make a pilgrimage to

Worpswede by bicycle. The circle, the symbol of perfection in the arts, was discovered by mathematicians and than used for the construction of the wheel and the bicycle. We feel united in spirit with the artists of Worpswede and consider their stamp also as ours.

Before the opening of the annual meeting there are a few announcements: The 1991 DMV meeting will be held in Bielefeld. At the business meeting it will be proposed to hold the 1992 annual meeting in Berlin at the Humboldt University. The members of the MGDDR already have accepted this proposal. Presumably the new German Mathematical Union will invite the International Congress of Mathematicians to Berlin for the year 1998. Georg Cantor was very much in support of establishing an international congress. The first one was in 1897 in Zürich, the second in 1900 in Paris, the third was in 1904 in Heidelberg. This was up to now the only time the International Congress has been held in Germany.

Now it is time to start our work. Herewith I inaugurate the anniversary meeting of the DMV with its 19 main lectures and numerous talks in the various sections.

Viel Freude und Erfolg![7]

---

[7] Much joy and success!

The First Woman Professor
and Her Male Colleague

In Sweden Sonja Kovalevsky is claimed to be the first woman anywhere to become a university professor. If this is exaggerated it is certainly true that she was the first in Sweden where the next woman professor was not appointed until 1938. Her appointment caused strong reactions in some quarters. In the second part of *Giftas (Getting married)*, published in 1886, August Strindberg wrote[1]:

"When the University of Stockholm divided the salary of the male mathematics professor to give half of it to a woman, it was a crime – against justice. And the men rejoiced."

Strindberg's book, written as a reaction to contemporary feminist aspirations, was considered scandalous by many. He was prosecuted, not for the main theme but for blasphemy because of an irreverent reference to transsubstantiation, but he was finally acquitted. The editors of the new national edition of Strindberg's works, published in 1982, comment on the passage above:

"In 1884 the University of Stockholm created a special professorship in mathematics for the Russian born Sonja Kovalevsky, in addition to the ordinary professorship; the salary was in no way 'divided', but Sonja Kovalevsky was paid far less than her male colleague."

There are several excellent recent biographies of Sonja Kovalevsky [1, 2, 5], but apart from two passages in [5] they do not seem to have made use of the diaries of the "male professor", Gösta Mittag-Leffler, which are now kept in the archives of the Royal Library in Stockholm. The Mittag-Leffler papers there fill tens of shelf meters. In order to draw attention to these untapped sources I would like to quote some of the diaries from 1887 to 1890 which illuminate the relationship between the two professors of mathematics in Stockholm, at least as seen from one side. However, I shall first recall some of the background. (For more details see [1] and [2].)

---

[1] All translations from Swedish are mine.

After receiving his doctorate in Uppsala in 1872 Gösta Mittag-Leffler spent the years 1873–1876 studying in Germany and France. He made many lasting friendships during this period, and in particular he became a devoted student and admirer of Karl Weierstrass in Berlin. It made a great impression on him to hear Weierstrass talk about Sonja Kovalevsky as his best student ever. After four years of study with Weierstrass she had returned to Russia in 1874 and it was there that Mittag-Leffler met her in 1876. In a letter quoted in [4] he described this encounter:

Ce qui m'a le plus vivement interessé à St. Pétersbourg, a été de faire la connaissance de Madame Kowalewsky. Aujourd'hui (10 février 1876) j'ai passé plusieurs heures chez elle. Comme femme elle est délicieuse. Elle est belle et, quand elle parle, son visage s'éclaire d'une expression de bonté féminine et d'intelligence supérieure qu'on ne soutient pas sans éblouissement. Ses manières sont simples et naturelles, sans aucune trace de pedantisme ou de savoir affecté. Du reste en tous points "dame du grand monde". Comme savante elle se distingue par une clarté et par une précision d'expression peu commune, ainsi que par une conception singulièrement prompte. On s'aperçoit aisément aussi du degré de profondeur où elle a poussé ses études, et je comprends parfaitement que Weierstrass la regarde comme le mieux doué de ses disciples.

Shortly afterwards Mittag-Leffler became a professor at Helsingfors University. He tried to arrange a position for Sonja Kovalevsky there but ran into opposition. The Finns were worried that as a Russian nihilist she would foment revolutionary ideas and so provoke the Russians controlling Finland at that time. However, Mittag-Leffler did not give up. After moving to Stockholm in 1881 as the first professor at the new university there, he renewed his efforts on her behalf. Already in June 1881 he wrote in a letter quoted in [2, p. 133]: *"As far as I am concerned, I shall be extremely happy if I have a chance to invite you to Stockholm as a colleague, and I do not doubt that with you in Stockholm our faculty will be one of the most advanced in the mathematical world."* After the death of Sonja Kovalevsky's husband Vladimir in 1883 it was finally arranged that Sonja would come to Stockholm as docent. She taught a course in partial differential equations there during the Spring term of 1884. In the meantime Mittag-Leffler managed to raise enough money so that she could be offered a five year appointment as extraordinary professor at the University, beginning during the fall term 1884. Half way through the appointment a crisis appears in the diaries of Mittag-Leffler:

JANUARY 11 [1887]

... Letter from Sonja. She says that she will remain here during the Spring but will accept a position either in Paris or in England. She even asks me to help her to get one. The reason: that she would only get 4000 Cr. in salary.[2] *Not a word about the fact that she has committed herself to stay for two more years, so that she has let me promise in her name that she will stay for five years. Then she just asked for a salary of 2000 Cr., but I raised 4000 Cr. I could only raise the money by declaring everywhere that she would stay for 5 years. Now I will look as a liar to all these people, who must pay their money although it will no longer go to her salary. And then I am supposed to keep all this a secret for Hermite and Sylvester so that she then some time will deceive them as she has deceived me. I had hardly deserved to be treated so by her.*

... I shall ask Ach.[3] *to try a last appeal to her heart. Feelings of honor, duty, fidelity to given promises she regards as true prejudices. To let oneself be led by them when self interest goes in another direction, a despicable weakness.*

JANUARY 17

Ach had come home last night. Visited her in the morning. She promised to try to appeal to Sonja's conscience, so that she does not make more trouble but stays here if I also get her the mechanics course so that the salary becomes 6000 in all.

---

[2] At this time the salary of a full professor was 6000 Cr. at the state universities but 7000 Cr. in Stockholm where the teaching load was just two hours a week, half of that at the state universities. The university had an annual grant of 40000 Cr. from the city of Stockholm, but the rules required that not more than half of a professor's salary could be taken from this grant. The rest had to be raised from private sources. The financial situation at the university was all the time very strained, particularly in the end of the 1880's when several retiring professors could not be replaced.

[3] Ann-Charlotte Leffler was Mittag-Leffler's sister, a successful author and playwright.

JANUARY 27

Ach had had a conversation with Sonja about her plans to leave Sth.[4] *Ach had told her that this would be impossible before the 5 years had passed, for which she had promised to stay, that all those who had contributed money to her position would now despise her deeply if she broke her promise, and that this would damage me. She had become very angry with me that she was tied in this way. She had just planned to turn to Ugglas and Lindhagen*[5] *to ask for a resignation, which she well knew that they would accept with pleasure. She had not had the slightest thought of how it would turn out for me, whom I could get as her successor and so on. However, she now said that she realized that she had to stay the remaining two years for my sake. She asked Ach to say nothing to me about their conversation.*

Mittag-Leffler was widely known as a successful and ruthless businessman, but it is clear that he lost out in this bargain although the raise in pay for Sonja Kovalevsky was accompanied by increased teaching duties.

The spring term of 1887 was largely spent in collaboration with Ann-Charlotte on a play "The struggle for happiness", and Mittag-Leffler was worried about the lack of dedication to the later so famous paper on rotating bodies which she was supposed to work on. Also here the financial aspect comes in:

MARCH 26

... I talked to Sonja about her paper, which she has entirely neglected although she is certain to get the prize of the Institute[6] *if it just becomes reasonably good. She answered this time that she has no time since she must write something which brings in money, for she cannot live on 6000 Cr. May she write during the summer. There will be no work then anyway. That she has not worked during the winter is entirely a consequence of the drama.*

---

[4] Stockholm.

[5] Chairman and secretary of the board, respectively.

[6] This refers to the competition for the Bordin Prize of the French Academy of Sciences.

Concerning the 6000 Cr., the original agreement, based on her own suggestion, was that she would have no salary at all. However, I soon discovered that this would not be possible, for the income from Russia seemed to be small or non existent. She then declared that 2000 Cr would be quite adequate. I doubled the sum and raised 4000 Cr. Then I got her an appointment to be in charge of mechanics with 3000 Cr extra in pay. To bring the salary up to 6000 this year I had to give 1000 Cr. myself, and I cannot manage more, the University gave 900 Cr., and Mrs Malmsten 100 Cr.

However, Miss Ellen Key[7] has declared the new drama undramatical, now the courage is close to zero and the mood miserable, until the meeting with the Russian inventor can create a new interest.

The Russian inventor mentioned here occurs in an earlier diary entry:

MARCH 25

... Sonja is now dreaming about a young Russian inventor, designer of an airship and living in Kaluga. Friends there have written, that he is dreaming of her. More is not required to start her dreaming about him. What is important now is to get away from here as soon as possible. She has asked the man to meet her in Petersburg ... Her imagination is always filled by erotic dreams, in all innocence of course, as in a 16 year old boarding school girl. She bids goodbye to science immediately if she just finds someone to devote herself entirely to.

Mittag-Leffler continued to worry about his colleague's work during the whole semester:

MAY 1

... I tried to talk with Sonja about her prize paper for the French Institute and to press her to work seriously on it. She told me definitely, and she was no doubt serious, that she cannot work on the paper with force and success without first being in love with a mathematician. It did not help to stress how narrowmindedly female this trait

---

[7] Author, critic and feminist.

was and that hardly since the creation of the world a male scientist felt and argued in this way. I fear seriously, that she will never finish. If she now falls in love again with somebody, the chances are not good for a mathematician.

During the Summer of 1887 Mittag-Leffler travelled to the continent. First he visited his old teacher Weierstrass in Berlin, where he writes in his diary:

[Weierstrass] was sceptical about Mrs Kovalevsky's work on the rotation problem. He was sad over her nihilistic imprudence and waited with anxiety for her permanent appointment in Sth. I did not have the heart to tell him that she no longer wants that but just dreams of coming somewhere else, where she has a better chance to meet the hero her imagination and heart are longing and yearning for, and that she spends her youth to no purpose in Stockholm. (Her own words to Ach.)

From Berlin the trip continued to Paris where Mittag-Leffler became seriously ill and wrote down his last will in the diary. One of his wishes was that Sonja Kovalevsky should be appointed temporarily on his position with a pay of 7000 Cr. on condition that she would also teach the mechanics course. However, he recovered so nothing came of that and the haggling about the salary continues during the fall although I shall not bore the reader further with that but pass to some passages from the diary of 1888 which shed some light on how Mittag-Leffler saw her personality.

FEBRUARY 25

... In the evening I visited Sonja. She intends to travel with Kovalevsky [8] from Stockholm to the Kaukasus where they will travel together. (Kov. does not know about this plan yet.) She will then see if there can be a match between them. That she is not and will not be in love with Kov. is very certain but I think she will persuade herself that it will be the case. And everything will then depend on her. I hope that the match will take place. He has more than any other man I have met the properties which would be required to

---

[8] Max Kovalevsky, a sociologist, was a distant relative of her late husband.

make her as happy as she can become with her temperament. And if this erotic longing could only be eliminated from her life, she would then again be able to produce splendid papers.

These discussions were pursued further in Wernigerode (in the Harz mountains) where both Sonja Kovalevsky and Mittag-Leffler were visiting Weierstrass in August.

AUGUST 10

... Took a walk in the evening with Sonja. During our walk the 7th she said, that she does not want to marry Kovalevsky, but wants them to live secretly together, so that this is only known to their most intimate friends. He does not want to, wants to marry, and finds another relationship degrading for her. She hopes rather to have a stronger grip of him if they are not married and hopes to be able to keep him then as the devoted lover she demands that he remains. ...

AUGUST 13

In the afternoon visited the Misses Weierstrass. Sonja, Cantor and I took a guide to climb the Brocken. On the way various confidences from Sonja. She does not want to marry, fears that Kovalevsky by vanity wants to have her as wife and will have other mistresses. ... Besides, she wants to be mistress but not wife. She wants to have one or several lovers during the vacation, who are intellectually as outstanding as possible, but during the semesters she wants to be alone and work. – That is not how it sounds during semesters. – She thinks it is hard that the men are so impressed by her that nobody can get the idea that she would be mistress but not wife.

Unfortunately for her she is not such that a man would desire to get her as his mistress whereas many would want her for a wife. Men do require from a mistress primarily physical attractions, which she lacks, but in a wife they can well accept the absence of these if compensated by other qualities, which she has in such eminent degree.

Besides, unhappy would be the man who would get her for a wife. Her so exceptionally well developed selfishness and the cold indifference which is so well hidden by her lively and interested manner would soon bring him to despair. In her everything is head and imagination, but her heart has not yet vibrated. And her terrible temper and the scenes she would provoke all the time. And the completely openly proclaimed and real absence of every feeling for duty and responsibility. No she is certainly right that it is better for her not to marry.

Even during this visit Sonja's bitterness about conditions in Stockholm came out:

AUGUST 16

Breakfast, conversation, and a walk with H[urwitz] and C[antor]. At dinner Sonja started a discussion with H. and C. and the Misses Weierstrass (W. himself never eats at the dinner table) about the woman question. I did not take part except that I developed to H. my opinion that also women should have the right to vote. Suddenly Sonja caught the attention of the whole company to expound on how much more she "leistet" in Sth than I, how she has lectured doubly, written more papers and so on, never been on leave while I have often been, never been ill while I am all the time and so on. I had on my lips the remark, that among my occupations I also have to maintain the position of my female colleague. However, I just said that I shall not make a defense. Cantor became somewhat agitated and assumed that for example she forgot all the work my journal takes. She answered that I have a secretary and that the journal does not take more time than it takes her to run her household.

Most of the diary entries from this time are devoted to discussions with Weierstrass concerning Poincaré's paper for the King Oscar prize, Weierstrass' own hopes to prove stability of the n body problem, and a small scandal. H.A. Schwarz, who was called to Berlin to help in a discussion about Sonja Kovalevsky's remaining problems in her prize paper, had spread scandalous gossips about Mittag-Leffler's private life. Cantor informed Mittag-Leffler and fanned the fire as well as possible. After Mittag-Leffler had noted that as a Swede he could not duel, he threatened Schwarz with the courts, aided by Weierstrass who suggested at least 6 months of forced labor. The whole matter ended with apologies from Schwarz in front of Weierstrass, whereafter the mathematical walks resumed seemingly in amiable spirit, with Schwarz submitting a paper to

Acta Mathematica as a token of reconciliation. (Inspection of the journal shows that none was published though.)

But let us return to Sonja Kovalevsky's relations to Stockholm and to men.

AUGUST 24

... In the evening a conversation with W[eierstrass] about Sonja. He very much dislikes the plans for Paris. He deplores her tactlessness to talk badly about Sth and Sweden at all times and has drawn her attention to the fact that no country would have done for her what has been done in Sweden.

Last night I talked with Sonja about the plans that W[eierstrass] would come to Sweden next summer and that I would there surround him with a group of young mathematicians. She does not want that, for she does not want to be in Sweden herself during the summer, and he must not come when she is not there.

The jealousy about the attentions of Weierstrass among his students which had been an underlying theme during the whole visit became explicit during this final entry:

AUGUST 26

... In the afternoon Sonja and I walked to the bathhouse where we were both going to bathe. W. had said at the dinner table that he had found a proof of stability and she told him right away that she was jealous because he had told it to me. During the walk to the bath she was first in a good mood, but she darkened when I started to talk about the proof of stability. When I wanted to talk about the transformation introducing the quantities ξ she said it was not necessary because she knew. Somewhat surprised I asked if W. had already had time to tell her. She answered that she knew it since a long time, but blushed at the same time and was silent the rest of the way. That she lied is completely certain, for I know definitely that W. has never told her about it. The truth struck me suddenly, the explanation why she has been so unpleasant all this month. She has been jealous over the kindness W. has shown to me, over the information he has given me now and then. When she has laid hands on a person she cannot bear that he shows another person the slightest

*interest. And I had innocently, when she wrote that W. had asked
her to come here to help him and she hinted that she could not and
would not come because of the love story with Kov, at once offered
to come in her place and help W. as much as I only could. She
could now come a bit later. That I have been so blind not to realize
this before. It was the same way the whole of last winter. Then she
had concentrated on Ach and wanted to control her completely. She
knew how close we were and then she worked persistently to separate
us. Now I understand much which has been incomprehensible before.
It has happened now and then when I have wanted to talk about
my conversations with W. that she has not wanted to listen or answered that she wanted to ask W. herself. I had interpreted this as
a low opinion of my ability to report, but I was somewhat suprised
by this explanation since she had never before shown me such disre-
spect. My decision is taken, I shall try to achieve, that she becomes
for me personally if not indifferent then at least more distant than
before. Only then can I endure the annoyance she gives me. As long
as she is as close to me personally as up to now, she torments the
life out of me. I can suffer much from those, who are personally indifferent to me, but nothing from the few who are really close to me.*

In earlier entries in the diary Mittag-Leffler had been unhappy
that Sonja Kovalevsky had become so cool towards his sister that
she did not even write to her. The preceding entry continues as
follows:

*... In the evening a walk with Sonja. I told her a part of what I
thought about her relationship to Ach. She defended herself by saying
that she did not dare to correspond with her since she feared to
be infected by her passionate nature and that she herself now wanted
to be calm and not be carried away by feelings when she had to
make a decision about Kov. The explanation is clever enough, but
it has just the drawback of not being true.*

In the fall of 1888 Max Kovalevsky was invited to give a series
of lectures at the University of Stockholm. Mittag-Leffler comments
in his diary that Sonja shows her devotion openly. Of course he
continues to press her to finish the paper for the Bordin prize;
a reminder arrived from Hermite in November.

NOVEMBER 30

... Sent to Paris Sonja's supplement. Dinner at Gyldén's[9] house. Conversation with Max. He does not believe that Sonja's plans for Paris will succeed. And he thinks that even if they do, her position here is incomparably better. "She really just wants to go to Paris to associate with Russian nihilists." Here she has a better position than she can ever get elsewhere. And almost six months free every year ...

Sonja did get the Bordin prize in Paris on Christmas Eve 1888. She seems to have been exhausted by her work and took a leave during the Spring term 1889 to stay in Paris, much to the disappointment of Mittag-Leffler who had to fight alone for her reappointment in Stockholm. (See [2, pp. 257–267] for the correspondence between Kovalevsky and Mittag-Leffler during this period.) It was not an easy battle.

FEBRUARY 21 [1889]

Visited A... in the evening to report the result of the Kovalevsky fund raising. He seemed pleased with my plan.

FEBRUARY 22

Visited Hammarskjöld[10] in the morning. He was satisfied with my plan for the appointment of Mrs K. ...

FEBRUARY 23

Went to Söderström[11] with the letter to the board concerning the Kov. fund raising. The fox, who had previously agreed that only 15000 had to be raised, was now no longer satisfied with a fund of 50,000 Cr. He wanted to have 60,000 to guarantee 2000 [income from the capital]. Finally we agreed that those who had donated the 15,000 should allow, that also the capital could be used if the interest should decrease below 4%. ...

---

[9] Hugo Gyldén was professor of astronomy at the academy of sciences.

[10] Hammarskjöld succeeded Ugglas as chairman of the board in 1889.

[11] A businessman appointed to the board by the city council.

FEBRUARY 28

In the morning visited Hammarskjöld to hand over the donations for the Kov. fund. He took it for granted, that everything would work out. However, he was wrong. Retzius and Nilsson[12] asked that the item be tabled and Retzius opened a debate in which he rather carelessly disclosed his batteries. He talked about how superfluous the higher analysis really was, how one really needed teaching of the elements, how we now could get Mrs Kov. replaced by a Swede, Phragmén, and more in this spirit.

MARCH 1

Visit to Hammarskjöld in the morning. He related then all that had happened in the board. ...

MARCH 3

... In the afternoon letter from Hammarskjöld that Retzius had visited him and proposed that Mrs K. should only be appointed for as long as the city council's subvention to the University is paid and not for lifetime as I have proposed. The reason: If the city council withdraws its support, then the board would come in the embarassing situation of having a professor with just 2000 Cr in salary, or the salary coming from the fund which I have raised. ...

MARCH 4

Hammarskjöld in the morning. Told him that I absolutely refuse the proposal of Retzius on the simple ground, that she will certainly not apply for the professorship if the humiliating condition that she is just appointed for a certain time and not for life is stipulated. I emphasized the shamefulness in treating her badly because she is a woman, and the impropriety in not appointing her for life when

---

[12] Gustaf Retzius, at one time professor of histology at Karolinska Institutet, was a member of the board. He was married to Anna Hierta, generally regarded as the most dedicated guardian of morals in Stockholm at the time. This might have had something to do with the position of Retzius. Fredrik Nilsson was professor of chemistry at the academy of agriculture and represented the academy of sciences on the board.

funds are available although earlier when there were no funds one had appointed for life such inferior persons as Petterson and Leche. H. agreed with me completely and would give my answer to Retzius.

I looked up Hedin and asked him to talk to Söderström. Warm and noble as he always is when one asks him to talk on behalf of justice, he promised at once to take on the problem, even though it obviously is unpleasant for him to go to Söderström. Went to A... who was also completely on my side.

Then I went to Gyldén's and talked very frankly with them. Mrs G. is very angry with Retzius that he lets his hatred and jealousy of Mrs K. replace objective reasons, Gyldén defended him and would try to reason with him. And that is his duty for ever since Mrs K. got the prize he has done nothing but fan the flames of Retzius' hatred. ...

MARCH 5

Slept during the night for the first time since long ago. The worry for Mrs K. has kept me sleepless for a long time. In the morning I telephoned to Gyldén. He had talked last night to Retzius who insists on his proposal since this is a genteel way to get rid of Mrs Kovalevsky.

Mittag-Leffler continued the round of his friends Key, Hedin, Leche to plead for the Kovalevsky professorship, for the moment of decision was now close:

MARCH 7

The board decided today that the professorship in Higher mathematical analysis[13] *would be declared vacant with a salary of the interest of 50,000 Cr. up to 2000 Cr and 2000 Cr as long as the University receives from the City Council at least 40,000 Cr a year. It was first decided to receive the 19,000 Cr. which I had delivered! Retzius' proposal, that he would be asked by the board to contact the donors with a request that the donation be changed so that it would not*

---

[13] After Kovalevsky's death the chair was held by E. Phragmén, I. Bendixson, F. Carlson, L. Carleson and the author of this article; the name disappeared in the end of the 1950's.

be necessary to appoint Mrs Kov, for life was unanimously voted down (as particularly offensive to me). ...

The motivation given by Retzius and his absent ally Nilsson was the same as mentioned in the diary on February 28. Another reason appeared later:

MAY 11

Retzius had visited Hammarskjöld and entirely won him over by his attacks against the socialists. He had said that he opposed Mrs K. because she had close contact with Branting[14], had married Miss Kjellberg to Vollmar[15], and was very active as a socialist agitator here. All this had convinced H. that R. had not acted from personal motives when he opposed the appointment of Mrs K.

It is remarkable that Mittag-Leffler was able to overcome both the objections to Sonja Kovalevsky as a woman and as a dangerous socialist in addition to the scepticism about the utility of pure mathematics. – There is a gap in the diaries after this big battle, and Sonja Kovalevsky does not reappear in them until a year later.

MARCH 6, 1890

Visited Sonja in the morning to talk about the organization of the seminar. She told me quite abruptly that Max Kovalevsky is ill, it concerns his foot, but she does not quite know how it is. She has wired him with a question if she ought to come down to him. If he answers yes, she wants to leave already tonight. She has a lecture tomorrow. As reason for her trip she will say that her brother is ill. It is probably not true that M. K. is ill, or that this is at least only a part of the truth. The planned trip is due to other conflicts.

I told her that the story of the brother cannot be told in any case. It would compromise her for ever and irrevocably. Adviced her to go today to Benzelius and talk to him about her heart condition. Some idea might emerge from that. The difficulties in engaging bril-

---

[14] Leader of the social democratic party and 30 years later prime minister in Sweden.

[15] A German socialist.

liant *Russian ladies for a regular activity start to grow beyond my head.*

The irritation with Sonja did not last long:

MAY 12

The worst day of my life so far. On Wednesday Netzius had made a small operation on Mother ... At noon I got his verdict. Cancer. ... Visited Sonja in the evening. I shall never forget her emotion and her deep sympathy.

Two days later Sonja left for Petersburg. – In the summer Mittag-Leffler went to Berlin to visit Weierstrass who was in bad health. Sonja had been there earlier in the summer but hurried away:

JULY 10

... Sonja has told him that she travels with her brother-in-law Alexander Kov. and that it was after a telegram from him that she suddenly had to leave Berlin. In reality the telegram was from Max. According to Ach. it is really true that Max at the beginning wanted to marry her. But he is utterly vain and afraid to be his wife's husband, and he wanted her to leave everything and follow him. However, she did not want to give up her independent position. I still believe that what mattered more was, what she told me several times in the Harz, that by not marrying she hoped to keep him as an always devoted lover. But the situation has changed. Now she wants to sacrifice everything and marry if he wants to. She would still prefer first to get a professorship at the academy in Petersburg or a position in Paris. However, he does not want to any longer. The sad fact, according to rumors in Paris, seems to be that he is very débauché. He has always very questionable individuals as male friends. And he likes to amuse himself with des drôleries together with them. How will this end? Without doubt a definite break would be the best. But poor Sonja, she cannot live without a love intrigue.

In Petersburg she had been celebrated as a Patti or Nilsson and was very pleased with that. This has always been her ideal. In Berlin she had flirted wildly with Pascale[16]. Poor Weierstrass! He said

---

[16] Ann-Charlotte's friend whom she was about to marry.

"I did not see much of her, she was quite naturally so entirely occupied by your sister." He did not know that she has hardly exchanged two words with Ach but according to the unanimous testimony of Pascale and Ach exclusively flirted with P. Besides she had been utterly excited and exhilarated by the coming meeting with Max.

Chances at the Academy in Petersburg exist but are uncertain, since nobody there talks energetically for her while the opponents are so much more active. ...

Here end the diary entries of Mittag-Leffler which concern Sonja Kovalevsky. Half a year later she was dead.

Why did Sonja Kovalevsky stay in Stockholm after the expiration of the first five years to which she had committed herself? The diaries give no clear answer; the best clue is perhaps Max Kovalevsky's observation (Nov. 30, 1888) that her position there was incomparably better than what she could obtain elsewhere. And the reason for that was of course Mittag-Leffler's untiring efforts on her behalf – as he observed on July 10, 1890, there was nobody in Petersburg talking energetically for her.

This raises the question why Mittag-Leffler devoted himself so wholeheartedly to supporting Sonja Kovalevsky. After all the diaries give many expressions of his frustration and irritation with her. One reason was undoubtedly that it was essential for his effort to build up mathematics in Stockholm to have a candidate for a second professorship with a high international reputation. But behind it all there must also have been a high regard and deep affection for her, which were not put down in the diary but come out in the obituary [3] where the last paragraph balances the diary entries:

Sophie Kovalevsky gardera une place éminente dans l'histoire des mathématiques, et son œuvre posthume qui doit bientôt paraître, conservera son nom dans l'histoire de la littérature. Mais ce n'est peut-être ni comme mathématicien ni comme littérateur qu'il sied d'apprécier et de juger avant tout cette femme de tant d'esprit et d'originalité; comme personnalité, elle était plus remarquable qu'on ne pourrait le croire d'après ses travaux. Tous ceux qui l'ont connue et approchée, à quelque cercle, à quelque partie du monde qu'ils appartiennent, resteront constamment sous la vivante et forte impression que produisit sa personne.

REFERENCES

1. Koblitz, A.H. (1983): A convergence of lives. Boston, Basel, Stuttgart: Birkhäuser
2. Kochina, P. (1985): Love and mathematics. Translated from Russian edition published by Nauka in 1981. Moscow: Mir Publishers
3. Mittag-Leffler, G. (1892): Sophie Kovalevsky. Acta. Math. 16:385–392
4. Mittag-Leffler, G. (1923): Weierstrass et Sonja Kowalewsky. Acta. Math. 39:133–198
5. Randver, G.W. (1981): Sonja Kovalevsky. Bokförlaget Trevi

Fritz John

Memories of Student Days in Göttingen

This is an account of my personal experiences at Göttingen University during a critical period in Germany. I have no way of comparing my impressions with student life and the state of mathematics instruction at other German universities at that time. Not all of it is "besonnte Vergangenheit". Some memories still are too disturbing to be mentioned here.

First a few words about my earlier exposure to mathematics before coming to Göttingen. The decisive influence was my teacher's in secondary school. During the 1920's I attended a highschool ("Gymnasium") in a tiny country existing at the time, the "Free City of Danzig". There was already a lot of incipient nazism in and out of school, which affected me since my father was Jewish. Still, science was taught very competently by teachers with a sound academic education. Mathematics started for us with plane geometry and algebra at the age of 12 or 13. There was no attempt to teach set theory or other abstract mathematical notions to younger children.

Mathematics came as a revelation to me: so much intuitive beauty! I took to it with enthusiasm, and soon tried my hand at finding and proving theorems of my own, of course on a very elementary level. I wrote up some of the results as part of the requirements for graduation from high school. By that time we had covered analytic geometry and formal differential calculus, with some excursions into spherical trigonometry (perhaps a remnant of the needs of navigation of an earlier time).

Though in many respects not well adjusted to the school, I was encouraged by my principal mathematics teacher, E. Schmidtke. He suggested attending the University of Göttingen, reputedly the world's foremost center of mathematics, where he himself had studied. Thus it came about in the spring of 1929 that I enrolled as a student in Göttingen, where I stayed to the beginning of 1934.

The mathematical Göttingen of that period has been described admirably in the two books on Hilbert and Courant by Constance Reid. My own impressions were conditioned by the fact that I arrived there as an almost pennyless student, with money scraped together by my hard working widowed mother, and that I came from a family with no claim to social or intellectual prominence. Success was important, since my funds might give out at any time. There was the usual shock of adjusting to living away from home, to huge numbers of well informed competitors and to the high level of the beginning courses. I managed to survive due to the fact that some of the faculty took an active interest in at least some of the students, and extended a helping hand. To be so favored depended mainly on mathematical performance, though other factors, like charm ("female"), accomplishments ("music"), connections ("children of colleagues"), may occasionally have played a role.

There was no opportunity for students to "perform" during the lectures with their large audiences. This was possible, however, in the exercise sessions ("Übungen") attached to the basic courses, and the more general problem sessions ("Praktika"), run by a professor with the help of numerous assistants. These practice sessions always offered some challenging non-routine problems, whose solution and subsequent discussion gave students an opportunity to come to the attention of the faculty. I was fortunate in my problem solving during my first year. Courant and some of the Dozenten, Cohn-Vossen and Wegner, became supportive. With Courant's help I obtained a stipend from the Studienstiftung des Deutschen Volkes, which relieved me of my main financial worries for the rest of my student days.

Courant, destined to play a major role in my life, was the acting director of the Mathematics Institute (the exact administrative set-up was not clear to me). He had just succeeded in getting the Rockefeller Foundation to donate a new and spacious mathematics building. Courant had a solid reputation in analysis, principally in the calculus of variations. His main impact though was through his teaching and his books. These included his "Differential- und Integralrechnung" (the first mathematics book I acquired, and on a revision of which I collaborated with him many years later) and his famous "Methoden der Mathematischen Physik" (nominally with Hilbert as co-author). His many personal connections with publishers, like Ferdinand Springer, scientists, like Harald and Niels Bohr, and industrialists, like Carl Still, also benefitted the Institute. Courant's lack of pomposity and his concern with students set an example to the faculty. Students whom he considered promising could count on his help, were invited to his house and, if suitable, participated in the musical activities of his family. He unselfishly strove to advance the cause of mathematics, though he undoubtedly enjoyed being in the center of things. His way of doing things may have aroused resentment in the less favored students, or in colleagues who felt imposed upon.

Courant, the mathematician, affected me only later. Hilbert, though still around, was ailing and not really active any more. I took my introductory courses from Herglotz, Cohn-Vossen and Wegner. Of the three Herglotz was aloof and artistic, the other two, down-to-earth and friendly. Herglotz gave beautifully polished lectures, seemingly on any subject under the sun, from celestial mechanics to geometry of numbers. In the course of time I attended many of his lectures and fell under his spell. Among the things I learnt from him, and which proved useful for my own work later on, were the manipulation of multiple integrals, the theory of the propagation of singularities, Herglotz's solution of partial differential equations with constant coefficients and, what I came to call, the "Radon transform" (or decomposition into plane waves). Herglotz's lectures were full of surprising mathematical connections, each new to me and unexpected, the arguments drawn from his vast range of knowledge, in particular from a theory of multiple algebraic integrals, only obliquely referred to. Sometimes his ethereal way of proving things in his lectures (and also publications) completely hid some simple direct access to the same result, say by Fourier transformation, which he was fully aware of. One could only admire, but without hope of entering on one's own, this fantastic world of beauty, with its 19th century flavor, which might be called "analysis without estimates".

Courant's course on partial differential equations was the exact opposite, with its concrete, systematic, well motivated existence proofs, based on a priori estimates. Courant's lectures lacked the glamour of those of Herglotz and of Herman Weyl. They were, however, deeply stimulating, and offered a chance to participate in the creative process. My main interests became analysis and geometry, and I was drawn more and more to Herglotz, Courant and Weyl, and to some extent to Lewy, Rellich and Fenchel. In retrospect I regret not having taken advantage of all the other extraordinary things going on in Göttingen, like the lectures of

E. Noether and the new physics developed by Born and others. In the beginning I was attracted to number theory, and enjoyed reading Landau's book on prime numbers. But I was repelled by Landau's later formalized style, and never attended any of his lectures. Altogether my ability to digest material proved to be strictly limited. I cut down on courses, avoided abstraction and even missed out on developments in modern analysis, trying instead to build independently on the foundations I was familiar with.

There was another development affecting me. In the fall of 1931 my fiancée, Charlotte Woellmer, transferred as a mathematics student from Berlin to Göttingen. Charlotte's funds were just as straitened as mine. Her coming was only made possible by exchanging a room in her parental home with that of a person in Göttingen. When this arrangement broke down after one semester, it looked as if she could not continue in Göttingen. Fate intervened to save us from separating. Courant was going to spend the spring semester of 1932 in America. When he heard of Carlotte's plight, he very generously offered her to stay with his family in his house during his absence. At the same time Oswald Veblen of Princeton University came as guest professor to Göttingen, and I was assigned to be his assistant and to prepare notes of his course on projective relativity. I was not very good at that. Without previous acquaintance with tensor calculus, I had a great deal of difficulty with the concepts underlying the course (which I only learnt later from Veblen's book on quadratic differential forms). Nevertheless Veblen and his wife were very nice to Charlotte and me, taking us on extended car rides (our first ones) and telling us about the strange world of distant America.

At that time Charlotte and I supplemented our income somewhat by preparing lecture notes ("Ausarbeitungen"); I had already done that before her arrival. Between us we produced notes on a great variety of courses (by Herglotz, Cohn-Vossen, Courant, Weyl and Rellich). Essentially one took down the lectures in short hand, almost verbally, or at any rate in sufficient detail for reconstructing the arguments. They were then written out (the lecturers rarely bothered to check them) and subsequently typed by us, producing up to six copies of varying quality. Formulae and symbols had to be entered painfully by hand in every copy. The best copy went to the Göttingen Mathematical Institute, which financed the enterprise. Some of these Ausarbeitungen can still be found in its library (though without mentioning who prepared them). Other copies were sold. One of the regular customers was an American, Professor Archibald, acting for Brown University.

Through a seminar talk I became acquainted with the work of the American mathematician Marston Morse on relations between the number of critical points of a function. I managed to prove that no relations other than those given by Morse exist between these numbers. This gave rise to my first paper, published in the "Mathematische Annalen". With its strong topological flavor it pointed in a direction I did not pursue later on.

By this time in 1932 Charlotte and I were accepted as junior members of the mathematical community, with friendly relations with Courant, Weyl, Lewy, Fenchel and Rellich. We had, of course, in addition many associations with other students and with visitors. Among those surviving are Hans Schwerdtfeger and Hannah Mäder (his future wife), Heinz Meyer-Leibnitz, Stefan Warschawski, Saunders Mac Lane, Jim McShane, Y.W. Chen, Leifur Asgeirsson, Wolfgang Hahn and Wolfgang Wasow.

In the natural course of events I would have had to think about a topic for my doctoral dissertation. Once, at the public swimming pool, H. Lewy brought up the problem of determining a function from its averages on spheres with a fixed radius. I got interested in this question and started working on it. It represented a special integral equation of the first kind of convolution type. The solution involved a blend of Fourier and Radon transforms, the wave equation and properties of Bessel functions. [A more abstract approach to related question was pursued independently in France by J. Delsarte, under the name of "functions periodic in the mean".] I hoped that once my thesis was finished, and I had received my Ph.D., I would be able to stay at the university, and start an academic career. This was, of course, far from certain. In addition I had imprudently neglected the customary precaution of preparing simultaneously for becoming a teacher in secondary schools.

But all plans were made irrelevant by the political developments in Germany. Up to then politics did not penetrate our sheltered existence. We ignored the increasing frequency of street demonstrations with their shouted slogans and threats of violence. Then, at the end of January 1933, Hitler took over the government, and events moved rapidly. The universities were purged progressively of "non-Aryans". Some people sensed the meaning of the revolution early on, and left quietly. One of these was Hans Lewy, who walked with us on his last evening in Germany, explaining why he was going to emigrate. Others with deeper roots in the past or more illusions, still stayed on, in the vain hope for a return of normalcy. One of these was Courant, for whom the collapse of the world around him was especially traumatic. The remarkable thing was the extent to which, in spite of his own worries, he tried to help others.

I do not recall the exact sequence of events. There was no immediate physical danger, if you did not draw attention to yourself. At one point I was expelled from the Studienstiftung. In particular I had trouble raising the 200 Marks needed as fee for receiving the Ph.D. A friend of my family finally came to the rescue. The time did not seem far off when the university would refuse to confer a degree on me. With the mathematics faculty decimated, I found in Rellich a sponsor for my examination in July 1933. The rules required examinations on three subjects. Mathematical analysis and geometry counted as two, and I chose philosophy as the third one. Compared with what I have seen since in other countries, the procedure was extremely informal. There was no audience, and only one examiner at a time was present (and taking notes). My examiners in mathematics were Rellich and Weyl, and in philosophy, M. Geiger. I had chosen the latter because of his interest in the foundations of mathematics, and that is what he questioned me about exclusively. Even in these easy circumstances I did not perform as well as I would have liked to. Essentially the degree was conferred on the merits of my dissertation, which had been approved earlier by the faculty. There was informality also in the actual confering of the degree, which consisted in receiving a scroll in exchange for 200 Marks at some office window. Here there were no solemn procession, no caps and gowns or speeches by the Rektor.

Ten days after I received my Ph.D. Charlotte and I got married in a civil ceremony. We were worried whether it was still legal or safe to enter a "mixed" marriage. It could not be done in complete secrecy. The bans had to be posted for public inspection in the city hall two weeks ahead. There also had to be two witnesses at the ceremony. We found out, however, that among our friends there were other couples in the same predicament, and we served mutually as witnesses. Lack of funds and the need for secrecy precluded an elaborate celebration. It was also important to involve our families as little as possible.

We lived in the center of town, in an old water mill adjacent to the Leinekanal. Money was scarce. Through her marriage Charlotte lost her stipend from the Studienstiftung which she had just been awarded. Courant helped out, drawing on funds or contributions at his disposal. During the summer of 1933 we paid our last visit to my mother in the Free City of Danzig, traveling there as cheaply as possible by bicycle and passenger-ship. On our return to Göttingen things looked bleak. There was no possibility of future employment in Germany for me. The atmosphere became more oppressive. There were anonymous denunciations against us. One felt like in a trap with the door slowly closing. Courant pursued various leads, for example to Istanbul or Odessa, which came to nothing (fortunately, as it turned out). I would have welcomed escape to any place outside Germany. In the meantime I wrote up my thesis for publication, and helped Courant with volume II of the "Methoden der mathematischen Physik". Charlotte was busy with an Ausarbeitung for one of Rellich's courses.

Finally, in the fall of 1933, Courant accepted an invitation for a year's stay in Cambridge. He managed to secure a scholarship for me there which enabled me to emigrate to England and safety in January '34. At that time I did not have to overcome any official hurdles, in the form of special permits required. I still had a valid German passport; I had no posessions to take along, or any jobs to take leave from. Charlotte followed two months later, after finishing her notes for Rellich's course. We were fortunate in leaving so early. Other people with more to give up put off leaving, and then had a much harder time in both getting out of Germany, and being accepted by another country. We stayed almost two years in England, supported by a small university grant, and the generosity of the academic community. We were part of a growing influx of refugees into Britain, which it seemed difficult for the universities to absorb. In the fall of 1935 I was offered and accepted a position in the United States, which became my adopted country.

My first return after emigration to a very different Germany took place in 1952. There have been many more since. My visits to Göttingen became less frequent over the years, as the number of people I knew from my student days diminished. Charlotte and I still were able to visit Herglotz, Rellich and Deuring not long before their deaths. Of those who emigrated there are also not many left. The Göttingen I knew as a student is gone. It was a splendid place, which formed me as a mathematician and determined my future.

Max Koecher

Castel del Monte und das Oktogon

1. CASTEL DEL MONTE

Die Entwicklung der staufischen Architektur im Königreich beider Sizilien gipfelt nach H. Götze ([1] und [2]) in dem *Castel del Monte*, der *»Krone Apuliens«*, das in der Regierungszeit Friedrich II. von Hohenstaufen errichtet wurde. *» Dieses Bauwerk hat Aufmerksamkeit und Bewunderung erregt, aber auch Staunen und Verwunderung – wie der Kaiser selbst.«* Der Grundriß des Castels zeichnet sich durch eine besondere Symmetrie aus und basiert auf dem *Oktogon*, dem regulären Achteck. H. Götze gibt einen Versuch seiner Deutung und schlägt eine Konstruktion vor, die mit geringsten Annahmen eine plausible Herleitung dieses Grundrisses nachvollziehbar macht.

2. EIN KONSTRUKTIONSPRINZIP ODER DIE »INNERE ÄSTHETIK«

Im folgenden soll dieser Grundriß auf seine »innere Ästhetik« untersucht werden. Der Grundriß wird dabei auf die aus Innenhof und Türmen bestehende Kontur reduziert. Dabei will ich unter einer *geometrischen Konfiguration mit innerer Ästhetik* eine Konfigu-

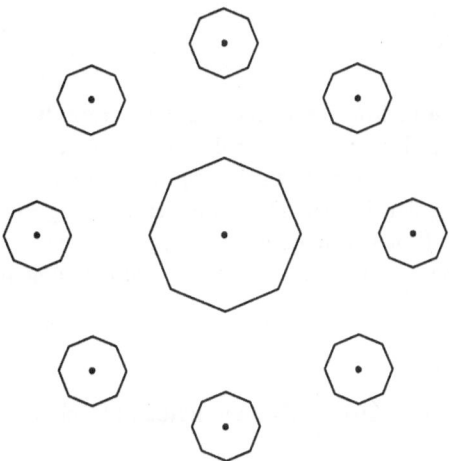

ration von Oktogonen (oder allgemeiner von regelmäßigen n-Ecken)
verstehen, die nach folgenden Prinzipien aufgebaut ist:

(i) Die Konfiguration besteht aus dem Mittelpunkt und den Ecken
   eines zentralen Oktogons $\mathcal{O}$ sowie dem Mittelpunkt und den
   Ecken von acht kleineren, aber gleich großen, parallel ver-
   schobenen Exemplaren $\mathcal{O}_\nu$ des zentralen Oktogons.
(ii) Die Mittelpunkte m_ν der $\mathcal{O}_\nu$ liegen auf den von dem Mittelpunkt
   m von $\mathcal{O}$ ausgehenden Strahlen durch die Ecken von $\mathcal{O}$.
(iii) Alle m_ν haben von m den gleichen Abstand.
(iv) Möglichst viele der

Eck- bzw. Mittelpunkte von $\mathcal{O}$ und von den $\mathcal{O}_\nu$

sind *kollinear*, d.h. liegen auf Geraden.

Evtl. kann man unter (iv) auch gewisse Mittelpunkte zweier unter
(i) und (ii) genannten Punkte zulassen.

Ein Beispiel (für Quadrate) soll dies erläutern:

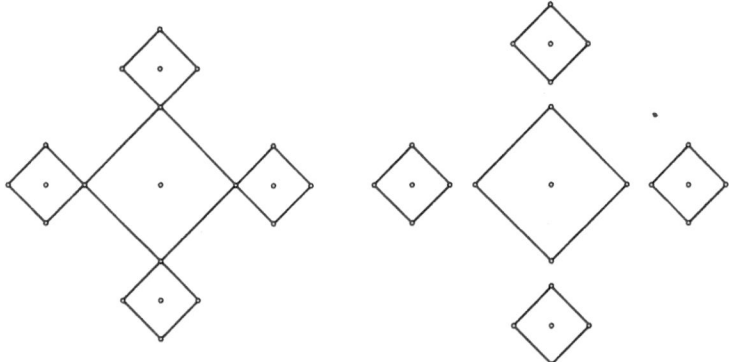

Hier ist die rechte Figur eine Konfiguration, für die (i), (ii) und
(iii) gilt, während die linke Figur die einzige ist, die noch (iv) erfüllt,
ohne daß auf Mittelpunkte zurückgegriffen wird.

Wenn man den halben Durchmesser von $\mathcal{O}$ gleich 1 setzt, also
die Konfiguration *normiert*, so hat jede Konfiguration, die (i), (ii)
und (iii) befriedigt, zwei Freiheitsgrade, nämlich erstens den

Abstand $\varkappa$ zwischen m und den m_ν

und zweitens den

Verkleinerungsfaktor λ, also den halben Durchmesser λ der $\mathcal{O}_\nu$.

Man kann dann von einer $(\varkappa, \lambda)$-Konfiguration sprechen. Gilt überdies (iv), so soll von einer *ästhetischen* $(\varkappa, \lambda)$-Konfiguration gesprochen werden. Es wird also im folgenden für ein Oktogon untersucht werden, welchen Einschränkungen die Zahlen $\varkappa$ und λ unterliegen, die zu ästhetischen Konfigurationen gehören. Im Anschluß daran wird der Fall eines Pentagons kurz diskutiert werden.

3. Die komplexen Zahlen als Hilfsmittel

Zur Diskussion von $(\varkappa, \lambda)$-Konfigurationen werden die Punkte der euklidischen Ebene als komplexe Zahlen, also als Elemente $z = x + iy$ des Körpers $\mathbb{C}$ der komplexen Zahlen interpretiert. Hier sind also

$$x = \operatorname{Re} z \ (\textit{Realteil}) \quad \text{und} \quad y = \operatorname{Im} z \ (\textit{Imaginärteil})$$

reell. Weiter ist die *konjugiert komplexe* Zahl $\bar{z}$ und der *Betrag* $|z|$ von z durch

$$\bar{z} := x - iy \quad \text{bzw. durch} \quad |z| := \sqrt{x^2 + y^2}, \quad \text{also durch} \quad |z|^2 = z \cdot \bar{z}$$

definiert. Damit gilt

$$\operatorname{Re} z = \tfrac{1}{2}(z + \bar{z}), \quad i \cdot \operatorname{Im} z = \tfrac{1}{2}(z - \bar{z}) \quad \text{und} \quad \overline{(\alpha z)} = \alpha \cdot \bar{z} \quad \text{für} \quad \alpha \in \mathbb{R}.$$

Mit der Abkürzung

$$[u, v, w] := \bar{u}v + \bar{v}w + \bar{w}u$$

hat man dann die

PROPOSITION. *Für $u, v, w \in \mathbb{C}$ sind äquivalent:*
(i) *u, v und w liegen auf einer Geraden.*
(ii) *$\operatorname{Im}[u, v, w] = 0$.*

BEWEIS. Durch Ausrechnen stellt man fest, daß

$$\operatorname{Im}[u, v, w] = \operatorname{Im}(\bar{u} - \bar{w})(v - w)$$

gilt.

Nun liegen 0, u und v genau dann auf einer Geraden, wenn es reelle α und β gibt mit $\alpha u = \beta v$ und $\alpha^2 + \beta^2 \neq 0$, wenn also $\operatorname{Im} \bar{u}v = 0$ gilt. Man wendet dies an auf $u - w$ bzw. $v - w$ anstelle von u bzw. v und sieht, daß (ii) genau dann gilt, wenn 0, $u - w$ und $v - w$ auf einer Geraden liegen. Dies ist aber gleichwertig mit (i). $\quad\square$

Für den interessierten Nicht-Mathematiker kann man die Proposition auf die Tatsache reduzieren, daß es Rechenvorschriften gibt, mit denen man stets entscheiden kann, ob vorgegebene Punkte auf einer Geraden liegen oder nicht.

4. DIE OKTOGON-KONFIGURATION

Es bezeichne $\mathcal{O}$ das in den Einheitskreis von $\mathbb{C}$ einbeschriebene Oktogon, also das reguläre Achteck, mit einer Ecke auf der reellen Achse im Punkt 1. Man betrachte die »primitive 8. Einheitswurzel«

$$j := \frac{1}{\sqrt{2}}(1+i) = e^{\frac{\pi i}{4}}$$

und die Ecken

$e_0 := j^0 = 1,$

$e_1 := j^1 = \frac{1}{\sqrt{2}}(1+i),$

$e_2 := j^2 = i,$

$e_3 := j^2 = \frac{1}{\sqrt{2}}(-1+i),$

$e_4 := j^4 = -1,$

$e_5 := j^5 = \frac{1}{\sqrt{2}}(-1-i),$

$e_6 := j^6 = -i,$

$e_7 := j^7 = \frac{1}{\sqrt{2}}(1-i)$

des in den Einheitskreis einbeschriebenen Oktogons $\mathcal{O}$.

Nach Vorgabe zweier positiver Zahlen $\varkappa$ und λ besteht die $(\varkappa, \lambda)$-Konfiguration des Oktogons also aus

dem Mittelpunkt 0 von $\mathcal{O}$,
den Ecken e_ν von $\mathcal{O}$,
den Mittelpunkten $m_\nu := \varkappa \cdot e_\nu$ von $\mathcal{O}_\nu$,
den Ecken $e_{\nu\mu} := \varkappa \cdot e_\nu + \lambda \cdot e_\mu$ der $\mathcal{O}_\nu$, $\mu = 0, 1, \ldots, 7$,

für $\nu = 0, 1, \ldots, 7$.

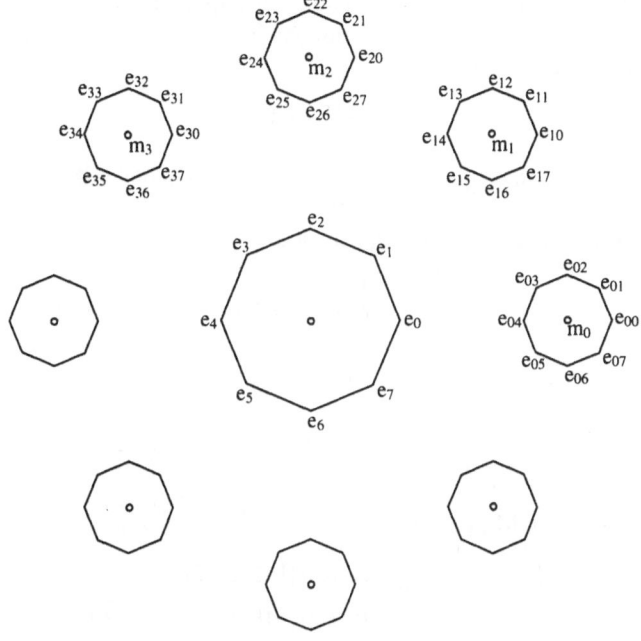

Offensichtlich liegen die Punkte

$$e_{01}, e_{02}, e_{31} \text{ und } e_{32} \quad \text{sowie} \quad m_0, e_{03}, e_{27}, e_{23} \text{ und } m_2$$

jeweils stets auf einer Geraden. Abweichend vom sonstigen mathematischen Sprachgebrauch sprechen wir von *Kollineationen*, wenn Punkte auf einer Geraden liegen.

5. ERSTE MÖGLICHE KOLLINEATIONEN

Erste mögliche Kollineationen sind solche, die nur die Punkte m_ν und $e_{\nu\mu}$ der kleinen Achtecke betreffen, wie

(A) $\qquad e_{00}, e_{15} \text{ und } e_{22}$ liegen auf einer Geraden

und

(B) $\qquad e_{01}, e_{16} \text{ und } e_{21}$ liegen auf einer Geraden.

Andere solche Bedingungen scheint es nicht zu geben!

Nach Konstruktion ist klar, daß eine solche Kollineation erhalten bleibt, wenn man $(\varkappa, \lambda)$ durch $(\alpha\varkappa, \alpha\lambda)$ mit positivem α ersetzt.

Die Bedingungen, die sich aus solchen Kollineationen an das Paar $(\varkappa, \lambda)$ ergeben, werden daher homogen in $\varkappa$ und λ sein. Eine mögliche solche Bedingung ist

$$(*) \qquad\qquad \varkappa(\sqrt{2}-1) = \lambda(\sqrt{2}+1).$$

LEMMA. *Die Bedingungen* (A), (B) *und* (*) *sind äquivalent.*

BEWEIS. (A) ⇔ (*): Es ist hier nach 4

$$e_{00} = \varkappa + \lambda, \quad e_{15} = \frac{1}{\sqrt{2}}(\varkappa - \lambda)(1+i), \quad e_{22} = (\varkappa + \lambda)i.$$

Im Hinblick auf Proposition 3 hat man also den Imaginärteil von

$$[e_{00}, e_{15}, e_{22}] = \frac{1}{\sqrt{2}}(\varkappa + \lambda)\{(\varkappa - \lambda)(1+i) + (\varkappa - \lambda)(1-i)i - \sqrt{2}(\varkappa + \lambda)\}$$

zu berechnen. Man erhält

$$\begin{aligned} \mathrm{Im}\,[e_{00}, e_{15}, e_{22}] &= (\varkappa + \lambda)\{\sqrt{2}(\varkappa - \lambda) - (\varkappa + \lambda)\} \\ &= (\varkappa + \lambda)\{\varkappa(\sqrt{2}-1) - \lambda(\sqrt{2}+1)\}. \end{aligned}$$

Nach Proposition 3 sind (A) und (*) äquivalent.

(A)⇔(B): Diese Äquivalenz sieht man elementar-geometrisch ein. Natürlich kann man die Äquivalenz »(B)⇔(*)« auch rechnerisch nachvollziehen. □

6. FLUCHTLINIENSCHNITTPUNKTE

Die Fluchtlinien des Innenhofes, also die Verlängerungen der Seiten des zentralen Oktogons, und ihre Schnittpunkte sind sicher ausgezeichnete Punkte der Konfiguration. Bis auf Drehungen handelt es sich dabei um den einen Schnittpunkt s der Geraden durch e_2 und e_1 sowie durch e_6 und e_7. Da die beiden Geraden durch Spiegelung an der reellen Achse ineinander übergehen, ist s gleichzeitig der Schnittpunkt der Geraden durch e_2 und e_1 mit der reellen Achse. In einer ästhetischen Konfiguration gibt es also die Möglichkeiten

(F) $s = e_{00} = \kappa + \lambda$, d.h., e_2, e_1 und e_{00} liegen auf einer Geraden,

(G) $s = m_0 = \kappa$, d.h., e_2, e_1 und m_0 liegen auf einer Geraden,

(H) $s = e_{04} = \kappa - \lambda$, d.h., e_2, e_1 und e_{04} liegen auf einer Geraden.

Als Schnittpunkt einer Geraden mit der reellen Achse ist s reell. Wegen

$$\mathrm{Im}\,[e_2, e_1, s] = \mathrm{Im}\left\{\frac{-1}{\sqrt{2}}\,i(1+i) + \frac{1}{\sqrt{2}}(1-i)\,s + si\right\}$$

$$= \frac{1}{\sqrt{2}}\{-1 - s + \sqrt{2}\,s\}$$

ergibt Proposition 3 schon $(\sqrt{2} - 1)\,s = 1$, also

$$s = \sqrt{2} + 1.$$

Wir greifen die von H. Götze (als Stufe »D«) vorgeschlagene Bedingung (G) heraus und formulieren das Ergebnis als die

PROPOSITION. *Die Bedingung* (G) *ist mit* $\varkappa = \sqrt{2} + 1$ *äquivalent.*

Bei dem entsprechenden Schnittpunkt der Fluchtlinien der Türme kann man sich wieder aus Symmetriegründen auf den Schnittpunkt s_0 der Geraden durch e_{00} und e_{01} bzw. durch e_{04} und e_{03} beschränken. Eine ästhetische Bedingung ist hier die Forderung, daß dieser Schnittpunkt s_0 in der Mitte der beiden betreffenden Türme liegt, daß also die Bedingung

(S) e_{04}, e_{03} und $\frac{1}{2}(e_{01} + e_{10})$ liegen auf einer Geraden

gilt. Überraschenderweise erscheint hier wieder die Bedingung (∗) aus 5:

LEMMA. *Die Bedingungen* (S) *und* (∗) *sind äquivalent.*

BEWEIS. Man hat

$$e_{04} = \varkappa - \lambda, \quad e_{03} = \varkappa - \lambda\bar{j}, \quad e_{01} = \varkappa + \lambda j, \quad e_{10} = \varkappa j + \lambda,$$
$$s_0 = \tfrac{1}{2}(\varkappa + \lambda)(1 + j).$$

Hier ist nun

$$\operatorname{Im}[e_{04}, e_{03}, s_0] = \operatorname{Im}\{(\varkappa - \lambda)(\varkappa - \lambda\bar{\jmath})\}$$
$$+ \tfrac{1}{2}(\varkappa + \lambda) \cdot \operatorname{Im}\{(\varkappa - \lambda j)(1 + j) + (1 + \bar{\jmath})(\varkappa - \lambda)\}$$
$$= \frac{1}{\sqrt{2}}(\varkappa - \lambda)\,\lambda + \tfrac{1}{2}(\varkappa + \lambda) \cdot \operatorname{Im}\{\varkappa(2 + j + \bar{\jmath}) - \lambda(1 + j + \bar{\jmath} + i)\}$$
$$= \tfrac{1}{2}\lambda\{\sqrt{2}(\varkappa - \lambda) - (\varkappa + \lambda)\} = \tfrac{1}{2}\lambda\{\varkappa(\sqrt{2} - 1) - \lambda(\sqrt{2} + 1)\}.$$

Mit Proposition 3 ist das Lemma bewiesen. □

Ist die Bedingung (S) erfüllt, dann kann man die entsprechend verkleinerte Oktogon-Konfiguration bei jedem der Türme wiederholen und dann dieses Verfahren wiederholen. Man erhält ein *Fraktal*, also etwa folgendes Bild:

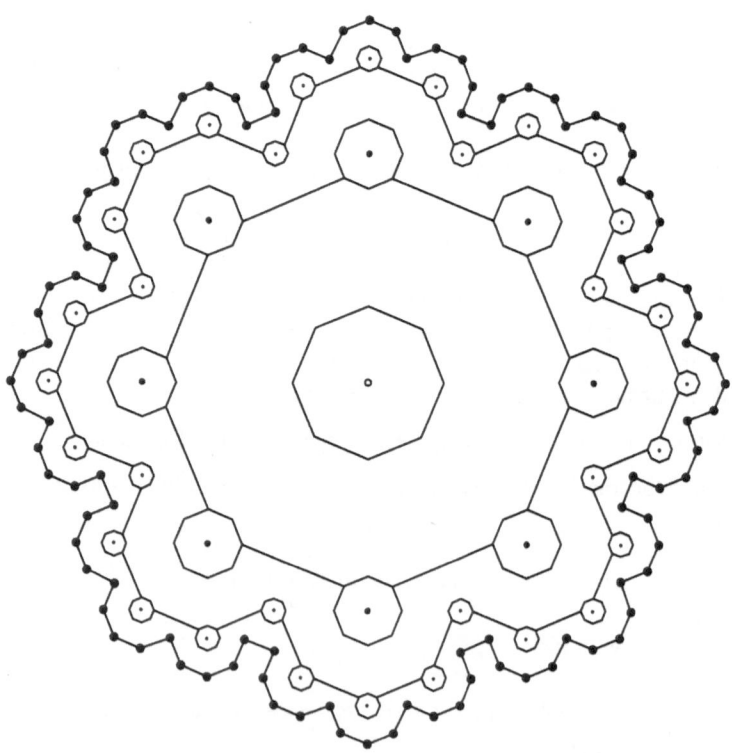

Ein Fraktal aus Quadraten und ein Pentagon-Fraktal findet man bei A. Beutelspacher und B. Petri ([3], Seite 88).

7. DIE FINALE BEDINGUNG

Neben der Bedingung (G), die sich wegen Proposition 6 auch als

(G) $$\varkappa = \sqrt{2} + 1$$

schreiben läßt, hat H. Götze (als » Stufe E «) vorgeschlagen, die Konstruktion dadurch festzulegen, daß die gemeinsame Fluchtlinie zweier Türme vom Mittelpunkt 0 des zentralen Oktogons einen doppelt so großen Abstand hat wie die Seiten des Innenhofes von 0. In der hier gewählten Bezeichnung und Normierung bedeutet dies einfach

(G') $$\varkappa - \lambda = 2.$$

Zusammen mit der Bedingung

(∗) $$\varkappa(\sqrt{2} - 1) = \lambda(\sqrt{2} + 1),$$

ergibt nun eine einfache Rechnung den

SATZ. *Je zwei der drei Bedingungen* (G), (G') *und* (∗) *implizieren die dritte Bedingung.*

KOROLLAR. *Eine normierte Oktogonkonfiguration ist genau dann ästhetisch, wenn die beiden Bedingungen* (G) *und* (G') *gelten. In diesem Falle gilt*

$$\varkappa = \sqrt{2} + 1 \ und \ \lambda = \sqrt{2} - 1.$$

8. ALGEBRAISCHE ASPEKTE

Im Hinblick auf die Definition der Punkte e_ν und $e_{\nu\mu}$ in 4 bildet man die Teilmenge

$$\mathscr{R} := \mathbb{Z} \oplus \mathbb{Z}\sqrt{2} \oplus \mathbb{Z}i \oplus \mathbb{Z}\sqrt{2}\,i$$
$$= \{a + b\sqrt{2} + ci + d\sqrt{2}\,i; a, b, c, d \in \mathbb{Z}\} = \mathbb{Z}[\sqrt{2}, i]$$

von $\mathbb{C}$ und überzeugt sich, daß die Punkte e_ν zu $\dfrac{1}{\sqrt{2}}\mathscr{R}$ gehören.

Der Definition entnimmt man weiter, daß $\mathscr{R}$ gegenüber Addition und Multiplikation abgeschlossen ist.

Mit der im Korollar 7 angegebenen Fixierung

$$\varkappa := \sqrt{2} + 1 \in \mathscr{R} \quad \text{und} \quad \lambda := \sqrt{2} - 1 \in \mathscr{R}$$

haben die nicht zu $\mathcal{O}$ gehörenden Punkte der Oktogon-Konfiguration die Form

$$m_\nu = (\sqrt{2} - 1) \cdot e_\nu \quad \text{bzw.} \quad e_{\nu\mu} = (\sqrt{2} + 1) e_\nu + (\sqrt{2} - 1) e_\mu.$$

Man faßt zusammen und bekommt die

PROPOSITION. *Alle Punkte der normierten ästhetischen Oktogon-Konfiguration liegen in* $\dfrac{1}{\sqrt{2}} \mathscr{R}$.

Wegen $\varkappa \cdot \lambda = 1$ sind $\varkappa$ und λ *Einheiten* von $\mathbb{Z}[\sqrt{2}]$, also auch *Einheiten* von $\mathscr{R}$. Die Frage nach allen Einheiten von $\mathscr{R}$ ist daher von Interessse:

LEMMA. *Die Einheiten von* $\mathscr{R}$ *sind genau die Zahlen der Form* $a + bi$, *wobei* a *und* b *Einheiten von* $\mathbb{Z}[\sqrt{2}]$, *also von der Form* $\pm(1 - \sqrt{2})^m, m \in \mathbb{Z}$, *sind.*

BEWEIS. Es sei x eine Einheit von $\mathscr{R}$ und $y \in \mathscr{R}$ mit $xy = 1$ gegeben. Für jeden Automorphismus $x \mapsto x'$ von $\mathscr{R}$ gilt dann $xx' \cdot yy' = 1$. Damit ist xx' eine Einheit des Fix-Ringes $\{x \in \mathscr{R}; x' = x\}$. Man wendet dies an auf die durch $\sqrt{2} \mapsto -\sqrt{2}$ und $i \mapsto -i$ erzeugten drei Automorphismen. $\square$

BEMERKUNG. Hätte man in 2 den halben Durchmesser von $\mathcal{O}$ zu $\sqrt{2}$ normiert, so würden die Punkte der ästhetischen Oktogon-Konfiguration sämtlich in $\mathscr{R}$ liegen.

9. DAS PENTAGON

Das Pentagon wird durch die 5. Einheitswurzeln beschrieben. Zur Vereinfachung der Schreibweise macht man zweckmäßig vom *goldenen Schnitt*

$$g := \tfrac{1}{2}(\sqrt{5} + 1) \quad \text{bzw.} \quad h := \tfrac{1}{2}(\sqrt{5} - 1)$$

Gebrauch. Es gilt hier offenbar $gh = 1$. Eine primitive 5. Einheitswurzel ist dann

$$\zeta := \tfrac{1}{2}(h + i\omega) \quad \text{mit} \quad \omega := \sqrt{\tfrac{1}{2}(5 + \sqrt{5})}$$

gegeben. Damit erhält man die Ecken des in den Einheitskreis einbe-schriebenen regulären Fünfecks bekanntlich in der Form

$$e_0 := \zeta^0 = 1,$$
$$e_1 := \zeta^1 = \zeta,$$
$$e_2 := \zeta^2 = h\zeta - 1,$$
$$e_3 := \zeta^3 = -h\zeta - h,$$
$$e_4 := \zeta^4 = -\zeta + h.$$

Eine Pentagon-Konfiguration
sieht dann wie folgt aus:

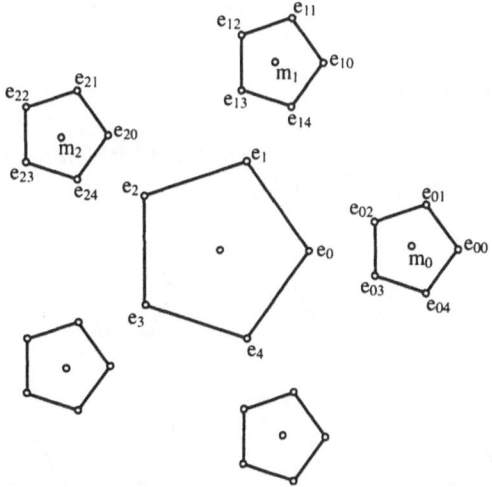

Für alle $\varkappa$ und λ liegen e_{00}, e_{01}, e_{10} und e_{11} sowie e_{04}, e_{02}, e_{14} und e_{12} auf je einer Geraden, während z.B. e_{00}, e_{01}, m_1 nie kollinear sind. Nicht-triviale Bedingungen sind hier

(A) e_{03}, e_{02} und e_{10} liegen auf einer Geraden,

sowie

(B) 0, e_{02} und e_{01} liegen auf einer Geraden.

Als weitere Bedingung kan man fordern, daß der Schnittpunkt der Fluchtlinien des zentralen Pentagons von 0 ebenfalls den Abstand $\varkappa$ hat. Dies bedeutet

(C) $-\varkappa, e_1$ und e_2 liegen auf einer Geraden.

Man zeigt unschwer das

LEMMA. a) *Die Bedingungen* (A), (B) *und*

$$\lambda = \varkappa(1-h)$$

sind äquivalent.

(b) *Die Bedingung* (C) *ist gleichwertig mit*

$$\varkappa = g^2 = 1 + g.$$

c) *Genau dann sind* (A), (B) *und* (C) *erfüllt, wenn*

$$\varkappa = 1 + g \ und \ \lambda = 1$$

gilt.

10. DAS HEXAGON

Das Hexagon soll als letztes Beispiel nicht fehlen, weil es sich hier stets um eine ästhetische Konfiguration handelt:

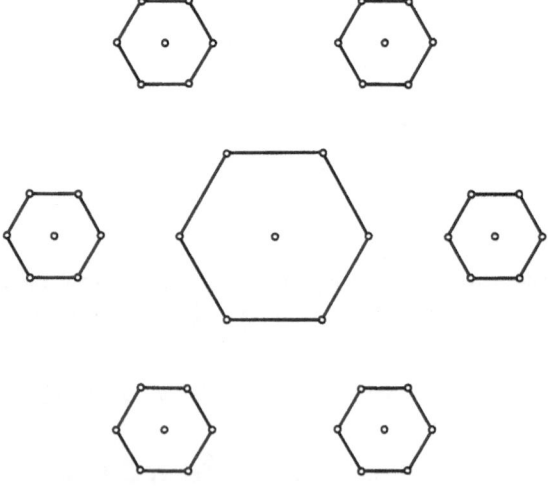

LITERATUR

1. Götze, H. (1984): Castel del Monte, Gestalt, Herkunft und Bedeutung. Sitzungsberichte der Heidelberger Akad. d. Wiss. Phil.-hist. Klasse 1984, Ber. 2, S. 9
2. Götze, H. (1984): Castel del Monte, Prestel
3. Beutelspacher, A., Petri, B. (1988): Der goldene Schnitt. BI Wissenschaftsverlag

The Coming of Age of Mathematics in India

To most people, the words "Indian Mathematics" probably suggest, first, the romantic figure of Srinivasa Ramanujan and then, perhaps, the contributions of the ancients (Aryabhatta, Bhaskara, Brahmagupta and so on). To a professional mathematician of today, they might also suggest the Tata Institute of Fundamental Research in Bombay and the many excellent mathematicians it has produced. However, during the period 1900–1950, there were several centres of mathematical activity in India about which people outside India, and even most Indians, know very little.

In what follows, I shall try to describe some of the figures who, often under very difficult circumstances, kept mathematical tradition alive in India. Their work may not be of great interest to a present day mathematician, but it was good work for the time and was very influential in attracting talented people to the subject. I am well aware that many more people merit inclusion than the ones I shall mention, but I hope that despite the shortcomings of my discussion, the general picture that emerges will reflect, at least with some accuracy, the state of affairs during the period 1900–1950.

India's position in the world of mathematics has become more important since 1950, due largely to the achievements of the School of Mathematics of the Tata Institute. I shall discuss the early days of the Tata Institute since they mark this change. A great deal of research comparable to the best work of the people discussed below was done at the Tata Institute, but I shall pass over this in silence. A full treatment would require an article longer than the present one and it might be difficult for me to be objective.

I shall not discuss scientists who worked for most of their productive lives outside India. Thus, I shall leave out of the discussion some of the best scientists of Indian origin; the most obvious of these are S. Chandrasekhar and Harish-Chandra.

Nor shall I deal with Ramanujan. A great deal has been written about him and his work recently, especially in connection with the "Centennial Year" (Ramanujan was born in 1887). Nevertheless, I believe that Ramanujan still awaits his mathematical biographer. It is very difficult to assess his work and put it in proper context in the mathematics of today.

Most aspects of life in India were controlled by the British in the nineteenth century and the first third of the twentieth; this certainly was the case with education. In 1857, the British established three universities, at Calcutta (which was the capital of India from 1833 to 1912), Bombay and Madras. Their aim was not the encouragement of intellectual activity. Rather, it was to produce a sizeable group of Indian "civil servants", Indians who would learn the English language and be sufficiently familiar with western ways and thought to be of real assistance to the British in administering territory of the size of the Indian subcontinent. The Civil Service offered the best paid jobs with the greatest security as well as a certain degree of authority over others. That this was attractive to relatively poor Indians who were often part of a large family is certainly understandable. But this system of education had no room for independent thinkers who might dedicate themselves to scientific research. The British did not want any challenge to their authority which independent thinkers might well have made. Thus the educational atmosphere was not conducive to a real academic career as we picture it now. Teaching was so abysmally paid that it could not hope to attract able people. The age of retirement was 55, and active people had to look for other positions when they were perhaps past their prime.

The university system which was set up was not centralised. It was not like the system of rather large centres (often state universities) that existed in Europe or that one finds in the United States today. The system in India was modeled on the University of London (according to Sir Charles Wood's Education Despatch of 1854), and consisted of a large number of small colleges affiliated to some university; some of these colleges, such as Presidency College in Madras and St. Stephen's in Delhi, predate the creation of the universities. Thus, when mention is made in what follows of a scientist being at some university, it should be remembered that this often meant that he was at one of the colleges where the teaching was onerous and the opportunity to influence students limited. There were some university positions outside the colleges, but except in Calcutta, the people who occupied them did little teaching, and their influence on the students was, at best, through their research and by example.

There was another way in which the British had a negative effect on mathematics in India. Mathematics in Britain was in a lamentable state in the second half of the nineteenth century. Despite the work of men such as Cayley and Sylvester, the British took almost no part in the great expansion of ideas which was occurring in Germany, France and Italy.

With the emergence of G.H. Hardy (1877–1947) and J.E. Little-wood (1885–1977), real analysis and certain aspects of number theory and of complex analysis were being practised at very high levels. But until the forties, there was no one in Britain who understood the work of, say, Poincaré or Hilbert. In his presidential address to the Mathematical Association at London in 1926, Hardy discussed the state of research in the country and characterised it as follows: "Occasional flashes of insight, isolated achievements sufficient to show that the ability is really there, but, for the most part, amateurism, ignorance, incompetence, and triviality."

The men who came from Britain to India to take charge of the universities were the less successful products of the British system. As far as I am aware, only one real British mathematician accepted a position in India; this was W.H. Young, who was at Calcutta for part of each year from 1913 to 1916, and I shall say something about him in a moment. But the large majority of the mathematicians who came to India did no research of their own, encouraged none in others, and transmitted a constipated view of mathematics to those with whom they came into contact. These same men were also often placed at the head of entire education programs for the provinces (large states into which India was then divided).

W.H. Young began to do serious research in mathematics rather late in life; he and his wife, G.C. Young, who was also an excellent mathematician, began to do fundamental work in real analysis in Göttingen where they lived from 1897 to 1908. Young's permanent residence was outside Britain from 1897 till his death in 1942 (in Switzerland). Young, Vitali and Lebesgue were led, independently of each other, to define the Lebesgue integral, but it was Lebesgue who proved the main convergence theorems and other basic properties which make this integral indispensable in real analysis. Young's work on Fourier series (the so-called Young-Hausdorff inequality comes readily to mind) was, in some ways, even more original.

Young was not appreciated in Britain; his candidacy for a professorship at Cambridge was not taken seriously and he accepted a chair at Calcutta in 1913. He spent the winter months of three years at Calcutta, leaving the university in 1916. He was also Professor of the Philosophy and History of Mathematics in Liverpool where he lectured during the summer months. He moved to Aberystwyth as Professor of Pure Mathematics in 1919.

To return to our discussion, I believe that the negative effects of the British system in India were made worse by the Indian mentality. There was a passiveness, not to say fatalism, among even the intellectual elite. However well Indians might follow a trail already blazed, there were few who were ready to take the risks of striking out on their own. One meets this attitude in India even today.

Fortunately, there were exceptions to what I have just said. (Sir) Asutosh Mookerjee (1864–1924), following the lead of Mahendra Lal Sircar, was among the first to propose the idea that a university should be a centre of independent intellectual activity, of research, and of teaching of high quality at all levels. Mookerjee came up through the British system and was appointed a judge of the Calcutta High Court in 1904. This remarkable man had always been interested in mathematics; he wrote some interesting papers dealing with aspects of plane algebraic curves and of differential equations in the years 1883–1890. He became Vice Chancellor of Calcutta University in 1906 and began systematically to improve the university. He brought talented people to Calcutta (S. Mukhopadhyay and N.R. Sen among them; see below). He founded the Calcutta Mathematical Society in 1908. He did his best to encourage talent and used the Society to bring the few researchers in India together and to publish their work (in the Bulletin of the Society).

The mathematical work of Asutosh Mookerjee is forgotten. His pursuit of the idea that Indians could, and should, undertake serious scientific investigations, that they should have, and raise, academic standards will not be forgotten so quickly.

Other universities were established from 1882 on.[1] Some of them began to attract individual mathematicians who did their best to do research and to improve the quality of teaching. In doing this, these men were certainly making personal sacrifices. They were generally able people and could easily have passed the examinations leading to the civil service, which, as already mentioned, meant good pay and an assured future. Instead, they chose to pursue an academic career which was ill paid, involved a great deal of drudgery and where advancement, even for the most talented, was problematic. Their dedication kept mathematical tradition alive and made conditions somewhat better for those who followed.

Among the places where some good work was being done (in the period 1900–1950) were the following: Calcutta (and Dacca), Benares, Lucknow, Allahabad, Aligarh, Punjab, Delhi, Poona and Bombay, Bangalore, Annamalai, Andhra, and Madras. I shall attempt to describe briefly some of the pure and applied mathematicians who functioned in these places. Calcutta and Madras were the most important of them, and I shall take them up at the end.

Before starting, let me try to locate these places geographically. I shall assume that the reader knows where Calcutta, Delhi, Bombay and Madras are (corresponding roughly to the east, north, west and south respectively). Benares, now known as Varanasi, is a holy city on the Ganges river about midway between Calcutta and Delhi. Allahabad is not far from Benares, to the west. Lucknow is some 200 miles northwest of Benares. Aligarh is a little southeast of Delhi. Punjab University was in Lahore which is now in Pakistan, just west of the Indo-Pakistan frontier. It was moved to Chandigarh after the partition of India. This city was designed and built by Jawaharlal Nehru, LeCorbusier, Maxwell Fry and the latter's wife Jane Drew; because of this, it is better known in the west than other Indian cities of its size. Chandigarh is about 150 miles north of Delhi. Poona is near Bombay, to the southeast. Bangalore is situated some 200 miles west of Madras. Annamalai University is on the outskirts of Chidambaram (an ancient centre of Hindu

---

[1] As mentioned earlier, the universities of Calcutta, Bombay, and Madras were founded in 1857. Punjab University was established in 1882, Allahabad University in 1887. The Benares Hindu University was founded in 1915, the Universities of Lucknow and Dacca in 1920, as was the Muslim University in Aligarh. Delhi University was created in 1922 by an Act of the Indian Legislature but consisted then simply of three existing colleges; funds were appropriated for a separate campus only in 1927. Andhra University was established in 1926, Annamalai University in 1929. Jamshedji Tata long wanted to create a research institute for technical training and donated land in Bangalore in 1898; the Indian Institute of Science was opened on this land in 1909 with funding from Jamshedji's sons, Dorabji and Ratanji, five years after Jamshedji's death.

religion) which is some 100 miles south of Madras. Andhra University is in Waltair on the east coast of India, about 350 miles northeast of Madras.

1. ALLAHABAD AND BENARES

One of the earliest influential figures in Indian mathematics was Ganesh Prasad (1876–1935). After early work in India, he went to Cambridge, England and then to Göttingen and came under the influence of Klein and Hilbert. He wrote some papers on differential geometry, mainly concerning surfaces of constant curvature, soon after his return to India. The bulk of his work is, however, on potential theory and the summability of Fourier series. After a year or so in Allahabad, he was in Benares from 1905–1914 and 1917–1923. He founded the Benares Mathematical Society. He went to Calcutta in 1923 and finished his career there. He also succeeded Asutosh Mookerjee as President of the Calcutta Mathematical Society.

Despite the fact that a large part of Ganesh Prasad's work (summability) was in the British tradition, he was one of the first Indian mathematicians to feel continental influence. He was a very powerful figure in mathematical circles. It is a little hard to escape the feeling that he could have had a more positive impact on Indian mathematics than he in fact did.

The career of B.N. Prasad (1899–1966) had some similarities to that of his teacher, Ganesh Prasad. He went to England (Liverpool) and then to Paris. Much of his work deals with the summability of Fourier series. But Prasad was not happy with the British system as it worked in India. He strongly advocated creating schools modeled on the Ecole Normale Supérieure in Paris. He lived to see the Tata Institute flourish, but his hopes of heading an institution of higher learning and research remained unfulfilled.

B.N. Prasad is a clear example of the sacrifices made by able people who went into teaching and research. His work was well thought of by such mathematicians as Titchmarsh and Denjoy. Nevertheless, he remained a lecturer at Allahabad for many years (1932–1946) and was then made a Reader. He became Professor on a temporary basis in 1958; the appointment was made permanent in 1960, shortly before his retirement. B.N. Prasad also founded the Allahabad Mathematical Society, which publishes its own journal even now.

Two others in Benares and Allahabad who must be mentioned are V.V. Narlikar (born 1908) and A.C. Banerji (1891–1968). Both were applied mathematicians. Narlikar was principally at Benares and was well known, at least in India, for his work on relativity. Banerji taught at Allahabad from 1930 to 1952 and became Vice Chancellor. He was a very broad-minded man and he worked hard to expose his students to new ideas.

2. DELHI, LUCKNOW, ALIGARH

Two men who were active at St. Stephen's College in Delhi were Ram Behari (1897–1981) and P.L. Bhatnagar (1912–1976).

Ram Behari's work was almost entirely in differential geometry. He studied at Cambridge and Dublin and was a student of J.L. Synge. His early work was on classical questions (rules surfaces and the like). In later years his work, done partly in collaboration with some of his students, became broader and more modern. He moved to the University of Delhi in 1947.

P.L. Bhatnagar worked both in pure and in applied mathematics. He studied with A.C. Banerji and B.N. Prasad. He was at Delhi from 1940 to 1955; he then moved to Bangalore to head the department of applied mathematics at the Indian Institute of Science. He, too, did some work on the summability of Fourier series, but the major part of his work was in such fields as astrophysics and fluid dynamics. He was also seriously interested in ancient Hindu mathematics. Bhatnagar was very influential in developing applied mathematics in India.

B.R. Seth (1907–1979) too spent several years in Delhi (1937–1949). His work was mainly on questions of elasticity and fluid dynamics. In 1950, he moved to the newly created Indian Institute of Technology at Kharagpur (near Calcutta). He founded something of a school of applied mathematics there.

A.N. Singh (1901–1954) spent most of his professional life at Lucknow. He worked hard to attract students to the subject; for example, he organised a mathematics exhibition regularly in the department in an attempt to make the subject more popular. Most of his work was on differentiability properties of functions of a real variable. He was also deeply interested in the history of Hindu mathematics, and even wrote a book on the subject.

Lucknow also had R.S. Varma (1905–1970) on its faculty from 1938 to 1946. He worked on classical analytic questions (integral transforms, special functions) till he moved to Delhi. He worked for the Defence Science Organisation from 1947 to 1962 and dealt with problems of ballistics and operations research.

The Muslim University at Aligarh had a very famous mathematician on its faculty from 1930 to 1932: André Weil. Sylvain Lévi, the well known Indologist, was approached by Syed Ross Masood (to whom E.M. Forster's *A Passage to India* is dedicated). Masood asked Lévi if he could recommend someone who was expert in Romance languages and in French culture and history. Weil was very interested in India and its culture, he knew Lévi, and was willing to spend some time in India. Masood had become Vice-Chancellor at Aligarh in October, 1929 and offered Weil a Chair in mathematics as being the only one available. Thus, Weil came to India.

At Aligarh, Weil came into contact with T. Vijayaraghavan and D.D. Kosambi; I shall discuss them later. He had a powerful impact on Vijayaraghavan, and helped to broaden his view of mathematics (as Vijayaraghavan himself told C.P. Ramanujam and me more than once).

Weil clearly understood what needed to be done in the colleges to improve the quality of the students and to make them better prepared for research. His outline of the lecture on this subject that he delivered to the Indian Mathematical Society in 1931 is printed in vol. 1 of his collected works. It is sad that Weil's remarks about the steps that need to be taken apply with as much force today as when he made them.

The three men mentioned above were at Aligarh for rather short periods. S.M. Shah (born 1905) was a junior colleague of Weil's in 1930 and remained in Aligarh till 1958. He then left for the United States, where he has been since.

Shah worked extensively on functions of a complex variable, in particular on entire functions. His work deals largely with the growth properties of entire functions and their Taylor coefficients. He and Ganapathy Iyer were among the few Indians who were aware of the revolutionary ideas of Rolf Nevanlinna concerning meromorphic functions and Picard's theorem.

3. PUNJAB

There are three well known number theorists to come out of Punjab.

S. Chowla (born 1907) had his early training at Lahore. He then studied in Cambridge and came under the influence of Little-wood. He taught for a while at Benares and at Andhra University in Waltair. He then became a professor at Lahore. After the partition of India, he went to the Institute for Advanced Study in Princeton in 1948 as a visiting member. He held professorships in Kansas, Colorado and Penn State, retired in 1976, and lives now in the United States.

Chowla's work was very extensive. He obtained very interesting results in analytic number theory (L-functions, Waring's problem and so on). He also worked on combinatorial problems. The best known of his results is part of work he did in collaboration with Atle Selberg on the so-called Epstein zeta function. It is of importance in the study of the class number of imaginary quadratic fields, and has led to further important work, for instance, by B. Gross.

Chowla was also very active in encouraging and promoting students. His best Indian student was R.P. Bambah (born 1925), who has done very interesting work on the geometry of numbers. Bambah has spent most of his life in Punjab; he is at present Vice Chancellor of the university at Chandigarh.

The other number theorist from the Punjab referred to above is Hansraj Gupta (1902–1988). Gupta was profilic, and has written on several aspects of number theory. The most important part of his work is on partitions. He did, for example, some numerical work extending tables of the number of partitions made by P.A. MacMahon which has proved to be very useful.

4. BOMBAY, POONA, BANGALORE, ANNAMALAI

The University of Bombay had no research department in mathematics, at least until rather recently (although it did in, for example, chemistry). However, D.D. Kosambi (1907–1966) was active at nearby Poona. He was at Fergusson College from 1932 to 1946 before he moved to the Tata Institute. He had spent a few years at Benares and at Aligarh before 1932.

His purely mathematical work was in differential geometry, in particular, on path spaces which are generalisations of spaces with an affine connection. He had studied at Harvard at both the school and the college level, and his view of mathematics was much broader than was common in India at the time.

He also became very interested in statistics, and in applying its methods to fields such as numismatics and Sanskrit literature, but particularly to Indian history; these subjects became his main interests.

He isolated himself almost completely from other mathematicians when he was at the Tata Institute. Despite his incisive intelligence, his influence on Indian mathematics was small.

The Indian Institute of Science was founded in Bangalore by Jamshedji Tata, one of the great industrialists of India. (Sir) C.V. Raman (1888–1970) spent some years (1933–1947) at this Institute before moving to an Institute named after himself. Raman had resigned a civil service position to pursue science in Calcutta. Asutosh Mookerjee appointed him to a professorship in physics in 1917; he held the position till 1933. It was here that he did the work in optics, partly in collaboration with K.S. Krishnan (1898–1961), for which he received the Nobel Prize. He and Krishnan were very active in many of the efforts being made to improve teaching and research in the sciences, including mathematics. It might be added that Mookerjee was also responsible for bringing Krishnan to Calcutta.

Another physicist who was to have a profound influence on the development of mathematics in India also spent some years at the Indian Institute of Science in Bangalore. This was Homi J. Bhabha (1909–1966). He had been working in Cambridge with Dirac and was in India in 1939, meaning to return to Cambridge, when the second world war broke out. He went to Bangalore, and while he was there, he founded the Tata Institute of Fundamental Research (TIFR) in Bombay. Bhabha was prophetic in his recognition of the inevitability of the development of nuclear energy. He wanted to assure the existence of Indians sufficiently close to the cutting edge of modern work in physics to take advantage of this development. He also realised that this required supporting mathematical research at the highest level. With financial help, again, from the Tata family and from the provincial government of Bombay, TIFR was created in 1945.

B.S. Madhava Rao (1900–1987) spent much of his professional life at Central College in Bangalore. His own work was in applied mathematics, but his view of the subject was very enlightened and he encouraged students to work in a wide variety of fields. He sent several of his students to TIFR in the late fifties, some of whom have done outstanding work. When he retired, he went to the Defence Science Organisation in Poona where he is reported to have been exceptionally effective.

A. Narasinga Rao (1893–1967) was at Annamalai University from 1929 to 1946. He moved to Andhra University in 1946, and to the Indian Institute of Technology at Madras in 1951. He worked mainly on aspects of euclidean geometry, but was also interested in other fields (e.g. aerodynamics). He did his best to ecourage young mathematicians. He was responsible for attracting Ganapathy Iyer and S.S. Pillai to Annamalai. He was the first editor of *The Mathematics Student*, a journal (founded by the Indian Mathematical Society) to which mathematics teachers and students could send their contributions. [He also instituted a medal, named after himself, awarded through the Indian Mathematical Society to outstanding young people.]

V. Ganapathy Iyer (1906–1987) too was a professor at Annamalai. He was an excellent mathematician, and wrote papers on many aspects of both real and complex analysis. His best work was on the theory of entire functions; in some of this work he applied methods from functional analysis to spaces of entire functions. As mentioned in connection with S.M. Shah, he was among the few in India who understood Nevanlinna theory. Given the high level of his accomplishments, it is rather surprising that he did not have more of an impact on the younger generation.

5. CALCUTTA AND MADRAS

Calcutta had the first, and for a long time, the most important university in India. There was always a strong tradition of intellectual pursuit in Bengal. The poet Rabindranath Tagore, the botanist Jagdish Chandra Bose and the political leader Subhash Chandra Bose all came from Bengal. Among scientists, one of the best known is S.N. Bose (1894–1974) who worked at Calcutta and Dacca; K.S. Krishnan and T. Vijayaraghavan were colleagues of his at Dacca. He was a long-standing member of the Calcutta Mathematical

Society. He is best known for his discovery of the so-called Bose-Einstein statistics which describes the statistical behaviour of certain types of particles. [Particles with certain spin characteristics are now called bosons in his honour.]

Calcutta played host to many distinguished scientists. I have already mentioned W.H. Young, C.V. Raman, and K.S. Krishnan who all spent fruitful years at Calcutta; there were many others.

Of the mathematicians and statisticians who were more permanently associated with Calcutta, I shall cite only a few. They are: S. Mukhopadhyay (1866–1937), N.R. Sen (1894–1963), R.N. Sen (1896–1974), P.C. Mahalanobis (1893–1972), R.C. Bose (1901–1987), and the German mathematician F.W. Levi (1887–1966).

Levi fled Nazi Germany and came to Calcutta as professor of mathematics in 1935 where he remained until 1948. He then retired from Calcutta and moved to the Tata Institute.

His influence on mathematics in India was considerable. He introduced Indians to algebra (which used to be called modern algebra). He was an active participant in mathematical meetings. Levi played an important part in the acceptance of algebra in university curricula all over India.

Levi seems to have been less effective when he was at the Tata Institute (he was there only about three years); perhaps a university milieu suited him better. He retired in the early fifties and returned to Germany.

Syamdas Mukhopadhyay was brought to the university in Calcutta by Asutosh Mookerjee to organise the mathematics department. He was clearly an independent man, and was not much influenced by the British view of mathematics. He wrote on the differential geometry of curves, both in the plane and in n-space; he considered global properties of ovals in the plane. His work was much better known in France than at home. Hadamard is supposed to have had a high opinion of his work on plane curves. [Note. The name Mukhopadhyay is derived from Sanskrit; its Bengali equivalent is Mookerjee, with variations in its English spelling. Thus our mathematician is referred to sometimes as Mukherji.]

Nikhilranjan (= N.R.) Sen too was brought to Calcutta by Asutosh Mookerjee. He worked mainly in applied mathematics and physics (relativity, cosmogony, fluid dynamics) but he also did useful work in potential theory and probability. He was very helpful to younger people.

Rabindranath (= R.N.) Sen came to Calcutta in 1933. His work was mainly on differential geometry, and especially on parallelism.

P.C. Mahalanobis was a very influential statistician. He did important theoretical work to which he was led by analysing actual data obtained in the field. This connection between theory and application was important to him and he tried to foster the relationship in the work of his colleagues. He did important statistical work for both the state and the central government.

Mahalanobis founded the Indian Statistical Institute (ISI) in 1931. He brought excellent scientists to the institute, for instance, R.C. Bose. One of the best statisticians to come out of ISI is C.R. Rao who was at the institute from 1944 to 1979. He then left India to go to the United States; he is currently at Pennsylvania State University.

R.C. Bose is very well known internationally for his work in statistics and combinatorics. He was a student of Mukhopadhyay, and his early work reflects this influence; it is largely on geometric questions, including hyperbolic geometry. Mahalanobis brought him to ISI and induced him to work on statistical questions. Bose responded by applying geometric ideas to statistics.

Bose is perhaps most famous for having disproved a conjecture of Euler on "mutually orthogonal latin squares" in collaboration with his student S.S. Shrikhande (who headed the mathematics department at the University of Bombay during the sixties). Bose went to the United States in 1949, where he remained till his death. He seems to have had more Indian students there than in India.

ISI produced four excellent pure mathematicians V.S. Varadarajan, R. Ranga Rao, K.R. Parthasarathy, and V.S. Varadhan. They have done impressive work on measure theory, stochastic differential equations, and especially on the representation theory of Lie groups. This was done on their own initiative.

Madras has never had a centre of mathematics of the quality of the department at Calcutta University. This applies not only to the university itself, but also to two other institutions meant to foster research, the Ramanujan Institute and Matscience. I shall say something about the first of these later. No serious work was produced at the second until recently.

Nevertheless, there was always a strong tradition of scholarship and learning in and around Madras. Certainly, the best pure mathematicians to emerge in India during the period 1900–1950, even leaving Srinivasa Ramanujan out of consideration, came from the Province of Madras.

The men I shall now discuss are: K. Ananda Rau (1893–1966), R. Vaidyanathaswamy (1894–1960), T. Vijayaraghavan (1902–1955), S.S. Pillai (1901–1950), and S. Minakshisundaram (1913–1968).

Ananda Rau's early work was done in Cambridge. He was a student of Hardy, and there is no mistaking the influence of Hardy and Littlewood on this work. Ananda Rau had a critical and independent mind and he was conscious of the vast extent of mathematics and the lack of interest in India in many of its branches. He knew Ramanujan in Cambridge and spoke of him with great admiration, but without a trace of the mysticism or romanticism of many others. On his return to India, he took up a position at Presidency College, where he remained until his retirement. He was in the Indian Educational Service, and was thus better off than most other mathematicians. (The IES was abolished shortly after his appointment.)

Ananda Rau's first successes concerned summability by general Dirichlet series (including power series). He established very general "tauberian theorems" which had resisted the efforts of some of the best analysts. One of the methods of summation to which he contributed some of his most original ideas, Lambert summability as it is called, is intimately connected with the distribution of prime numbers. In later years, Ananda Rau was occupied with modular functions and the representation of integers as sums of squares. It has always seemed to me a great pity that neither Ramanujan nor Ananda Rau ever came into contact with someone like Erich Hecke when they were young. Hecke was a master at combining hard analysis with the arithmetic behaviour of modular (and automorphic) functions, exactly the two topics central in the work of these two Indians.

Ananda Rau's work is of great depth and elegance. He had a strong influence on his students: Vijayaraghavan, Pillai, Ganapathy Iyer, Minakshisundaram, Chandrasekharan and others. He taught them, besides mathematics, the value of intellectual independence.

Vaidyanathaswamy studied in Edinburgh with E.T. Whittaker, then visited Cambridge and worked with the geometer, H.F. Baker. When he returned to India, he was something very new on the scence. He was greatly interested in "structures". He emphasised general principles over special results, however deep. Even in his work on properties of plane algebraic curves, he was fascinated by general properties of homogeneous forms and of birational transformations. He also worked on number theory, but was more interested in the general behaviour of multiplicative arithmetical functions than in special properties of even the most important among them.

Vaidyanathaswamy was the first Indian to study mathematical logic, set theory and general topology; he even wrote a book on this last subject. He was very influential. For over two decades, he was the motive force behind the Indian Mathematical Society. Even if one cannot point in his work to individual results of the depth of those obtained by Ananda Rau, his opening up of a different view of mathematics was most beneficial. He avoided the danger of empty generalities, just as Ananda Rau focused on special problems that would illustrate general behaviour. He lectured on topics in pure mathematics at the Indian Statistical Institute in Calcutta for about two years after his retirement from Madras.

It was Ananda Rau who recognised Vijayaraghavan's talents; he even had to intervene to get him admitted to Presidency College. Vijayaraghavan began to work independently even as a student and, following Ramanujan's example, sent some of his work to Hardy. This led, albeit with some delay, to his proceeding to Oxford to work with Hardy. On his return to India, he was a colleague of André Weil's for about a year, taught at Dacca (near Calcutta; now in Bangladesh) for a few years and then went to Andhra University.

In 1949, Alagappa Chettiar created the Ramanujan Institute in Madras, and brought Vijayaraghavan there as Director. Based on some fairly extensive contacts with this institute since 1954, my impression was that it existed mainly as a place of work for Vijayaraghavan and his younger colleague, C.T. Rajagopal, at least till the latter's retirement in 1969. Vijayaraghavan would have liked having at his disposal the funds necessary to bring young researchers and outside scholars to the Ramanujan Institute. In this, he was disappointed. But he was very helpful and kind to the students who did go to him. Thus, for well over a year, up to the time of his death in 1955, Vijayaraghavan received C.P. Ramanujam (about whom more later) and me regularly. He helped us with the material we were trying to study, gave us a few lectures, and encouraged us to work on our own. I need hardly add that these activities were not part of his normal duties at the Institute.

Vijayaraghavan's work, like that of Ananda Rau, began with an impressive theorem about summability, in this case, Borel summability; the theorem was suggested by work of Ananda Rau as Vijayaraghavan himself has said. He then did a piece of work which is typical of his abilities, disproving a conjecture of E. Borel about the growth of solutions of non-linear ordinary differential equations. This brought him to the attention of the mathematical world; G.D. Birkhoff, for example, was impressed and was instrumental in having Vijayaraghavan invited to the United States as Visiting Lecturer of the American Mathematical Society in 1936.

His third major investigation was the distribution of the fractional parts of the numbers $\alpha \cdot \xi^n$ (α and ξ being fixed real numbers) as $n \to \infty$. This led him very naturally to a class of numbers which, for a short time, were called Pisot-Vijayaraghavan numbers. The distribution is far from being understood even now, despite major advances in diophantine approximation.

When Vijayaraghavan died, C.T. Rajagopal became the director of the Ramanujan Institute. For many years, the institute housed the library of the Indian Mathematical Society, and Rajagopal was Librarian. Shortly before his retirement, the Ramanujan Institute was absorbed into the University of Madras as an Institute attached to the Department of Mathematics.

S.S. Pillai studied with Ananda Rau. He also came into contact with Vaidyanathaswamy, but his own inclination was towards important specific problems and Ananda Rau's influence on him was the greater. He was a lecturer at Annamalai University, and for a brief period, at Calcutta where he went because of the interest taken in him by F.W. Levi. He obtained a stipend from the Institute for Advanced Study in Princeton (for the year 1950/51). He was to go to the Institute and, from there, take part in the International Congress of Mathematicians at Harvard. Tragically, the plane carrying him crashed near Cairo.

The work for which Pillai will always be remembered concerns Waring's Problem. The question itself is easily understood: Given an integer $k \geq 2$, what is the smallest integer $g(k)$ with the property that every natural number n can be written as a sum $n = x_1^k + \ldots + x_s^k$ of s k-th powers of natural numbers x_j with $s \leq g(k)$? It is, of course, far from obvious that this can be done at all, viz. that $g(k)$ is finite. A famous classical theorem of Lagrange tells us that $g(2)=4$.

In 1909, David Hilbert proved the existence of $g(k)$ for all integers $k \geq 2$. By a remarkable analytic argument, Hilbert first proved a beautiful algebraic identity stated by his friend A. Hurwitz; Hurwitz himself had only been able to check this identity in a few special cases. Hilbert then added an interpolation argument to complete the proof. Hardy and Littlewood, using, in part, techniques introduced by H. Weyl, brought powerful analytic methods to bear on this problem. By improving Weyl's techniques, I.M. Vinogradov sharpened these results very considerably; his methods have been central in analytic number theory since he introduced them.

But none of these methods enables one to actually compute $g(k)$. What Pillai did was to evaluate $g(k)$, at least when $k \geq 7$; L.E. Dickson independently obtained very similar, and perhaps slightly stronger results at about the same time. The value depends on the behaviour of the fractional part of $(\frac{3}{2})^k$. It should be added that the problem is not completely solved; it is strongly suspected that the behaviour of this fractional part is such as to produce a single simple formula for all values of $k \neq 4$, but this remains conjectural. Further, the value of $g(4)$ was only determined recently (in 1986) by R. Balasubramanian, J.-M. Deshouillers and F. Dress. In principle, $g(k)$ can now be computed for any given value of k.

Pillai wrote many papers on the theory of numbers. Each of them shows analytic power and originality, but his work on Waring's problem remains his greatest achievement.

Minakshisundaram began as a research student at Madras University. After his doctorate, when he could not find a job, Father Racine (about whom I shall say something later) helped him earn a living coaching students. Minakshisundaram was then appointed Lecturer in Mathematical Physics at Andhra University. Thanks to the interest taken in him by Marshall Stone, he was able to spend a few years (1946–1948) at the Princeton Institute for Advanced Study; he retained his position at Andhra University during this time. When he returned to India, he could still not find a job better suited to him. Although he was promoted to a professorship in Mathematical Physics, circumstances at the university made it impossible for him to strike out on his own. This is another clear example of the sacrifices that were necessary in pursuing a career of intellectual achievement in India.

Minakshisundaram's early work, influenced by Ananda Rau, was on the summability of Dirichlet series and eigenfunction expansions. Then he met Fr. Racine and M.R. Siddiqui. Siddiqui, who was in Hyderabad (some 300 miles north of Bangalore), had been a student of Lichtenstein in Leipzig and had done good work on the initial value problem for parabolic equations, extending results of Lichtenstein. (He became Vice Chancellor of the University of the Northwest Frontier Province in Pakistan, and later, President of the Pakistan Academy of Sciences.) Minakshisundaram thus became interested in partial differential equations, and was led to study the zeta-function associated to boundary value problems. This culminated in his fundamental joint work with Å. Pleijel (done when they were both in Princeton) on the eigenvalue problem for the Laplace operator on a compact Riemann manifold. His idea of using the heat equation in this study has proved very fruitful and extremely powerful. It led to work of Atiyah, Bott and Patodi on a new approach to the so-called index theorem for elliptic operators, and is being used even today in what has come to be known as the index theorem for families of elliptic operators.

Minakshisundaram did quite a lot of other work. For example, he and K. Chandrasekharan worked extensively on subtle analytic properties of the so-called Riesz mean and applications to multiple Fourier series. It is sad that Minakshisundaram's exceptional abilities did not have the impact on Indian mathematics that they should have had.

Fr. C. Racine (1897–1976) came to India in 1937 with a Jesuit Mission. He had studied analysis in Paris with such people as E. Cartan and J. Hadamard. He was a friend of many of the outstanding French mathematicians: Leray, Weil, Delsarte, Lichnerowicz, H. Cartan and others. These men were making fundamental discoveries; they were also making fundamental changes in the way mathematics is taught and learned. In fact, some of them were in the process of creating Bourbaki. Thus Fr. Racine came to India with a view of mathematics completely unlike the prevailing one. In the classroom, but especially in personal contacts outside, he communicated this dynamic view of the subject. He sent students interested in mathematics to the Tata Institute to experience this kind of mathematics for themselves when that institution was beginning to function well.

Fr. Racine was deeply committed to India. He went back to France on his retirement as was expected of him, but returned to Loyola College, Madras to live out his life among the people for whom he had done so much.

C.T. Rajagopal (1903–1978) was mentioned in connection with Vijayaraghavan. He too was a student of Ananda Rau. He taught at Madras Christian College just outside the city till he went to the Ramanujan Institute in 1951. Rajagopal's work was almost entirely devoted to summability and related questions in analysis, although he wrote some papers on functions of a complex variable. In his last years, inspired by work of K. Balagangadharan, he became interested in the history of mathematics in medieval Kerala (southwestern coastal state of India). Rajagopal's work, while not as deep as that of Ananda Rau, had some outward similarity to it.

The last person from Madras I shall mention is V. Ramaswami Aiyar (1871–1936). He did some teaching in different colleges and then became a civil servant, but he was a true amateur (in the French as well as the English sense of the word) of mathematics. It was Ramaswami Aiyar who founded the Indian Mathematical Society in 1907, and started publication of its Journal. He worked hard to foster mathematical research in the country. Ramaswami Aiyar was also the first of several Indians to recognise the exceptional talents of Ramanujan before Ramanujan wrote to Hardy; he tried to get modest financial support for Ramanujan to enable him to continue his work, but in this he was only partially successful.

I have not treated Andhra University separately, but it will undoubtedly have been remarked that several of the mathematicians mentioned above spent some time there. This was no accident. Andhra had two very enlightened Vice Chancellors in S. Radhakrishnan (the philosopher and statesman) and his successor, C.R. Reddy. In acting to help good scientists who were in some difficulty, they followed the advice of C.V. Raman. At Waltair, as at Calcutta, the qualities of the Vice Chancellor were crucial in enhancing those of the university. The situation in Madras illustrates the opposite side of the coin. But for the prejudices of its long-time Vice Chancellor Lakshmanaswami Mudaliar, Madras might well have had one of the better research departments in mathematics in the country.

The picture of mathematics in India during 1900–1950 that I have tried to describe is the following. Indians were gradually shak-

ing off the effects of an outmoded system and a narrow view of
mathematics. There were a few truly outstanding scientists, and
many more good ones, in various centres distributed all over India.
They formed scientific societies and published journals which
formed a good outlet for work being done in the country. These
societies often had an international membership, and foreign
members contributed papers to these journals. All this was done
under difficult circumstances; for instance, the publication of journals was continued even when the second world war created terrible
paper shortages.

The contributions of the people I have mentioned, and those
of their many colleagues whom I have not, should not be completely
forgotten.

It will not have escaped the reader's notice that all the Indians
who have been mentioned so far were men. Social conditions in
most parts of the world form one of the main reasons that it is
difficult for women to become seriously interested in mathematics
at an early age. The conditions are even more difficult in India
which has a system of arranged marriages and near ostracism of
unmarried or divorced women and widows. Nevertheless, there
have been women mathematicians in India. S. Pankajam and
K. Padmavally are among the earliest of whom I am aware. The
former was active in the forties, the latter in the fifties. They were
both students of Vaidyanathaswamy. There have been several wom-
en at the Tata Institute. One of the best women mathematicians
the country has produced is Bhama Srinivasan, who is currently
at the University of Illinois at Chicago. Her work on group representations is well known.

We come now to the Tata Institute of Fundamental Research
(TIFR). As stated earlier, the Institute was the brainchild of
H.J. Bhabha and was founded in 1945. Originally, it received no
financial support from the central government, but this support
increased gradually, and the Institute is now financed almost entirely by the Government of India. However the Institute continues
to function with a large degree of independence.

Bhabha brought F.W. Levi and D.D. Kosambi to Bombay; neither of them was particularly effective there. Upon consultation
with John von Neumann, André Weil, and Hermann Weyl, he
approached K. Chandrasekharan in Princeton to try and bring him
to Bombay to develop mathematics at TIFR. Bhabha had already

heard about Chandrasekharan several years earlier, probably from Kosambi.

Chandrasekharan was a first rate analyst. He was a student of both Ananda Rau and Vaidyanathaswamy; he also came into close contact with Vijayaraghavan later. His early work was on intuitionistic logic and on functions of a complex variable. He then began, on his own initiative, important work on multiple Fourier series and related questions of analysis.

In 1944, Marshall Stone was in Madras and wanted to meet the best young mathematicians there, and especially Chandrasekharan and Minakshisundaram of whom he had heard through Kosambi. With Stone's help, Chandrasekharan went to the Institute for Advanced Study in Princeton. This was, for him, a turning point. Hermann Weyl had a profound influence on him which, I think, would be difficult to exaggerate. He was, in fact, Weyl's assistant for one year. He also met, and became friendly with, many of the outstanding figures in the subject; I shall mention only C.L. Siegel, A. Selberg and S.S. Chern as having had the most powerful influence on his thinking. His mathematical output was very extensive. He began a fruitful collaboration with S. Bochner, with whom he studied Fourier series and transforms in several variables, and Dirichlet series.

Bhabha offered Chandrasekharan a position at TIFR with the understanding that he would have Bhabha's full support in building up the school of mathematics. Chandrasekharan insisted that he must have a free hand in running the school, and to a very large extent, he did.

The school of mathematics at TIFR was not modeled on any existing institution, although I am sure that the Institute in Princeton, and the vast experience of Oswald Veblen and Hermann Weyl provided Chandrasekharan with ideas for what the shape should be. Within a few years of his arrival at TIFR in 1949, he managed to form a sizeable group of people and to create a remarkable atmosphere. Let me make some more comments on this.

Members of the school were free to learn and to do research with no distractions (except of their own making). There were no restrictions on the field of work, no unfashionable subjects. This was specially important to people newly arrived from a restrictive university environment. These newcomers were given a stipend sufficient to cover necessities and encouraged to test their mettle. The older members (those who had been there longer) provided advice, information, and encouragement to the newer ones. Members of the school did not have to give lectures, nor did they have to undertake administrative duties, unless, of course, they wanted to.

There was continuous contact with the best minds in mathematics. Outstanding mathematicians from all over the world, including the Soviet Union, came regularly to TIFR and lectured on the most diverse branches of the subject, presenting connected accounts of topics of current interest. These courses were written for publication by one or two members of the school. This often led to independent work by the "notes-takers", and sometimes, to collaborations with the lecturers.

Material success and advancement in TIFR followed ones work; it was fair and rapid, based solely on the merits of the work done.

When I joined the school of mathematics at TIFR in 1957, the atmosphere there was heady. Nothing seemed as important or as exciting as mathematics. New subjects were being talked about constantly and trying to learn them was a challenge. Listening to a colleague try out his ideas and attempting to understand and improve on them was the best instruction one could have. And then there was the excitement of working on problems oneself.

One quality of the greatest importance in a leader of such a group is the ability to recognise significant mathematics and important problems even in fields far removed from his own areas of specialisation. In my opinion, Chandrasekharan had this quality to an extraordinary degree; otherwise the Institute would have been very different.

Chandrasekharan was helped in all the work that creating this school entailed by K.G. Ramanathan. (Ramanathan was a number theorist who worked for a few years with Siegel.) He made the major decisions himself, as, for instance, the decision to organise periodic International Colloquia on important subjects. There is little doubt that the running of the school involved personal sacrifices on the part of Chandrasekharan. Let me mention only the most obvious of them. He continued to do excellent mathematics when he was directing the school; he even broadened his interests and began a series of papers of some importance (in collaboration with others) on analytic number theory. But there can be little doubt that he would have done much more personal research if he had not spent a great deal of time on "trivial details" so that others could work uninterrupted.

The school of mathematics became famous internationally. In many fields such as algebraic geometry and complex analysis, Lie groups and discrete subgroups, and number theory, it has produced a body of work of lasting value.

Chandrasekharan left India for Switzerland in September, 1965; he still lives in Zürich. Shortly after his departure, Bhabha died in a plane crash in January, 1966. It is my belief that the Tata Institute has been unable to absorb the loss of its two most visionary members, and that this has changed the atmosphere.

The reputation of the school of mathematics of TIFR is, of course, based on the outstanding work done by many of its members. With one exception, I shall not try to describe these members or their work; I do not know recent developments and conditions at TIFR very well, and the work with which I am familiar would take too much space to describe. I make an exception for C.P. Ramanujam because his was certainly one of the most powerful mathematical minds to emerge in India since the mid-fifties, and he was, in many ways, a singular figure.

Ramanujam came to TIFR in 1957 with a great deal of knowledge of deep mathematics. This would be unusual anywhere; in India, it was indeed exceptional. He was one of those rare people who feel completely at home in all branches of mathematics. Thus, he understood sophisticated analysis as deeply as he did Grothendieck's view of algebraic geometry which appears very abstract, but illuminates the fundamental relation between geometry and arithmetic. Ramanujam helped many of the people at TIFR to understand difficult subjects. It was natural to turn to him when one reached an impasse in ones work.

In his short mathematical life, Ramanujam did some very profound work. This included definitive solutions of well known problems (as with his solution of Waring's problem for number fields) as well as the introduction of methods and results which formed the basis of progress by others (the Kodaira-Ramanujam vanishing theorem, characterisation of the affine plane, ...).

He was diagnosed as being schizophrenic when he was 26. This meant that the time he could devote to mathematics in the ten years that remained to him was severely curtailed. He collected an impressive library of books dealing with the malady and read them carefully. He came to the conclusion that the condition was incurable and took his own life when he was at the height of his intellectual powers (at the age of 36 in 1974).

When Chandrasekharan left India in 1965, the state of mathematics in India was very different from the situation in 1950. There was substantial government support for mathematics, not only at TIFR, but through several other agencies as well. There was a very strong group of mathematicians, mainly at the Tata Institute, whose work was internationally acclaimed. Famous mathematicians from all over the world were ready to come to India and establish beneficial contacts.

It is hard to assess the situation today. Government support has certainly continued and even increased; a National Board of Mathematics exists with very substantial financial resources and a free hand to operate all over India. TIFR still attracts some outstanding young talent from throughout the country.

The colleges in India have, however, not improved very much. The curricula today are more modern than they used to be, but the quality of the teachers in these colleges has not improved substantially, and this is unlikely to change as long as they continue to be so poorly paid and to have so much uninteresting teaching to do. Where the civil service attracted the most talented, it is now industry, some of it private, much of it run by government, which competes with the lure of intellectual pursuit for the commitment of the best people.

It is to be hoped that these things will change sufficiently for the future of mathematics in India to be truly distinguished.

Hans Lewy · 1904–1988

Hans Lewy was one of the first of that great group of mathematicians who, leaving Germany after Hitler came to power in 1933, substantially influenced the development of mathematics in the United States.

I became acquainted with him only some thirty years later. I had been told that in his youth he had been in Göttingen and had personally known many of the great mathematicians of the early twentieth century. It was largely as a result of my conversations with him that I wrote a life of David Hilbert, the first of my books to be published by Springer-Verlag.

Although Lewy loved to tell stories that illustrated the characters and personalities of others, he relentlessly removed any personal details from published accounts of his own life. The following is a typical Lewy entry:

Born Breslau, Ger. [now Wrocław, Poland], Oct. 20, 1904. Ph.D., U. Göttingen 1926. Married. Naturalized U.S. citizen. Priv. Doz., Göttingen 1927–33. Associate, Brown University 1933–35. Lecturer, University of California (Berkeley) 1935–37, Asst. Prof. 1937–39, Assoc. Prof. 1937–39, Prof. 1945.

No mention of honors, honorary academies or honorary degrees. He did not choose, he said, to participate in the contemporary cult of the personality, which he dismissed as a "star" system. And yet he was a star. And now that he is gone it seems that in addition to his more than six decades of influential mathematical work, something about the man and his life should also be in print.

Lewy was the first son and second child of Max and Greta (Rösel) Lewy. His father was a merchant who dealt in accessories for women's millinery, and his mother before her marriage had been a teacher of German in an expatriate enclave in Hungary. Her interest and ability in languages were inherited by her son. In addition to a thorough grounding in Greek and Latin, which he had received in his *Gymnasium* days, he was so fluent in French and Italian that he was frequently mistaken for a native on the streets of Paris and Rome. He could converse in Russian, once delivering a lecture in that language, and he actively studied Chinese up to the time of his death. On several occasions, marooned and without the solace of his piano, he set out to teach himself a new language – once it was Spanish, which he attacked by plunging into *Don Quixote*.

Both of Lewy's parents were Jewish, but only the father was religious. Lewy himself did not observe the rituals of Judaism and, as he once told a reporter for a Jewish newspaper, did not believe that the Jews were chosen by God. Needled by a friend about whether he would don a yarmulke when he was awarded the $100,000 Israeli-based Wolf Prize, which he would share with the distinguished Japanese mathematician Kunihiko Kodaira, he parried gracefully, "In matters of dress I will be guided by my co-recipient."

The boy Lewy so loved mathematics that his mother felt she had to hide his mathematics books from him, but he also loved music and was a gifted violinist. His parents, not wanting his gift to be exploited, did not permit him to perform in public until his fifteenth year, when he took part in a concert in the resort town of Bautzen. There is in existence a letter that he wrote to his family after that event: "Saturday was THE day: an orchestra, a male chorus, a woman soloist, [and] I – *Chaconne* by Bach, *Devil's Trills* by Tartini .... During the first piece I still felt a little tight, but the second I could hardly wait to start playing." One local critic praised his technical expertise and another his mature understanding of the music. A few months later, as soloist with the 46-piece Bautzen orchestra, playing Mozart's *Concerto in D Major*, he was praised for his "solid technique" and hailed as "a 'Wunderkind' with true Mozartian charm."

The choice between music and mathematics was not easy for Lewy. His parents favored mathematics as more secure, but he chose it for a different reason, seeing in the constant playing of often familiar pieces required by a musical career "a kind of enslavement." Although eventually he gave up the stringed instruments (the viola was actually his favorite), he continued to play the piano until the end of his life.

A mathematics teacher at the *Gymnasium* in Breslau prevailed upon Lewy's father to send the boy to the mathematically prestigious university in Göttingen rather than to the local university, which was "too stuffy and old-fashioned" in the teacher's opinion, although a number of distinguished mathematicians (Ernst Hellinger, Otto Toeplitz, Richard Courant) had begun their higher education there. So it happened that on the wave of the staggering inflation and hardship that followed the first World War, Lewy arrived in Göttingen. For a while he worked as a manual laborer on the railroad and (so he told me) thought more about food than about mathematics.

At eighteen he was exceptionally young for the mathematical society in which he found himself. He felt "quite overwhelmed" by the level of the lectures and the abilities of such fellow students as K.O. Friedrichs, who was a couple of years older than he. The director of the famous mathematics institute, which existed at that time only as a name on a letterhead, was Richard Courant, who had been a student of Hilbert and who had only that year (1922) taken over from Felix Klein. Hilbert, who was still lecturing, had aged prematurely as a result of an as yet undiagnosed case of pernicious anemia. In the first conversations I had with Lewy I was impressed by the fact that although he had been very young at the time he had been interested in observing the workings of the mind of this great mathematician of an earlier day, with whose disabilities other young mathematicians were more often simply annoyed.

As a student Lewy did not limit himself to mathematics. He regularly attended the lectures of Alfred Kühn, who was the head of the zoological institute – there was great interest among the Göttingen mathematicians at that time in "the new biology." He also attended the lectures of the physicist Max Born. A few years later, however, he found himself "repelled" by the quantum mechanics of Born and other physicists in Göttingen:

"Often if you would corner them, ask them to explain precisely, they would be evasive or else you would find out that they didn't quite understand what they were saying themselves. They obviously had some physical intuition which I didn't have, but their mathematics was objectionable. The type of personality that was needed at that time in physics was not to my liking. Well, what happened teaches one humility, and it showed that different situations in science require different types of personalities. So one cannot make up one's mind what is *the right way* of being for a scientist."

In Lewy's development as a mathematician both Courant and Friedrichs played significant roles, one as teacher, the other as collaborator. It was Courant who guided the younger men into analysis and – especially important in the early years – suggested that they study the application of finite difference equations in mathematical physics. In 1926, still only twenty-two, Lewy took his Ph.D. under Courant with a dissertation that used the equivalent finite difference problem to solve the question of existence and construction for the simplest variational problem. The following year Courant recommended his *Habilitation* as a *Privatdozent* to the Göttingen faculty.

The dissertation, Courant wrote, "offers something essentially new as far as methods and results are concerned, and reveals in its approach a completely original and independent mind." The success of the *Habilitation* thesis stemmed "from a novel turn that he gave to the theory of characteristics by which he completely solved a problem hitherto considered unsolvable, i.e., the reduction of integration of partial differential equations of second order to the differential equations of characteristics."

By the time Courant wrote the above, the groundbreaking Courant-Friedrichs-Lewy paper "On the partial difference equations of mathematical physics" had also been submitted for publication. It was to be a dramatic anticipation. Forty years later, "the ideas exposed still prevail[ing]," IBM was to reprint it (in English) in its entirety, describing it as "one of the most prophetically stimulating developments in numerical analysis [and] the basis of modern investigations into the numerical analysis of partial differential equations."

Although the paper was an extension of earlier publications by Courant, his young collaborators had made extensive independent contributions, using finite differences to develop an existence theory of hyperbolic equations of the first order. The result that was to be so significant for numerical analysis was Lewy's observation that a hyperbolic differential equation cannot be replaced by a difference equation in an arbitrary manner without a severe restriction on the ratio of the time difference to the space difference.

Friedrichs told me that at the time he was "absolutely amazed" by Lewy's "uncanny insights." Many mathematicians in future years were to echo Friedrichs's astonishment. Louis Nirenberg, a next-generation worker in the field, going over Lewy's papers in preparation for a talk on his work, said he found himself asking, again and again, how did he ever think of that?

"He seemed to think in a different way from most mathematicians I have met. He would have, somehow, fantastic ideas, and sometimes these ideas would come out in a paper of three or four pages. And these papers would have enormous influence on later development."

Courant wrote his letter of recommendation only a month after Lewy's twenty-third birthday; yet in it he was able to cite, in addition to the dissertation, the thesis and "C–F–L" (as it came to be known), three other significant papers, including one recently submitted for publication "of great importance" that solved "an important problem with which many mathematicians have struggled unsuccessfully" by means of "a novel and very original idea" that would "open up future investigation" of related problems.

"All in all, the obvious scientific mathematical development of this candidate presents a picture of an extremely productive and penetrating scientific mind, rich in independent ideas and possessing the technical ability to work them out."

On the personal side, Courant mentioned "a certain reticence and abruptness" in the candidate's manner, which he was sure would disappear in time since these were "associated with an inner modesty." He emphasized to the Faculty that while pursuing his mathematical studies, Lewy had also continued his music, playing quartets each week, performing with the symphony orchestra and composing chamber music: "One may indeed be amazed at the energy and ability of this man who does not appear impressive or at all robust." (This description will startle anyone who knew Lewy's expressive and interesting face in age and the physical energy that he radiated up until his death.)

As a Privatdozent, whose meager fees were paid by his students, usually few in number, Lewy jokingly reported in 1928 a "lucrative" semester: "Now when people ask why I went into mathematics I can answer, 'For the money!'"

In addition to teaching, he served with Friedrichs as one of Courant's many assistants. Although in the future Courant was often to be criticized for exploiting such younger men, Lewy always considered the time he spent as assistant exceedingly valuable.

In 1929, again on Courant's recommendation, Lewy obtained a year's fellowship from the Rockefeller Foundation. He spent the first semester in Rome – the beginning of a lifelong love affair with Italy, to which he was to return on countless occasions – and the second semester in Paris. He has described his reactions to these two centers and to the mathematicians there in an interview that will appear in the second volume of *Mathematical People* so I shall not repeat them.

If Adolf Hitler had not come to power, Lewy might well have spent the rest of his life in Göttingen; but, unlike many others, he recognized instantly that the situation in Germany could do nothing but worsen. That spring, while Courant was struggling to retain his position with letters and petitions to the Government, Lewy left Germany, passing on to a fellow Jewish student the Marks that he was not permitted to take out of the country. With support from the Duggan Foundation he obtained a position as an associate at Brown University in the fall of 1933. Two years later, when that stipend expired, Brown could not find money in its own budget to keep him. Offers came from Ohio and Kentucky and an indication of possible interest from the University of California, which had recently engaged Griffith Evans to upgrade its mathematics department. Lewy, attracted by the lure of the American west, "followed the hope" rather than the certainty – he was typically optimistic – and in the late summer of 1935 received a contract from Evans. The more than half century of his subsequent tenure at Berkeley was to span the transformation of that institution from a provincial state university to one of the great intellectual centers of the world.

Lewy grew to feel very much at home in America, becoming a citizen in 1940, but he also continued to be at home in Europe. Unlike many scholars, he took great pleasure in moving his household, even with a small child, for a number of extended stays abroad.

His mathematical work in the United States, as in Europe, was in analysis, a branch of mathematics that he once described to me as consisting of "kind of hesitant statements":

"In some ways analysis is more like life than certain other parts of mathematics ...," he said. "There are some fields in mathematics where the statements are very clear, the hypotheses are clear, the conclusions are clear. The drift of the subject is also quite clear. But not in analysis. To present analysis in this form does violence to the subject in my opinion. The conditions are or should be considered temporary, also the conclusions, sort of temporary. As soon as you try to lay down exact conditions, you artificially restrict the subject."

It was a source of regret to him that analysis was not a popular field of research in the United States.

Lewy's mathematical work was all connected in one way or another with partial differential equations. It falls into five groups, according to Nirenberg: (1) partial differential equations themselves, (2) the calculus of variations and the applications of variational methods to nonlinear partial differential equations, (3) the theory of several complex variables (a group of three famous papers for which he received the Steele Prize of the American Mathematical Society), (4) fluid dynamics, and (5) offbeat properties of solutions of certain classes of partial differential equations. Although he did some work on analytic functions themselves, in the main he used analytic function theory as his principal tool – always, according to Nirenberg, "in an extremely clever, ingenious, in fact I would say almost *devilish* way."

In spite of the playfulness that many have noted in his work, Lewy approached mathematics with a deep seriousness, seeing the "desire to fill in the details" as an ethical matter – what he described to me as "a certain fanatic adherence to the truth that is necessary in mathematics." In his foreword to the collected works of Riemann, he wrote that the mathematician must not be satisfied with the "blandishments of brilliance [but must] search for deeper truths, hidden beneath the reflection of the surface of facts."

His collected papers, which number approximately sixty over a period of more than sixty years (a relatively small number when one considers the influence they had), are being edited by David Kinderlehrer, a former student.

"Mathematics is like a Riemann surface," Lewy used to tell his students, Kinderlehrer recalls. "There are ordinary points and singular points. You have to work on the singular points."

In September 1933, as classes took up at Berkeley, Hitler invaded Poland, and the long expected Second World War began in Europe. Lewy was eager to get into combat, and the following year he began to take flying lessons. Although he was already in his mid-thirties, he hoped that by having a pilot's license he might be accepted by the Army Air Corps in spite of his small stature. Instead he spent the war years at the Aberdeen Proving Grounds, where he may have been more useful.

In his forty-third year (1947) Lewy married Helen Crosby, an artist, writer and translator who had been with the American Office of Strategic Services in Europe during the war. The couple almost immediately set off on a year-long journey around the wartorn planet, returning to Göttingen, Paris and Rome to do what they could to help old friends. The last three months they spent in the far west of China at a mathematics institute headed by a colleague from Göttingen days, Wei Si Luan. Lewy became very fond of the Chinese people, whom he liked to describe as "the Italians of the Orient."

The Lewys had one son, Michael, who is now also a mathematician.

Once he became a family man and a householder, Lewy discovered the joys of home repair, woodwork and gardening. He took pride in the fact that, unlike professors he had known in his youth, he could – and would – do things with his hands. In domestic activities he exhibited the same freedom from constraints that characterized his mathematical work. To the despair of his wife and son, he refused to read the directions that came with household appliances but insisted on discovering for himself how they worked. Because of his propensity for seeing both sides of a question, he could not make life and death decisions even for plants and left undisturbed "volunteers" that other gardeners would have eradicated. As a result the Lewy garden now contains an elm tree that first made its appearance in young Mike's sandbox.

He was often joyously unrestrained and enthusiastic. At these times he might suddenly swing up his arms and clap his hands – or even shout – in sheer delight. He was fun, his friends agree, "just enormous fun."

The abiding passion of his life was the desire for freedom that had caused him to choose a career in mathematics over one in music. While he was at the Aberdeen Proving Grounds, crowded into wartime housing, he found in sailing on Chesapeake Bay the space his soul required. But he thrilled especially to the wildness of great mountains, feeling perhaps most free there. Only a few months before his death he was hiking the Crest Trail in the Tahoe National Forest, leaping exultantly from rock to rock.

The most dramatic event of his more than fifty years at Berkeley occurred during the McCarthy era when the Regents of the University decided that members of the faculty should sign a loyalty oath – widely considered redundant because they had already sworn to uphold the Constitution and the laws of the State of California in their teaching contracts. Lewy recognized the threat to academic freedom and refused to sign, one of the three members of the Berkeley mathematics department who refused and were subsequently dismissed by the University. He was not, however, doctrinaire on the subject and counseled others, especially younger colleagues, that they must look at their weapons before they decided to fight. He himself was well armed, having resolved when he had had to leave Göttingen that he would save a year's salary as soon as possible in case he ever had to leave another job. In the tense months that followed, as the non-signers took their case to the courts, he was one of the few who managed to remain on friendly terms with those who had different views. Ultimately the California Supreme Court declared the oath unconstitutional and ordered the University to reinstate the non-signers with back pay and privileges. In spite of this difficult period in their relationship, Lewy remained grateful to the University of California "for taking him in and treating him royally" and provided in his will for a post-doctoral fellowship at Berkeley.

Lewy became emeritus in 1972, but he did not stop doing mathematics. Even in the summer of his death he gave a talk at a meeting in Cortona, Italy, on new work that involved attacking the Carathéodory conjecture from a different, variational angle. He had hoped to finish off the problem, but the complete solution was not to be granted to him. While in Europe, following a strenuous schedule in order to see as many as possible of his European friends, he caught a cold that developed into pleurisy. On his return to Berkeley he was hospitalized, fatally ill. He died on August 23, 1988, two months before his eighty-fourth birthday.

One of the last times I talked with him was in connection with the interview for *More Mathematical People* mentioned above. We were in his office in Evans Hall, surrounded by a half century's accumulation of books, journals, letters and reprints. Although he had stipulated "no personal details," recalling the great mathematical centers and the great mathematicians he had known had resulted to a certain extent in a review of his own life.

"I must say," he concluded with wry appreciation of life's ironies, "that as far as I strictly on a personal basis am concerned, I owe Hitler a much more interesting life than I would have had otherwise."

Reinhold Remmert

INVENTIONES MATHEMATICAE: Die ersten Jahre

> »Man hat der Historie das Amt, ..., die Mitwelt
> zum Nutzen zukünftiger Jahre zu belehren,
> beigemessen: So hoher Aemter unterwindet sich
> gegenwärtiger Versuch nicht: er will blos zeigen,
> wie es eigentlich gewesen.«
> (Leopold von Ranke, 1824)

Die *Inventiones mathematicae* werden heuer 25 Jahre alt. Sie entstanden in den 60er Jahren als *europäische* Zeitschrift und wurden in kurzer Zeit ein anerkanntes Publikationsorgan für Mathematik. Heinz Götze nennt gelegentlich die *Inventiones* ein Flaggschiff unter den Journalen des Springer-Verlags, die der Mathematik zugewandt sind. Im folgenden wird über die *Inventiones mathematicae* von ihren Anfängen bis zum Erscheinen der ersten Bände berichtet, wie ich es erlebt habe.

1963/64 · EINE DEUTSCH-FRANZÖSISCHE ANREGUNG

In den fünfziger Jahren bricht Europa zu neuen Ufern auf[1]. Was lag näher, als von einer *europäischen* Zeitschrift für Mathematik zu träumen. Dieser Traum wurde erstmals im Herbst 1963 während der Jahrestagung der Deutschen Mathematiker-Vereinigung in Frankfurt offen diskutiert: in einem Gespräch zwischen Charles Ehresmann und Alexander Dinghas. Zu den nationalen mathematischen Journalen soll ein international verbindendes *europäisches* Organ kommen, das für alle Gebiete der Mathematik offen ist (und die überall langen Veröffentlichungsfristen verkürzt). Herausgeber sollen *zwei bis drei* Mathematiker aus Frankreich und Deutschland werden; zulässige Sprachen sollen *Englisch, Französisch, Deutsch* und *Italienisch* sein. Enrico Bompiani (Sekretär der International Mathematical Union von 1952–1965) wird in den Plan eingeweiht und zeigt sich aufgeschlossen.

---

[1] Bei einer Nachsitzung zu einem Colloquium in Münster im Jahre 1951 bringt Henri Cartan einen Toast »à l'Europe« aus, was uns Studenten damals verwunderte.

Ehresmann macht am 26. Juni 1964 Götze bei dessen Besuch im *Institut Henri Poincaré* mit der Idee vertraut. Götze ist sofort motiviert und sagt die grundsätzliche Bereitschaft des Springer-Verlags zu, die Herausgabe einer europäischen mathematischen Zeitschrift zu übernehmen. Der ursprüngliche Gedanke, das neue Journal durch Zusammenarbeit mehrerer europäischer Verlage zu fördern, wird rasch verworfen, da schnelle Publikation so nicht zu erzielen ist. Man denkt auch daran, über Freunde in Straßburg finanzielle Unterstützung vom Europarat zu erhalten; doch wegen etwaiger Einflußnahme durch Kameralisten wird diese Möglichkeit nicht ernsthaft verfolgt.

1964 · Beratung mit » Mathematische Annalen « und » Mathematische Zeitschrift «

Erste Diskussionen. Am 10. Juli 1964 werden Heinrich Behnke und Helmut Wielandt, die federführenden Herausgeber der MA und der MZ, von Götze informiert: » Auf keinen Fall wollen wir Entscheidungen treffen, ohne die Herausgeber unserer angesehenen mathematischen Zeitschriften zu hören.« Beide Herren antworten postwendend. Behnke gibt sogleich gute Ratschläge; Wielandt sagt lapidar ja zur neuen Zeitschrift, » wenn diese nicht mit weit überlegenen Mitteln ausgestattet wird«. Jahre danach hat Behnke mir gesagt, daß er dem Plan keine Chancen eingeräumt habe und daß seine Skepsis noch bestärkt wurde, als er die Namen der ersten Herausgeber hörte.

Schon im Herbst 1964 finden Beratungen über Grundsatzfragen statt. Götze delegiert die organisatorischen Probleme an Klaus Peters, der gerade als junger Doktor aus Erlangen zum Springer-Verlag gekommen ist. Es ist klar, daß die neue Zeitschrift die junge Generation ansprechen muß und daher die Herausgeber nicht bereits im Establishment verwurzelt sein dürfen. So werden von Beginn an *keine Kollegen mit einschlägiger Erfahrung* gesucht. Aus Straßburg wird Peter Gabriel gewonnen, der sofort mit seiner Dynamik in die Geschehnisse eingreift. Am Rande des Bayerischen Mathematischen Kolloquiums 1965 in Dinkelsbühl führen wir erste konspirative Gespräche. Da die Zeitschrift europäisch werden und keine Nation herausragen soll[2], müssen mehr Herausgeber als ursprünglich gedacht gefunden werden, insgesamt sollen es *höchstens zwölf* sein. Die Suche nach Mitstreitern ist bald erfolgreich, so zeichnen 1966 beim Erscheinen des ersten Bandes Mathematiker aus den Niederlanden und der Schweiz, aus Belgien, Deutschland, Frankreich, Großbritannien, Italien und der Sowjetunion verantwortlich:

Vladimir Arnold; Moskau	Nicholaas Kuiper; Amsterdam
Heinz Bauer; Erlangen	Bernard Malgrange; Paris
Bryan Birch; Manchester	Yuri Manin; Moskau
Pierre Cartier; Straßburg	Reinhold Remmert; Göttingen
Peter Gabriel; Straßburg	Jacques Tits; Bonn
André Haefliger; Genf	Eduardo Vesentini; Pisa

Die *Inventiones* konnten damals mithelfen, das Tor zwischen Ost und West etwas zu öffnen: Wir waren sofort eine gesamteuropäische Zeitschrift.

Bis 1974 gibt es nur in Europa wirkende Herausgeber. Doch eine europäische Zeitschrift muß auch in den USA und in Asien präsent sein. »We need a bridgehead in the United States to compete with the Annals.« So wird 1974 die europäische Isolation durchbrochen und der erste Editor in den Vereinigten Staaten, Jürgen Moser, New York gewählt.

DIE CONSTITUTION. Die Zeitschrift erhält eine Verfassung und einen »Modus operandi«. Diese für ein Publikationsorgan der Mathematik ungewöhnliche Vorstellung machen sich sogleich alle

---

[2] August Leopold Crelle und Joseph Liouville heben in der Vorrede bzw. im Avertissement zum ersten Band ihrer Journale nationale Gesichtspunkte hervor. Crelle schreibt 1826: »Es giebt kaum einen Gegenstand des Wissens, der nicht auch eine Deutsche Zeitschrift hätte. Nur die weite, unbegrenzte Mathematik ... hat dermalen keins. ... Dieses scheint nicht billig zu seyn; denn die ... Deutschen haben ... gerade für die Mathematik einen vorzüglichen Beruf.«

Und Liouville klagt 1836: »La chute d'un Journal utile qu'ils [les mathématiciens français] auraient refusé de soutenir ne serait honorable ni pour eux ni pour la France.«

1882 gründete Gösta Mittag-Leffler die erste *internationale* Zeitschrift *Acta Mathematica*. In der Vorrede im Band 1 liest man in deutsch und französisch: »Hervorragende Mathematiker aller Länder haben, indem sie ihre Mitwirkung zusagten, uns einen Beweis ihrer Theilnahme gegeben, ..., welchem wir durch den Eifer und die Sorgfalt zu entsprechen hoffen, die wir unseren Veröffentlichungen widmen werden.«

Herausgeber in spe zu eigen. Punkt 1 der 1965 verabschiedeten und seither nicht geänderten Constitution lautet:

A mathematical journal »Inventiones mathematicae« is edited by a committee of at most twelve mathematicians who share collectively the whole responsibility concerning editorial policy.

Die Constitution schließt die Wiederwahl der editores quondam nicht aus; von dieser Möglichkeit wird 1982 Gebrauch gemacht.

Der zentrale Punkt, wie über Annahme oder Ablehnung von Arbeiten zu entscheiden ist, wird lange diskutiert. *Mindestens zwei* Referenten sollen jedes Manuskript kritisch beurteilen, wenigstens einer soll eine detaillierte Beurteilung vorlegen. Eine Zeitschrift ist so gut wie ihre Referenten; neben *Gutachten* wünscht man auch objektive *Schlechtachten*. Wir träumen vom *iudex severus*.

Constitution, Modus operandi und Quisquilien werden mit großem Zeitaufwand zu Papier gebracht. Trotz aller Sorgfalt begehen wir bei den Annahmekarten eine läßliche Sünde: Da keine Frau mitredet, schleicht sich die Anrede »Dear Sir« ein. Das geht sogar lange gut. Doch schließlich empört sich eine Autorin: »I am disturbed by the implications of the ›Dear Sir‹, as it conveys the implicit assumption that all research papers are written by men.« Wir streuen Asche auf unsere Häupter und schreiben ab sofort »Dear colleague«.

DIE THESE VOM »SYSTÈME NON-GAULLISTE«. Was sind und was sollen Herausgeber? Darüber reden sich die selbsternannten Editores der ersten Stunde die Köpfe heiß. Dürfen Herausgeber auch Referenten sein, dürfen sie auch in der Zeitschrift veröffentlichen? Diese Möglichkeiten werden als Ausnahmefälle zugelassen. In der Regel soll der Herausgeber sich mit einem *team of subreferees* beraten. Die vornehmste Aufgabe soll indessen sein, Arbeiten *anzuwerben*, wie es bei den *Mathematischen Annalen* zu Hilberts Zeiten üblich war. Dieses *»soliciting principle«* hat sich hervorragend bewährt.

In unserer Unbedarftheit stellen wir uns vor, daß eine Arbeit erst dann angenommen ist, wenn *alle* Herausgeber zugestimmt haben. »Einsame Entscheidungen darf es nicht geben«; der managing editor ist durch das gesamte Herausgeber-Kollegium dauernd zu kontrollieren. Wir sprechen von »système non-gaulliste« in Anspielung auf die damalige politische Situation in Frankreich. In der späteren Praxis hat sich nicht alles so entwickelt, wie es erdacht wurde; der Alltag rückte die Dinge zurecht. Doch viele Autoren reichen gerade deshalb ihre Manuskripte bei den Inventiones ein, um fundierte Stellungnahmen zu erhalten. Manchmal lassen sich allerdings auch renommierte Kollegen zu ätzender Kritik hinreißen, so schließt ein Referent mit dem Satz: »Une suite finie de mots n'est pas encore un théorème.«

Neben dem Problem des Referierens hat uns damals die Frage, wie lange man Herausgeber sein könne, sehr beschäftigt. Wir schrieben explizite Daten nieder; dies führte dazu, daß es schon nach wenigen Jahren eine stattliche Zahl von »editores quondam« gab. Derzeit sind es achtzehn, ihr Rat ist wie eh und je gefragt.

Die Zusammenarbeit in der Redaktion war von Anfang an gut. Gelegentlich gibt es verschiedene Ansichten; einmal schreibt ein Herausgeber: »Herewith I withdraw my paper and myself from Inventiones.« Es gelang, den Kollegen zu besänftigen; er blieb weitere Jahre lang ein engagierter Herausgeber.

1965 · TAUFE

Die Zeitschrift braucht einen attraktiven Namen. Darüber sind wir uns alle einig, wenngleich keiner an »marketing« oder »units« denkt. *C'est le ton qui fait la chanson.* Gesucht wird eine nicht zu seriöse, eher spielerische Bezeichnung. Der Mathematiker als Glasperlenspieler soll angesprochen werden, der *mathematicus ludens.* Die glückliche Eingebung hat Marie-Jeanne Tits im Sommer 1965. Ihr Vorschlag *Inventiones mathematicae* wird sofort – auch vom Verlag – akzeptiert. Auf der Bonner Arbeitstagung 1965 wird die Taufe des noch ungeborenen Kindes gefeiert.

1966 · CARL LUDWIG SIEGEL UND DIE INVENTIONES

Im Herbst 1966 spreche ich in Göttingen mit Siegel über den Plan der neuen Zeitschrift. Er war – abgesehen von einer kurzen Zeit im Advisory Board der *Annals of Mathematics* – nie Herausgeber gewesen und erläutert dies durch ein Schockerlebnis in seiner Jugend: In den zwanziger Jahren bat ihn Hilbert, eine Arbeit zu referieren. Siegel fand das Manuskript für die Mathematischen

Annalen nicht geeignet. Worauf Hilbert entgegnete: »Das weiß ich.
Ihre Aufgabe ist, diese Arbeit eines alten Freundes Annalen-würdig
zu machen.«

Etwas Mut macht Siegel mir dennoch durch die Zusage, dem
neuen Journal eine Arbeit zu überlassen. Und in der Tat schickt
er am 10. Februar 1968 »eine kleine Arbeit mit der Bitte, sie für
eine Veröffentlichung in Ihrer [sic] Zeitschrift in Betracht zu ziehen.
Das behandelte Thema ist nicht unzeitgemäß, und für die Richtig-
keit glaube ich mich einsetzen zu können.« Das Manuskript war
»nicht mit der Schreibmaschine hergestellt. Es ist aber sauber mit
der Hand geschrieben, und ich habe bisher immer in dieser Weise
veröffentlicht. Sollten Sie eine grundsätzliche Schwierigkeit sehen,
so zögern Sie bitte nicht, mir die Arbeit zurückzugeben; ich würde
sie dann bei den Göttinger Nachrichten anbringen können.«

Siegels Arbeit wird *sofort* angenommen; binnen 16 Tagen hat
er Korrekturen. Das aber mißfällt ihm; es sei guter Brauch, Autoren
eine Frist für eventuelle Änderungen einzuräumen (was die Inventiones seitdem auch tun). Die Arbeit erscheint 1968 in Band 5.

1966/68 · ERSTE BÄNDE

Heft 1 von Band 1 erscheint am 22. Februar 1966. *Inventiones* hatte
das »fortune«, sofort mit guten Arbeiten hochangesehener Kollegen
auf sich aufmerksam zu machen. Hinzu kam – wie sich später her-
ausstellte – ein »Glücksfall«, der allerdings zunächst schreckte: Ich
mußte eine Arbeit von anerkannten und einflußreichen Kollegen
ablehnen. Der junge Referent schrieb knallhart und unbekümmert,
die Arbeit zeige, »daß die Autoren seit Jahren ihre Gedankenwelt
nicht erneuert haben«. Das war die klassische Situation: Wie lehnt
man Arbeiten ab, deren Verfasser man sehr schätzt und von denen
man selbst viel gelernt hat? Nun, ich handelte nach dem von einem
erfahrenen Freund gegebenen Motto »an editor must not have a
conscience« und schrieb einen Brief »fortiter in re, suaviter in
modo«, dessen Abfassung Stunden kostete. Durch eine redaktio-
nelle Panne sickerte durch, daß die Inventiones es wagten, Arbeiten
gestandener Herren abzulehnen. Die junge Generation war begei-
stert. Und die Betroffenen waren nicht beleidigt, einer von ihnen
erzählte mir Jahre später, daß er von diesem Augenblick an die
Inventiones immer als *sehr* kritische Zeitschrift empfohlen habe.

Natürlich sind Autoren selten einverstanden, wenn ihre Arbeiten nicht angenommen werden. Dann wird es manchmal polemisch, so empörte sich einst ein Kollege, daß es keinen *court of appeal* gibt *to nail down the arrogance of power*. Doch hört man auch versöhnliche Töne, z.B. »We are grateful for your attention, even though the outcome was not what we had hoped.«

Die Redaktion läßt sich auch heute noch von den Visionen der Gründer leiten. Bei allem idealistischen Tun hatten wir von Beginn an die volle Unterstützung des Verlages. Er war stets kooperativ und hat ganz wesentlich zum Erfolg der *Inventiones mathematicae* beigetragen. Bei Herausgeberbesprechungen ist der Verlag immer dabei; es kommt niemandem in den Sinn, bei der Besprechung der sogenannten wichtigen Punkte die Vertreter des Verlages auszuschließen (wie es früher einmal bei einer anderen Zeitschrift geschah). Mir machte in den ersten Jahren – es war die Zeit der *angry young men* eines John Osborne – das vielleicht allzu forsche Auftreten gegenüber der älteren Generation Sorge. Als ich Heinz Götze das einmal vortrug, beruhigte er mich feinsinnig:

»Wenn sich der Most auch ganz absurd gebärdet,
Es giebt zuletzt doch noch'n Wein.«

Jean-Pierre Serre

Les petits cousins

Als Dokument aus den Anfängen der Garben- und Cohomologie-
theorie veröffentlichen wir hier Briefe von J-P. Serre an H. Cartan
aus den Jahren 1952/53. In diesen Noten werden erstmals die gerade
von Cartan entwickelten algebraischen Methoden auf klassische
Fragen der Funktionentheorie mehrerer Veränderlicher angewen-
det, u.a. auf die Probleme von P. Cousin (1895). Mit diesen Pionier-
studien beginnt die Perestrojka der komplexen Analysis, sie hatte
fortan ein anderes Gesicht. Die Ergebnisse aus I, II und IV wurden
1953 von Serre im Brüsseler Colloquium vorgestellt: *Quelques problèmes globaux relatifs aux variétés de Stein*, Œuvres I, 259–270.
Ein deutscher Teilnehmer urteilte: »Wir haben Pfeil und Bogen,
die Franzosen haben Panzer«.

Die französische Revolution regte eine Fülle von wichtigen
Arbeiten an. Das allgemeine Überlagerungsproblem der Note III
wurde 1956 von K. Stein gelöst: *Überlagerungen holomorph-voll-
ständiger komplexer Räume*, Arch. Math. 8, 354–361. Zum Themen-
kreis der Note VI gehören Arbeiten von Y. Matsushima und A.
Morimoto: *Sur certains espaces fibrés holomorphes sur une variété
de Stein*, Bull. Soc. Math. France 88, 137–155 (1960) und von
H. Skoda: *Fibrés holomorphes à base et à fibre de Stein*, Inv. Math.
43, 97–107 (1977).

Die kleinen Cousins von Serre werden hier im Original wiederge-
geben, lediglich Schreibfehler wurden vom Autor korrigiert.

R. Remmert

LES PETITS COUSINS · 30-4-52

Soit X une variété analytique complexe; on va introduire un certain nombre de faisceaux sur X:

$\underline{F}_a$: faisceau des éléments de fonctions holomorphes en tout point de X (ce faisceau est muni de l'*addition* des f. hol.)

$\underline{F}_m$: faisceau multiplicatif des éléments de fonctions holomorphes inversibles.

$\underline{G}_a$: même définition que $\underline{F}_a$, les fonctions *méromorphes* remplaçant les fonctions holomorphes.

$\underline{G}_m$: même définition que $\underline{F}_m$, les fonctions méromorphes non nulles remplacant les fonctions holomorphes non nulles.

On notera que $\mathbb{C} \subset \underline{F}_a \subset \underline{G}_a$ et $\mathbb{C}^* \subset \underline{F}_m \subset \underline{G}_m$, $\mathbb{C}$ et $\mathbb{C}^*$ notant des faisceaux *constants*.

On peut donc parler des groupes de *cohomologie* de l'espace X à valeurs dans ces différents faisceaux. On vérifie d'abord ceci:

Pour que le pb. de Cousin additif soit toujours résoluble dans X, il faut et il suffit que $H^1(X, \underline{F}_a) \to H^1(X, \underline{G}_a)$ soit biunivoque; pour que le pb. multiplicatif le soit, il faut et il suffit que $H^1(X, \underline{F}_m) \to H^1(X, \underline{G}_m)$ soit biunivoque.

En particulier, pour que le pb. de Cousin additif soit résoluble, *il suffit que $H^1(X, \underline{F}_a) = 0$*, et de même pour le multiplicatif.

Rapports entre $H^1(X, \underline{F}_a)$ et $H^1(X, \underline{F}_m)$.

Si, à tout $f \in \underline{F}_a$, on fait correspondre $e^f \in \underline{F}_m$, on définit un homomorphisme e du faisceau $\underline{F}_a$ sur le faisceau $\underline{F}_m$, dont le noyau est le faisceau constant isomorphe à $\mathbb{Z}$. On a ainsi un diagramme où les lignes sont exactes:

$$
\begin{array}{ccccccccc}
0 & \to & \mathbb{Z} & \to & \underline{F}_a & \to & \underline{F}_m & \to & 1 \\
& & \uparrow & & \uparrow & & \uparrow & & \uparrow \\
0 & \to & \mathbb{Z} & \to & \mathbb{C} & \to & \mathbb{C}^* & \to & 1.
\end{array}
$$

Chacune des suites exactes précédentes donne en cohomologie une suite exacte, et l'on obtient le diagramme suivant, que je complète par les $H^1(X, \underline{G}_a)$ et $H^1(X, \underline{G}_m)$:

$$
\begin{array}{ccccccccc}
& & H^1(X, \underline{G}_a) & & H^1(X, \underline{G}_m) & & & & \\
& & \uparrow & & \uparrow & & & & \\
H^1(X, \mathbb{Z}) & \to & H^1(X, \underline{F}_a) & \to & H^1(X, \underline{F}_m) & \to & H^2(X, \mathbb{Z}) & \to & \dots \\
\uparrow & & \uparrow & & \uparrow & & \uparrow & & \\
H^1(X, \mathbb{Z}) & \to & H^1(X, \mathbb{C}) & \to & H^1(X, \mathbb{C}^*) & \to & H^2(X, \mathbb{Z}) & \to & \dots
\end{array}
$$

Ce diagramme contient quelques renseignements sur les relations liant les deux problèmes de Cousin. Par exemple, supposons que $H^1(X, \underline{F}_a) = 0$ (cas des domaines d'holomorphie), et que $H^2(X, \mathbb{Z}) = 0$, alors la seconde ligne montre que $H^1(X, \underline{F}_m) = 0$, et comme conséquence le pb. multiplicatif est résoluble.

On peut également se demander quelle est l'image de $H^1(X, \underline{F}_m)$ dans $H^2(X, \mathbb{Z})$ si l'on ne suppose plus ce dernier groupe nul. A cause de la 3° ligne du diagramme, on voit que l'on obtient en tout cas comme image les éléments d'ordre fini de $H^2(X, \mathbb{Z})$ (ceci sans hypothèse sur X). Mais il est plus intéressant de se demander quelle est l'image dans $H^2(X, \mathbb{Z})$ des «données de Cousin multiplicatives», c'est-à-dire de noyau $H^1(X, \underline{F}_m) \to H^1(X, \underline{G}_m)$; il n'est plus du tout évident (ni même vrai en général – cf. X = variété kählérienne compacte) que l'on trouve les éléments d'ordre fini de $H^2(X, \mathbb{Z})$; c'est cependant le cas si l'injection: $H^1(X, \mathbb{C}^*) \to H^1(X, \underline{G}_m)$ est nulle, c'est-à-dire si l'on peut toujours trouver sur X une fonction méromorphe multiforme ayant des multiplicateurs donnés. (Ce qui est le cas, d'après Stein, si X est un domaine d'holomorphie?)

Questions naïves.

X étant un domaine d'holomorphie, a-t-on $H^1(X, \underline{G}_a) = 0$?

X étant un domaine d'holomorphie, a-t-on $H^n(X, \underline{F}_a) = 0$ pour $n \geq 2$?

Peut-on calculer $H^1(X \times Y, \underline{F}_a)$ (et aussi H^n :) connaissant les H^1 de X et de Y? Même question pour un espace fibré …

Etc.

Les petits cousins II · 20-5-52

On garde les notations du papier précédent. Mais on considère une variété X, étalée sur $\mathbb{C}^N$, et telle que:

(A) *On a $H^i(X, \underline{F}_a) = 0$ pour tout $i > 0$.*

Imposer cette condition n'est pas ridicule a priori, puisque on sait qu'elle est vérifiée par les *pavés*, et sans doute par bien d'autres espaces.

On va introduire un nouveau faisceau sur X: le faisceau $\underline{D}^p$ des formes différentielles *holomorphes* de degré p sur X. On notera que ce n'est un faisceau qu'au sens de la *Topologie*, et pas au sens des «faisceaux de modules» de H.C. Cependant, *ce faisceau est iso-*

morphe (toujours au sens topologique) à $(\underline{F}_a)^q$, où $q = \binom{N}{p}$; il suit
de là que $H^i(X, \underline{D}^p) = 0$ pour tout $p \geq 0$ et tout $i > 0$. Ceci a d'importantes conséquences:

1. Soit $\underline{Z}^p$ le sous-faisceau de $\underline{D}^p$ formé des formes dif. *fermées*.
Ce faisceau a été considéré par Dolbeault dans sa Note aux *C-R*.
Je dis que: $H^i(X, \underline{Z}^p) = H^{i+p}(X, \mathbb{C})$ (où $\mathbb{C}$ désigne le corps des complexes qui se rappelle ainsi à notre bon souvenir!).

Dém. Considérons la suite exacte de faisceaux:

$$0 \to \underline{Z}^{p-1} \to \underline{D}^{p-1} \to \underline{Z}^p \to 0,$$

où le premier homomorphisme est l'injection et le second est la différentiation extérieure d. Puisque $H^i(\underline{D}^{p-1}) = 0$ pour $i > 0$, la suite exacte de cohomologie donne $H^i(X, \underline{Z}^p) = H^{i+1}(X, \underline{Z}^{p-1})$ pour $i > 0$; si l'on remarque que $\underline{Z}^0$ n'est rien d'autre que le faisceau constant isomorphe à $\mathbb{C}$, on en déduit notre résultat.

Note. On pourrait aussi examiner par cette méthode le cas où $i = 0$; mais il est inutile de se fatiguer, vu qu'il y a dans le Sém. 50–51 exactement le th. voulu:

2. Considérons la suite exacte de faisceaux:

$$0 \to \mathbb{C} \to \underline{D}^0 \xrightarrow{\ d\ } \underline{D}^1 \xrightarrow{\ d\ } \cdots$$

où toutes les flèches (sauf les deux premières) représentent la différentiation extérieure. Cette suite exacte définit un *faisceau gradué à cobord* que nous noterons F.

Appliquons maintenant au faisceau gradué F le Th. 3 de XIX-5, 50–51. La condition (1) est vérifiée, puisque on a déjà dit que $H^i(X, \underline{D}^p) = 0$ lorsque $i > 0$; la condition (2) l'est aussi pour des tas de raisons.

La suite spectrale dont parle le th. 3 a pour terme $E_2 \approx H(X, H(F))$; mais le faisceau $H(F)$ est nul en toute dimension, sauf en dimension 0, où c'est $\mathbb{C}$; il suit de là que ce terme E_2 se réduit à $H(X, \mathbb{C})$ et qu'en outre toutes les différentielles leraytiques sont nulles.

D'après le th. 3, on trouve pour E_∞ la cohomologie de l'espace des sections de F; ici cet espace n'est pas autre chose que *l'espace des formes différentielles analytiques sur X, muni du cobord de la différentielle extérieure*. En d'autres termes:

On peut calculer la cohomologie de X (à coef. complexes) *en utilisant les formes différentielles holomorphes.*

(Ce résultat était [bien] connu pour la dimension 1.)

Les petits cousins. III · Juil. 52

1. *Soit* $\bar{X} \to X$ *un revêtement fini* (galoisien ou pas), *X et $\bar{X}$ étant des var. anal. comp. et la projection étant analytique. On a*:

$$X \ \text{est une var. de Stein} \ \Leftrightarrow \ \bar{X} \ \text{est une var. de Stein.}$$

a) Supposons X de Stein, et soit F un faisceau analytique cohérent sur $\bar{X}$; définissons un faisceau G sur X en posant $G_x = \Sigma_y F_y$, où y parcourt l'ensemble (fini) des pts de $\bar{X}$ qui se projettent en x. On définit la topologie de ce faisceau de façon évidente, et c'est un faisceau analytique cohérent, car il est localement isomorphe à une somme directe finie de faisceaux analytiques cohérents. En outre, et c'est un pur résultat de topologie (analogue au th. de Shapiro en théorie des groupes), on a $H^i(\bar{X}, F) = H^i(X, G)$ pour tout G, d'où $H^i(\bar{X}, F) = 0$, $i > 0$ ce qui caractérise les variétés de Stein.

b) Supposons $\bar{X}$ de Stein, et montrons que X remplit les conditions bien connues (α), (β), (γ). Il est d'abord clair que l'enveloppe d'un compact est compacte (ceci parce que le revêtement est fini); montrons que si x et x' sont deux points distincts de X il y a une fonction holomorphe f telle que $f(x) \neq f(x')$: soient $y_1, \ldots, y_n$ et $y'_1, \ldots, y'_n$ les points de $\bar{X}$ qui se projettent sur x et x' respectivement. Puisque $\bar{X}$ est de Stein, il existe une fonction holomorphe g sur $\bar{X}$ telle que $g(y_1) = \ldots g(y_n) = a$ et $g(y'_1) = \ldots = g(y'_n) = b$, a et b étant des nombres complexes donnés. Soit f la fonction holomorphe sur X dont la valeur en tout point est la moyenne des valeurs de g sur les pts se projetant sur ce pt. Il est clair que $f(x) = a$ et $f(x') = b$, d'où $f(x) \neq f(x')$. Reste à montrer qu'il y a sur X, en tout point, un système de coordonnées locales; on raisonne comme précédemment; si $x \in X$ et si les y_i se projettent en x, on commence par prendre des fonctions $g_1, \ldots, g_m$, toutes nulles aux pts y_i, et tangentes en ces pts aux images réciproques d'un système de coordonnées locales autour de x. On fait alors la moyenne des g_i, et ça marche.

(On pourrait raisonner de la façon «plus savante» mais équivalente suivante: on prend un faisceau analytique cohérent F sur X, et son image réciproque $\bar{F}$ sur $\bar{X}$; supposant $\bar{X} \to X$ galoisien de groupe de Galois G (ce qui est licite, car si le revêtement n'était pas galoisien on «monterait» à un revêtement galoisien et on redescendrait par la partie (a)), on a une suite spectrale partant de $H^i(G, H^j(\bar{X}, \bar{F}))$ et aboutissant à $H^1(X, F)$. Mais G est fini, et $H^j(\bar{X}, \bar{F})$ est un espace vectoriel sur $\mathbb{C}$ (d'ailleurs nul si $j > 0$). Il s'ensuit que l'on a un terme E_2 nul en dimensions > 0, d'où $H^i(X, F) = 0$ pour $i > 0$ et X est de Stein.)

2. *Soit* $\bar{X} \to X$ *un revêtement galoisien, le groupe étant* $\mathbb{Z}$. *Si* X *est de Stein,* $\bar{X}$ *est de Stein.*

(Ici l'implication inverse est fausse comme le montre l'exemple de $\mathbb{C}^* \to T^2$.)

L'existence en tout point d'un système de coordonnées locales est claire. Montrons que si $x \neq y$ sont deux points de $\bar{X}$ il y a une fonction holomorphe g telle que $g(x) \neq g(y)$. On peut évidemment supposer que x et y se projettent au même point de X, donc que $y = T^n . x$, T étant un générateur du groupe de Galois. D'après le th. des fonctions additivement automorphes il existe une fonction f sur $\bar{X}$ telle que $f(T . x) - f(x) = 1$ pour tout $x \in \bar{X}$; on a alors $f(y) = f(x) + n$, d'où $f(x) \neq f(y)$. Montrons maintenant que sur toute suite discrète de points, soit y_n, il y a une fonction holomorphe non bornée. Soit x_n l'image de cette suite dans X. Si les x_n sont discrets, on prend une fonction de X non bornée sur les x_n et son image réciproque dans $\bar{X}$ répond à notre condition. Sinon, extrayons des x_n une suite qui converge vers $x \in X$ (nous la noterons encore x_n). Je dis que la fonction f construite précédemment est non bornée sur les y_n. En effet, relevons x en un point $z \in \bar{X}$, et pour tout n assez grand, soit z_n le point du feuillet de z qui se projette en x_n (ceci a un sens, puisque x_n tend vers x); on a donc $y_n = T^{s_n} . z_n$, et puisque les y_n forment un ensemble discret, on a $s_n \to \infty$. Mais $f(y_n) = f(z_n) + s_n$, et $f(z_n) \to f(z)$. Il s'ensuit que $f(y_n)$ tend vers l'infini, cqfd.

(Le rédacteur est tout honteux, mais n'arrive pas à mettre le raisonnement précédent sous forme «savante»; il ne sait pas non plus pour quels groupes, autres que $\mathbb{Z}$, son raisonnement marche.)

LES PETITS COUSINS. IV · OCT. 52

(Le n° III de la série a trait aux revêtements des variétés de Stein
– il n'a d'ailleurs aucun intérêt.)

Soient X une variété de Stein, P un espace fibré principal de
base X et de fibre un groupe de Lie complexe G. Soit R une représentation linéaire de G, l'espace de représentation étant $\mathbb{C}^N$; la donnée de P et celle de R déterminent de façon unique un espace
fibré associé à P, soit E, dont la fibre est homéomorphe à $\mathbb{C}^N$ (et
même isomorphe au sens ana. compl.). En outre, puisque G opère
linéairement sur $\mathbb{C}^N$, chaque fibre est munie de façon canonique
d'une structure d'espace vectoriel de dimension N sur $\mathbb{C}$. En tout
point $x \in X$ les germes de section locales analytiques forment un
O_x-module F_x, et la collection des F_x forme un *faisceau analytique*
F, localement isomorphe au faisceau des applications holomorphes
de X dans $\mathbb{C}^N$, donc *cohérent*.

On peut définir F cartolocalement de la façon suivante: on
recouvre X par des U_i, dans $U_i \cap U_j$ on prend des $N \times N$-matrices
$R_{ij}(x)$ vérifiant la relation standard $R_{ij} R_{jk} R_{ki} = 1$, et on définit F
comme étant le faisceau qui coincide sur chaque U_i avec le faisceau
des fonctions holomorphes à val. dans $\mathbb{C}^N$, en convenant dans
$U_i \cap U_j$ d'identifier f et $R_{ij} f$ (f étant considérée comme associée
à U_i, et $R_{ij} f$ à U_j).

Ceci étant le Th. A dit que les *sections* de F, i.e. les sections
de E, engendrent (au sens de O_x) les sections locales. En particulier,
il y a toujours une section de E qui n'est pas identiquement 0 (et
même qui prolonge une section donnée sur une sous-variété).

APPLICATION AU THÉORÈME DE STEIN : *Tout espace fibré de fibre*
$\mathbb{C}^*$ *est défini par un diviseur positif* (démontré par Stein quand X
est un produit de domaines à 1 variable).

Soit P l'espace, et faisons opérer $\mathbb{C}^*$ sur $\mathbb{C}$ à la façon usuelle;
soit E l'espace associé, s une section holomorphe de E distincte
de O; soit $D = (s)$ le diviseur formé des zéros de s (c'est évidemment
un diviseur positif). Montrons que D définit P.

Le plus simple est peut-être de prendre des U_i; soit v_{ij} à
val. $\in \mathbb{C}^*$ les fonctions qui définissent P, et soient s_i les fonctions
qui définissent s. Puisque les s_j se «recollent», i.e. se prolongent,
on a $s_j = r_{ij} \cdot s_i$ dans $U_i \cap U_j$, d'où $r_{ij} = s_i^{-1} \cdot s_j$ et on reconnait là
le procédé standard pour définir un espace fibré de fibre $\mathbb{C}^*$ a
partir d'un diviseur.

COROLLAIRE. *Si* $u \in H^2(X, \mathbb{Z})$ *est donnée, il existe un diviseur* $D \geq 0$ *correspondant à* u.

En effet, on sait (petits cousins, passim) qu'il y a un espace fibré P correspondant à u, de fibre $\mathbb{C}^*$. D'où le fourbi.

APPLICATION AUX ESPACES FIBRÉS DE FIBRE $\mathbb{C}^*$. Soit n la dimension de X; nous allons voir qu'il y a un N assez grand pour que tout espace fibré principal P de fibre $\mathbb{C}^*$ et base X soit image réciproque de $\mathbb{C}^N - \{0\}$, fibré de base $P_{N-1}(\mathbb{C})$, espace projectif complexe.

On posera $S = \mathbb{C}^N - \{0\}$; d'après les fourbis traditionnels il nous faut montrer que l'espace (P, S), fibré par S, base X a une section holomorphe (cf. Sém. II, exp. VII, début). Or, on peut associer à cet espace, par le procédé vu plus haut, un espace fibré de fibre $\mathbb{C}^N$, et tout revient à trouver une section de ce dernier espace qui soit *partout* $\neq 0$. Si N est assez grand on doit pouvoir y arriver en prenant N sections bien indépendantes et en en faisant des combinaisons linéaires (il est clair que ça doit marcher pour des raisons de dimension-admettons-le).

On peut certainement faire marcher de même le th. de classification de Steenrod au grand complet – il faudrait voir le rapport de tout ça avec une note récente d'un américain dans les Proc. USA, présentée par Bochner.

Il serait plus intéressant d'étendre ce qui précède aux espaces fibrés de groupe structural le groupe linéaire (par exemple), l'espace universel étant alors la variété des systèmes de n vecteurs lin. ind. de $\mathbb{C}_N$, N grand. On plongerait cette variété dans $\mathbb{C}^{nN}$, et on devrait chercher des sections ne rencontrant pas toute une famille de sections. Il faut espérer qu'une raison de dimension rend ça possible. Peut-être cela aurait-il une application au «problème de Frenkel»? Etc.

COMPLÉMENT (Notes manuscrites de H. Cartan)

Soit donné, sur X de Stein, un diviseur positif D. Alors il existe une *section* de l'espace fibré E (associé à D) dont la variété de zéros est *exactement* D. (Prolonger la section nulle sur la sous-variété du diviseur; s'il y a d'autres zéros, leur diviseur est l'ensemble des zéros d'une *vraie* fonction holomorphe, donc on peut l'enlever.)

Soit V une sous-variété analytique (sans point singulier) de X. Soit donné un diviseur Δ sur V. Pour que Δ soit l'intersection de V et d'un diviseur de X, il faut et il suffit que l'élément de $H^2(X, \mathbb{Z})$ défini par $\varDelta$ soit l'image d'un élément α de $H^2(X, \mathbb{Z})$. [C'est évidemment nécessaire. C'est suffisant: considérons l'espace fibré sur X défini par l'élément α de $H^2(X, \mathbb{Z})$; l'espace fibré induit au-dessus de V possède une section ayant pour zéros le diviseur $\varDelta$. Prolongeons cette section au dessus de X: on obtient dans X un diviseur ...]

Les petits Cousins. V · Oct 52

On va s'occuper du problème des fonctions additivement automorphes, sous la forme que lui a donnée Stein. Soit donc X une variété anal. complexe, $\bar{X}$ un revêtement galoisien de X, le groupe de Galois étant G; dans tout ce qui suit, on notera $\bar{F}$ (resp. F) le faisceau des germes de fonctions holomorphes sur $\bar{X}$ (resp. X), et $\bar{A}$ (resp. A) l'algèbre des fonctions holomorphes sur $\bar{X}$ (resp. X). On a donc: $\bar{A} = \Gamma(\bar{F})$ et $A = \Gamma(F)$.

Soit maintenant $g \to f_g$ un homomorphisme de G dans A, identifié au sous-espace des pts fixes de $\bar{A}$; il s'agit de montrer qu'il y a (si X est de Stein) un élément $f \in \bar{A}$ tel que $f_g = f - g.f$. C'est évidemment un cas particulier du:

THÉORÈME 1. *Si X est de Stein, $H^1(G, \bar{A}) = (0)$.*
On peut démontrer ce théorème, c'est bien connu, en associant à tout cocycle de G à valeurs dans $\bar{A}$ un espace fibré de base X et de fibre $\mathbb{C}$, et utilisant le fait qu'un tel espace est trivial. Cf. Séminaire où ce sera sans doute fait.

Mais il est plus intéressant de raisonner comme ceci:

Appliquons au revêtement $\bar{X} \to X$, et au faisceau $\bar{F}$ la théorie des revêtements; on obtient une suite spectrale E_r telle que le terme $E_2^{p,q}$ soit isomorphe à $H^p(G, H^q(\bar{X}, \bar{F}))$ et que le terme E_∞ soit isomorphe au fourbi gradué associé à $H^i(X, F)$ (donc 0 si $i > 0$, puisque X est de Stein).

Je n'insiste pas sur cette partie topologique qui ne présente certainement pas de difficulté si on la prend bien. Il faut utiliser le fait que $\bar{F}$ est stable par G, et que F est localement isomorphe à $\bar{F}$; en particulier les sections de F s'identifient aux sections de $\bar{F}$ fixes par G, comme on l'a d'ailleurs déjà vu.

Regardons alors les termes de degré 1 dans la suite spectrale précédente. Dans E_2 on trouve notamment $H^1(G, H^0(\bar{X}, \bar{F})) = H^1(G, \bar{A})$, et pour des raisons de degré tous les éléments de ce groupe sont

des cycles et seul 0 est un bord. Comme E_∞ est identiquement
nul, c'est donc que $H^1(G, \bar{A}) = 0$, cqfd.

Supposons maintenant que X et $\bar{X}$ soient des variétés de Stein,
supposition qui n'est point trop absurde comme on le verra plus
loin. Alors $H^q(\bar{X}, \bar{F}) = (0)$ pour $q > 0$, et dans la suite spectrale précédente on a $E_2^{p,q} = (0)$ pour $q > 0$, et $E_2^{p,0} = H^p(G, \bar{A})$. Il est clair que
avec un pareil terme E_2 («réduit à la base») toutes les différentielles
spectrales sont nulles et comme E_∞ est nul, on a finalement:

THÉORÈME 2. *Si X et $\bar{X}$ sont de Stein $H^p(G, \bar{A}) = (0)$ pour tout $p > 0$.*
L'hypothèse du th. 2 est vérifiée dans pas mal de cas particuliers
(produits de domaines d'une variable complexe par exemple). En
outre on sait (cf. petits cousins III et Séminaire) que si G est un
groupe fini, ou un groupe $\mathbb{Z}$, ou une extension (multiple) de groupes
$\mathbb{Z}$ et de groupes finis, alors X de Stein $\Rightarrow \bar{X}$ de Stein (si G est
fini, l'implication inverse est également vrai). Il est donc assez natu-
rel de conjecturer que tout revêtement d'une variété de Stein est
également une variété de Stein, ce qui renforcerait l'intérêt du th. 2.

Il serait intéressant de généraliser ceci aux espaces fibrés principaux de base un X de Stein, la cohomologie du classifiant du groupe
structural remplaçant la cohomologie du groupe G.

LES PETITS COUSINS. VI · 2 DÉC. 52

ESPACES FIBRÉS À FIBRES VECTORIELLES ET APPLICATIONS
Soit P un espace fibré principal de groupe structural G, base X;
on suppose que G opère à droite sur P. Soit V un e.v. de dimension
finie sur $\mathbb{C}$, sur lequel G opère à gauche par des transformations
linéaires. Le couple (P, V) définit un espace associé P_V de base X
et fibre vectorielle isomorphe à V; c'est le quotient de $P \times V$ par
la relation: $(p\sigma, v) \approx (p, \sigma . v)$ si $\sigma \in G$.

On rappelle que les germes de sections de P_V forment un faisceau
analytique cohérent sur X. Si X est de Stein, on peut donc appliquer
à ces sections les mêmes raisonnements qu'aux fonctions holomorphes à valeurs «non tordues», et on a en particulier, comme on
sait:

*Soit Y une sous-variété sans singularités de X, r une section holo-
morphe de P_V au-dessus de Y. Il existe une section au-dessus de X
qui prolonge r.*

En particulier, prenant $Y = \{x\}$, $x \in X$, on voit qu'il y a une section holomorphe de P_V qui prend une valeur donnée en x, et plus particulièrement une section non nulle en x (on admirera comment cette conséquence classique du th. A est ici démontrée avec le th. B!). Prenant pour G le groupe $\mathbb{C}^*$, pour V l'espace $\mathbb{C}$ où $\mathbb{C}^*$ opère par homothéties, on en tire:

Soit $d \in H^2(X, \mathbb{Z})$ une classe de diviseurs. Il existe un diviseur positif $\mathcal{D} \in d$ qui ne contient pas un point donné $x \in X$ (et plus généralement qui ne rencontre pas une sous-variété Y donnée, pourvu que d induise 0 dans $H^2(Y, \mathbb{Z})$).

On va tirer de là:

Soient X une variété de Stein. $\mathcal{D}$ un diviseur de X; la variété $W = X - \mathcal{D}$ est une variété de Stein.

(Noter que si X est un domaine d'holomorphie, cela résulte de la caractérisation locale des domaines d'holomorphie, due à Oka.)

On suppose évidemment $\mathcal{D} \geq 0$, et on va vérifier (α), (β), (γ). Seul (α) offre une difficulté. Soient alors K compact $\subset W$, $\bar{K}$ son enveloppe dans W; puisque $\bar{K}$ est contenue dans un compact de X, il suffit de voir que tout $x \in \mathcal{D}$ a un voisinage qui ne rencontre pas $\bar{K}$. Or, d'après ce qui précède il existe $\mathcal{D}' \geq 0$, linéairement équivalent à $\mathcal{D}$, $x \notin \mathcal{D}'$. Il existe alors une fonction g méromorphe sur X, ayant $\mathcal{D}$ pour variété polaire, $\mathcal{D}'$ pour variété des zéros. Cette fonction est donc holomorphe sur W, et tend vers l'infini lorsque $y \in W$ tend vers x. Comme elle est bornée sur $\bar{K}$, cela signifie bien qu'il y a un voisinage de x qui ne rencontre pas $\bar{K}$, cqfd.

Revenons maintenant aux espaces fibrés à fibre vectorielle arbitraire V, groupe structural G. Soit V^* l'espace vectoriel dual de V, et faisons opérer G sur V par la représentation contragrédiente, de telle sorte que $\langle v, v^* \rangle = \langle \sigma . v, \sigma . v^* \rangle$, $v \in V$, $v^* \in V^*$. Le produit scalaire $\langle v, v^* \rangle$ est alors compatible avec le passage au quotient qui définit P_V et P_{V^*}, et on voit ainsi que *deux fibres de P_V et de P_{V^*} correspondant au même point $x \in X$ sont des espaces vectoriels en dualité*. Il s'ensuit en particulier que *toute section* holomorphe s de P_{V^*} définit une *fonction* holomorphe f_s sur P_V par: $f_s(y) = \langle y, s \rangle$ si $y \in P_V$. La fonction f_s est *linéaire* sur chaque fibre de P_V, et on pourrait voir facilement que cette propriété caractérise les f_s (on n'utilisera pas ce fait).

Nous allons maintenant utiliser les fonctions f_s pour prouver:

Si X est une variété de Stein, la variété P_V est une variété de Stein.

Remarquons que, d'après ce qui a été dit au début (appliqué à P_{V^*}), si l'on se donne une fibre V_x de P_V, et une forme linéaire sur V_x, il existe une fonction f_s qui coincide sur V_x avec cette forme linéaire. En particulier, il existe une f_s prenant des valeurs distinctes en deux points donnés distincts de V_x, ce qui montre que la condition (β) est remplie; démonstration analogue pour (γ). Pour (α), on prend un compact $K \subset P_V$, soit $\bar{K}$ son enveloppe; la projection de $\bar{K}$ sur X est contenue dans un compact, et il suffit donc de prouver que tout $x \in X$ a un voisinage U tel que $p^{-1}(U) \cap \bar{K}$ soit relativement compact. Pour cela, on choisit p fonctions $f_s^1, \ldots, f_s^p$ (dim. $V = p$) dont les restrictions à V_x soient linéairement indépendantes, et on prend U assez petit pour que ceci subsiste pour tout $x' \in U$. Comme les f_s^i sont bornées sur $\bar{K}$, il est alors bien clair que $p^{-1}(U) \cap \bar{K}$ est relativement compact, cqfd.

Il est clair que la démonstration précédente a une portée plus générale: si P_F est l'espace associé à P, de fibre F où opère G, on pourra affirmer que P_F est de Stein dès qu'on aura trouvé un espace vectoriel *de dimension finie $\mathscr{E}$* de fonctions holomorphes de F, stable par G, et tel que F vérifie (α), (β), (γ) *vis-à-vis des fonctions de $\mathscr{E}$*. Dans le cas considéré plus haut, on prenait pour $\mathscr{E}$ les formes linéaires sur l'espace V, i.e. V^*.

Nous allons maintenant tirer de là:

Si G est un sous-groupe fermé d'un groupe linéaire complexe $GL(n, \mathbb{C})$, et si X est une variété de Stein, tout espace fibré principal P de base X et de fibre G est une variété de Stein.

Supposons d'abord que $G = GL(n, \mathbb{C})$, et considérons l'espace fibré P_V associé à P, où l'on prend pour V l'algèbre de *toutes* les matrices carrées d'ordre n. D'après ce qui précède, P_V est de Stein; mais P s'identifie au sous-espace ouvert de P_V formé des matrices inversibles; c'est donc le complémentaire d'un *diviseur* de P_V (c'est bien d'un diviseur qu'il s'agit, puisqu'on doit annuler le déterminant d'une matrice générique), et d'après ce qui a été dit plus haut, c'est une variété de Stein.

Si maintenant G est plongé dans $GL(n, \mathbb{C})$, on étend le groupe structural à $GL(n, \mathbb{C})$, et P apparaît comme une sous-variété bien plongée d'un espace P' principal pour $GL(n, \mathbb{C})$, donc P est une variété de Stein.

Remarquons que l'on pourrait aussi bien appliquer la généralisation indiquée précédemment: on prendrait pour espace vectoriel de fonctions sur G, les restrictions à G des fonctions linéaires sur l'espace de toutes les matrices carrées d'ordre n augmenté de l'inverse du déterminant; G vérifie bien (α), (β), (γ) pour cet espace $\mathscr{E}$. Au fond, cela revient identiquement au même.

Notons un corollaire:
Tout espace fibré principal de fibre $\mathbb{C}^*$, base de Stein, est lui même de Stein.

Un méchant cousin · Nov. 52

Je vais donner un exemple d'espace fibré analytique, dont la base est une variété de Stein X, dont la fibre est une variété de Stein analytiquement rétractile, et qui n'a pas de section analytique.

Cet exemple met donc en défaut le principe d'Oka, puisqu'il existe visiblement une section continue. Plus précisément, il montre que les espaces justiciables du principe d'Oka doivent être d'un type assez particulier (fibrés principaux, par exemple).

EXEMPLE. On prend $X = \mathbb{C}^*$, $Y =$ couronne circulaire (à 1 variable complexe); on désigne par $\bar{X}$ et $\bar{Y}$ leurs revêtements universels (i.e. $\mathbb{C}$ et une bande dans le plan complexe); les groupes $\pi_1(X)$ et $\pi_1(Y)$ sont canoniquement isomorphes, et on les identifiera. Posons alors $Z = (\bar{X} \times \bar{Y})/\pi_1(X)$, $\pi_1(X)$ opérant des deux côtés à la fois. L'espace Z est fibré de base X et de fibre $\bar{Y}$, comme on sait. C'est l'espace fibré cherché.

En effet, X est de Stein, $\bar{Y}$ est de Stein et est analytiquement rétractile, et Z n'a pas de section analytique, car une telle section correspond biunivoquement à une application analytique $f : \bar{X} \to \bar{Y}$, commutant avec les transformations de $\pi_1(X)$, et une telle application étant une fonction entière à partie imaginaire bornée est constante!

REMARQUE. On a trouvé en même temps un couple X, Y, tous deux de Stein, tel que les classes d'applications holomorphes de X dans Y ne correspondent pas aux classes topologiques. Triste!

Espaces fibrés « à supports compacts » · Déc. 53
(d'après Grothendieck)

Soit X une variété de Stein connexe de dimension $n \geq 2$, E un espace fibré analytique à fibre vectorielle de dimension p, de groupe structural le groupe G des matrices inversibles à p lignes et p colonnes.

Nous noterons $\underline{E}$ le faisceau des germes de sections holomorphes de E; on sait que $H^q_*(X, \underline{E}) = 0$ pour $q \neq n$ (coh. à supports compacts-cela résulte par exemple du théorème de dualité analytique); en particulier on a $H^1_*(X, \underline{E}) = 0$. Un raisonnement immédiat de suite exacte montre alors (cf. conférence au Colloque de Bruxelles, n° 13):

Si s est une section holomorphe de E sur $X - K$, K compact, il existe une section s' de E sur X qui coïncide avec s sur $X - K'$, où K' est un compact contenant K.

Soit maintenant $\underline{G}$ le faisceau (non abélien) des germes d'applications holomorphes de X dans G. On sait que $H^1(X, \underline{G})$ (qui est un *ensemble*, avec point marqué) est en correspondance canonique avec les classes d'espaces fibrés analytiques de base X et groupe G. De la même façon, on montre que $H^1_*(X, \underline{G})$ correspond aux classes d'espaces fibrés principaux analytiques de base X, groupe G, qui ont une »section marquée« au voisinage de l'infini (i.e. dans le complémentaire d'un compact); bien entendu, un isomorphisme est assujetti à respecter ces sections.

THÉORÈME. *L'ensemble $H^1_*(X, G)$ est réduit à l'élément neutre* (autrement dit, tout fibré à support compact est trivial).

DÉMONSTRATION. Soit P un fibré à support compact, principal, et soit E le fibré à fibre vectorielle associé. Du fait que P possède une section marquée au voisinage de l'infini, il existe p sections holomorphes $s_1, \ldots, s_p$ de E sur $X - K$ (K compact) telles que les vecteurs $s_1(x), \ldots, s_p(x)$ soient linéairement indépendants en tout point $x \in X - K$; les s_i forment donc un *repère* sur $X - K$, et dire que P est trivial (au sens des classes d'espaces fibrés à support compact) équivaut à dire qu'il existe des $s'_1, \ldots, s'_p$, sections de E sur X, qui coïncident avec $s_1, \ldots, s_p$ sur $X - K'$ (K' compact contenant K), et qui sont linéairement indépendantes en tout point $x \in X$.

Le résultat indiqué plus haut montre tout d'abord l'existence des $s'_1, \ldots, s'_p$ prolongeant $s_1, \ldots, s_p$ sur $X - K'$. Reste à voir que les $s'_1, \ldots, s'_p$ sont bien linéairement indépendantes en tout point $x \in X$; voici deux démonstrations de ce fait:

a) L'ensemble D des points où les s'_i sont linéairement dépendantes s'obtient en annulant localement un déterminant, donc est un diviseur; comme ce diviseur est à support compact (puisque contenu dans K') et que $n \geq 2$, ce diviseur est vide.

b) On considère le fibré E' dont la fibre est l'espace dual de la fibre de E; sur $X - K$ il possède un repère $t_1, \ldots, t_p$ tel que $\langle t_i(x), s_j(x) \rangle = \delta_{ij}$ pour tout $x \in X - K$; d'après ce qu'on a vu plus haut, on peut prolonger les t_i en des t_i', et par prolongement analytique on aura encore $\langle t_i'(x), s_j'(x) \rangle = \delta_{ij}$ pour tout $x \in X$, ce qui montre bien que les $s_j'(x)$ sont linéairement indépendants en tout point $x \in X$, cqfd.

APPLICATION. Nous prendrons $p = 1$, d'où $G = \mathbb{C}^*$, $\underline{G} = \mathbb{C}^*$. Ecrivons la suite exacte traditionnelle:

$$0 \to \mathbb{Z} \to \mathbb{C} \to \mathbb{C}^* \to 0.$$

On en tire:

$$H_*^1(X, \mathbb{C}^*) \to H_*^2(X, \mathbb{Z}) \to H_*^2(X, \mathbb{C}),$$

et comme on vient de prouver que $H_*^1(X, \mathbb{C}^*) = 0$, il s'ensuit que $H_*^2(X, \mathbb{Z}) \to H_*^2(X, \mathbb{C})$ *est biunivoque.* D'où:

si $n = 2$, $H_*^2(X, \mathbb{Z}) = H_2(X, \mathbb{Z})$ *est un groupe sans torsion* (vu que $H_*^2(X, \mathbb{C})$ est un espace vectoriel complexe),

si $n \geq 3$, $H_*^2(X, \mathbb{Z}) = H_{2n-2}(X, \mathbb{Z})$ *est réduit à* 0 (car il en est de même de $H_*^2(X, \mathbb{C})$ d'après la dualité analytique, par exemple).

Plus généralement, si on savait prouver que $H_*^q(X, \mathbb{C}^*) = 0$ pour $q < n$, on en tirerait une démonstration de la conjecture $H_{2n-q}(X, \mathbb{Z}) = 0$ pour $q < n$.

Jacques Tits

Symmetrie[1]

Das Wort »Symmetrie« wird in der Mathematik ganz anders als in der Umgangssprache gebraucht. Verwendet man es im alltäglichen Leben, so denkt man meist an eine zweiseitige, rechts-links-Symmetrie; nicht so in der Mathematik. Freilich wird dem Wort manchmal auch in der Umgangssprache eine allgemeinere Bedeutung gegeben; jeder wird zum Beispiel anerkennen, daß die Figur der Abbildung 1 hochsymmetrisch ist, obwohl sie keine zweiseitige Symmetrie besitzt. Allerdings handelt es sich nur um einen seltenen Ausnahmefall. (Das angegebene Beispiel ruft eine Bemerkung hervor: bei der Vorbereitung eines Vortrags über dieses Thema ist mir aufgefallen, daß man leicht auf politische oder religiöse Symbole stößt, wenn man nach einem Beispiel eines gleichzeitig einfachen und hochsymmetrischen Gebildes sucht; das bezeugt, daß Symmetrien auf den Menschen immer eine große Wirkung gehabt haben.)

Ein zweiter Unterschied zwischen Mathematik und Umgangssprache hinsichtlich des Symmetriebegriffes besteht darin, daß *voll-*

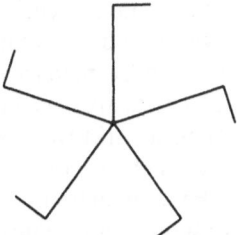

Abbildung 1

---

[1] Dies ist eine etwas überarbeitete Version des Textes eines am 28. November 1986 in Bonn gehaltenen Vortrags (siehe »Mathematische Betrachtungen«, Bouvier Verlag, Bonn, 1988, 32–44; ich danke dem Bouvier Verlag herzlich für die Erlaubnis, diesen Text teilweise abzudrucken). Die Hauptänderung besteht in dem Ersetzen der Abbildungen 3 und 4 durch eine neue Abbildung 3, wesentlich reicher in Symmetrieeigenschaften, und die ich ganz besonders Herrn Dr. H. Götze widme: siehe dazu das Nachwort. Die Kommentare zu den genannten Abbildungen (Seiten 34 und 35 von *loc. cit.*) wurden natürlich entsprechend geändert.

Abbildung 2

kommene Symmetrien nur in der Mathematik und nicht im Leben existieren. Damit spiele ich nicht nur auf die Tatsache an, die Hermann Weyl in seinem Buch »Symmetry« betont, daß in der abendländischen Kunst die Künstler nach Möglichkeit die vollkommene Symmetrie vermeiden und sie immer ein bißchen brechen. Hierzu gibt er schöne Beispiele, wie etwa die berühmten etruskischen Reiter vom Triklinischen Grabmal in Corneto (Abbildung 2).

Das Bild ist fast symmetrisch, aber eben nicht ganz: die vollkommene Symmetrie in der Kunst wirkt wohl oft ein wenig langweilig! In der Mathematik ist es nicht so (obwohl in letzter Zeit auch Mathematiker sich für »Fastsymmetrien« interessiert haben). Aber die Aussage, daß eine vollkommene Symmetrie in der Wirklichkeit nie vorkommt, ist an sich mehr grundlegend. Schauen wir uns zum Beispiel die Abbildung 3 an.

Grob gesehen zeigt sie Symmetrien, auf die wir gleich zurückkommen wollen, die aber verschwinden, wenn man das Bild näher anguckt: dies ist direkt klar, falls die den Eckpunkten angeknüpften Ziffern beachtet werden – die 30 Eckpunkte tragen nämlich 30 verschiedene Namen – aber auch wenn man dies nicht berücksichtigt, findet man leicht in der Zeichnung kleine Unregelmäßigkeiten, die alle augenscheinlichen Symmetrien zerstören.

Sieht man hingegen von den Ziffern und den kleinen Unregelmä-ßigkeiten ab, so erkennt man in der Figur sofort eine Rotationssymmetrie der Ordnung 5, ja sogar eine Symmetrie der Ordnung 10 wenn man beschließt, keine Unterschiede zwischen den »weißen« und den »schwarzen« Punkten zu machen. Aber eine viel höhere Symmetrie ist in diesem Bild verborgen und offenbart sich, wenn man die Figur lediglich als *Graph* betrachtet, womit gemeint ist, daß man in ihr nur 30 Eckpunkte (die 30 Punkte auf dem Randkreis) wahrnimmt, aus denen einige Paare verbunden sind und andere nicht: man denke etwa, daß die Eckpunkte 30 Leute darstellen, und daß eine Verbindung einer Bekanntschaft entspricht. Die Längen und Winkel der Verbindungsstrecken (genannt Kanten) sollen hier keine Rolle spielen. Diese verborgenen Symmetrien lassen sich folgendermaßen einsehen. Jeder weiße Punkt ist mit drei schwarzen

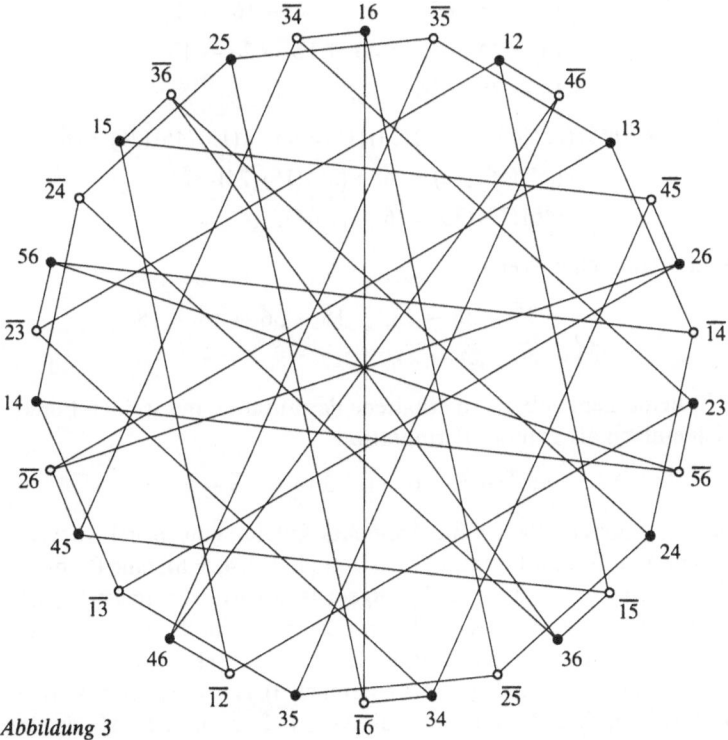

Abbildung 3

Punkten verbunden, und die diesen drei Punkten angeknüpften Ziffernpaare bilden eine sogenannte *Partition* der Menge $\{1, 2, 3, 4, 5, 6\}$; zum Beispiel ist der Punkt $\overline{12}$ mit 46, 15 und 23 verbunden, und wir erhalten die Partition (46) (15) (23) der obigen Menge. Umgekehrt, fängt man mit einer beliebigen Partition, wie etwa (12) (34) (56), an, so stellt man fest, daß die drei entsprechenden schwarzen Punkte 12, 34 und 56 mit genau einem weißen Punkt, in diesem Falle $\overline{23}$, verbunden sind. So sind den weißen Punkten die 15 Partitionen von $\{1, 2, 3, 4, 5, 6\}$ in drei Paaren eineindeutig zugeordnet. Nun liefert jede Permutation σ der Menge $\{1, 2, 3, 4, 5, 6\}$ eine Permutation der schwarzen Punkte (sie entsprechen ja den Paaren von Elementen dieser Menge) und eine Permutation der weißen Punkte (als Partitionen angesehen), die offenbar zusammen eine Symmetrie des ganzen (als Graph betrachteten) Gebildes darstellen. Zum Beispiel liefert die Permutation $1 \to 2 \to 3 \to 4 \to 5 \to 6 \to 1$ die Symmetrie

$$12 \to 23 \to 34 \to 45 \to 56 \to 16 \to 12,$$
$$13 \to 24 \to 35 \to 46 \to 15 \to 26 \to 13,$$
$$14 \to 25 \to 36 \to 14,$$

$$\overline{12} = (15)(23)(46) \to (26)(34)(15) = \overline{45} \to (13)(45)(26) = \overline{26}$$
$$\to (24)(56)(13) = \overline{14} \to (35)(16)(24) = \overline{25}$$
$$\to (46)(12)(35) = \overline{46} \to (15)(23)(46) = \overline{12},$$

und in ähnlicher Weise

$$\overline{13} \to \overline{35} \to \overline{36} \to \overline{13}, \quad \overline{15} \to \overline{56} \to \overline{16} \to \overline{15},$$
$$\overline{23} \leftrightarrow \overline{34}, \quad \overline{24} \to \overline{24}.$$

Wir bemerken, daß die so erhaltene Permutation der weißen Punkte sich einfach als von der Permutation

$$\bar\sigma: \overline{1} \to \overline{5} \to \overline{6} \to \overline{1}, \quad \overline{2} \leftrightarrow \overline{4}, \quad \overline{3} \to \overline{3}$$

induziert beschreiben läßt. Ähnliches gilt für eine beliebige Wahl von σ, so daß jeder Permutation σ von $\{1, 2, 3, 4, 5, 6\}$, eine Permutation $\bar\sigma$ von $\{\overline{1}, \overline{2}, \overline{3}, \overline{4}, \overline{5}, \overline{6}\}$ zugeordnet wird, die meistens sehr verschieden von σ aussieht (man bemerke zum Beispiel, daß die oben gewählte σ die sechs Ziffern, 1, ..., 6, zyklisch permutiert, während $\bar\sigma$ einen »Fixpunkt« besitzt, nämlich $\overline{3}$). Die 720 Permutationen von 1, 2, 3, 4, 5, 6 (und auch die von $\overline{1}, \overline{2}, \overline{3}, \overline{4}, \overline{5}, \overline{6}$) bilden eine

Gruppe, die sogenannte *symmetrische Gruppe* S_6. Die Zuordnung $\sigma \rightarrow \bar{\sigma}$ ist eine Symmetrie – oder, wie man in der mathematischen Sprache sagt, ein *Automorphismus* – der Gruppe S_6. Die Existenz dieses sog. äußeren Automorphismus von S_6 ist ein wohlbekanntes und bemerkenswertes Phänomen, das kein Analogon hat, wenn 6 durch eine andere ganze Zahl ersetzt wird. Mit Hilfe der obigen Methode haben wir 720 Symmetrien der als Graph betrachteten Abbildung 3 erhalten; es werden 1440, wenn man sie auch noch mit der zentralen Symmetrie $12 \leftrightarrow \overline{12}$, $13 \leftrightarrow \overline{13}$, ..., $56 \leftrightarrow \overline{56}$ kombiniert, die allerdings nur dann eine Symmetrie ist, wenn man weiße und schwarze Punkte nicht unterscheidet.

Man sieht, daß die Symmetrieeigenschaften einer Figur sehr davon abhängen, wie man diese Figur anschaut, e.g. im obigen Beispiel, ob man auf Unregelmäßigkeiten der Zeichnung, auf Längen von Strecken, auf Unterschiede zwischen weißen und schwarzen Punkten usw. achtet oder nicht. Das führt auf natürliche Weise zu einer Vorstellung von dem, was man ein *mathematisches Objekt* nennen könnte, nämlich ein Ding, bei dem man von vornherein entscheidet, welche Eigenschaften genau man berücksichtigen will. Solche Objekte können echte Symmetrien besitzen. Im alltäglichen Leben hingegen ist es üblich und oft notwendig, alle Aspekte der Dinge soweit als möglich zu berücksichtigen, was natürlich jede Symmetrie zerstört.

In der Mathematik und Physik kommen oft bemerkenswerte Symmetrien in ganz versteckter Weise als sogenannte »verborgene Symmetrien« vor. Die Abbildung 3 hat uns zwei Beispiele davon gegeben: es sind einerseits die 720 Symmetrien des Graphen, die aus den Permutationen von $\{1, 2, 3, 4, 5, 6\}$ herkommen und die außer der Rotation um 72 Grad und deren Vielfache gar nicht augenscheinlich sind, und andererseits der äußere Automorphismus der symmetrischen Gruppe S_6. Es gehört zu den interessanten Aufgaben der Mathematiker, solche verborgene Symmetrien zu entdecken. Wir wollen noch weitere Beispiele geben.

Um das nächste Beispiel einzuführen, stellen wir uns zwei scheinbar sehr elementare Probleme. Zum ersten: auf wieviel verschiedene Weisen läßt sich eine natürliche Zahl N in Summen von ungeraden Zahlen zerlegen? Man bilde sich etwa ein, daß man einen Brief mit N D-Mark frankieren muß, und zwar mit Briefmarken von 1 DM, 3 DM, 5 DM usw., und frage sich, auf wieviele Weisen dies möglich ist. Für $N = 6$ gibt es zum Beispiel vier Lösungen. Die

Anzahl M der Möglichkeiten wächst mit N nach einem Gesetz, das zunächst nicht leicht überschaubar ist (siehe Abbildung 4).

Das zweite in der Abbildung 4 formulierte Problem läßt sich nicht so leicht mit Briefmarken erklären, aber versuchen wir es trotzdem: diesmal stehen zwei Typen von Briefmarken zur Verfügung, »normale« Briefmarken, deren Werte die geraden Zahlen (2, 4, 6, …) sind, und »spezielle« Briefmarken, deren Werte die soge-

---

PROBLEM 1. Auf wieviele verschiedene Weisen läßt sich eine vorgegebene positive ganze Zahl N als Summe einer monoton absteigenden Folge von positiven ungeraden Zahlen ausdrükken?

Beispiel:: $\begin{aligned} 6 &= 1+1+1+1+1+1 \\ &= 3+1+1+1 \\ &= 3+3 \\ &= 5+1 \end{aligned}$

PROBLEM 2. Auf wieviele verschiedene Weisen läßt sich die Zahl N in eine geordnete Summe zerlegen, deren erstes Glied eine Zahl der Gestalt $\frac{1}{2}n(n+1)$ ist, während die anderen Glieder eine monoton absteigende Folge von positiven geraden Zahlen bilden?

Beispiel: $\begin{aligned} 6 &= 0+2+2+2 \\ &= 0+2+4 \\ &= 0+6 \\ &= 6 \end{aligned}$

Beide Probleme haben dieselbe Antwort M:

N	1	2	3	4	5	6	7	8	9	10	11	12	13	…
M	1	1	2	2	3	4	5	6	8	10	12	14	17	…

Abbildung 4

Die Äquivalenz der Probleme 1 und 2 entspricht der Gaußschen Formel

$$\frac{\eta(q^2)^2}{\eta(q)} = \sum_{n=-\infty}^{+\infty} q^{2(n+\frac{1}{4})^2}$$

wobei

$$\eta(q) = q^{1/24} \cdot \prod_{n=1}^{\infty} (1-q^n)$$

Abbildung 5

nannten Dreieckszahlen (1, 3, 6, 10, ...) sind, und es wird die Zusatzbedingung gestellt, daß höchstens eine spezielle Briefmarke benutzt werden darf. Es wird wieder nach der Anzahl der möglichen Kombinationen solcher Briefmarken, die sich zu einer Gesamtsumme von N D-Mark addieren, gefragt.

Merkwürdigerweise haben diese zwei sehr verschieden aussehenden Probleme dieselbe Antwort: für gegebenes N ist die Anzahl der möglichen Kombinationen in beiden Problemen gleich (Abbildung 4). Das entspricht einer wohlbekannten und tiefliegenden Formel von Gauß (Abbildung 5).

Würde man nun sagen, daß die ganze Sachlage völlig verstanden ist, sobald irgendein Beweis der Gaußschen Formel vorliegt? Ich meine, es ist nicht so. Eine tiefere Einsicht gewinnt man, indem man ein mathematisches Objekt aufbaut, in welchem sich die zwei gestellten Probleme widerspiegeln und zwar so, daß die Äquivalenz der Probleme (i.e. die Gleichheit ihrer Lösungsanzahlen) einer – wohl verborgenen – Symmetrie des Objektes entspricht; die Existenz dieser Symmetrie erleuchtet dann sowohl die Äquivalenz der zwei Probleme als auch die Formel von Gauß. Solch ein Objekt findet man in der Tat nach Frenkel, Kac, Lepowsky et al. in der Darstellungstheorie gewisser Kac-Moody-Lie-Algebren.

Um unsere letzten Beispiele verborgener Symmetrien einzuführen, betrachten wir die auf den Abbildungen 6 und 7 gezeigten Gitter, deren Symmetrien wir kurz untersuchen wollen. Beide haben sogenannte Translationssymmetrien, von denen wir hier absehen möchten; um sie auszuschließen, genügt es zum Beispiel, jeweils

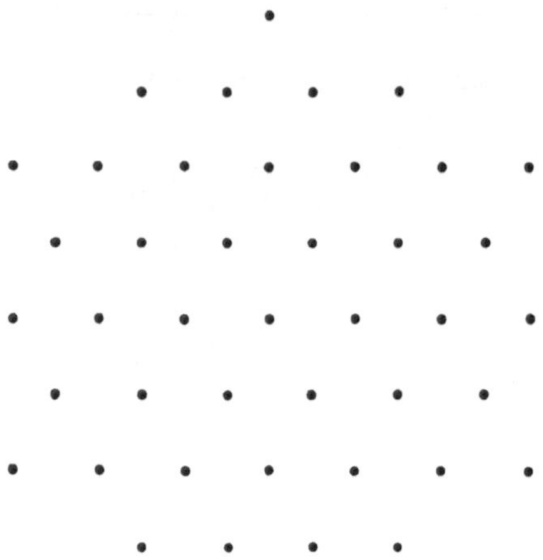

Abbildung 6

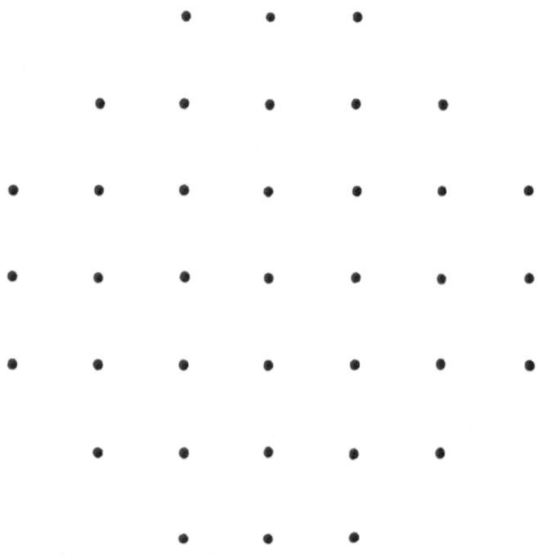

Abbildung 7

einen Punkt des Gitters festzuhalten. Dann wird die Symmetriegruppe endlich, und zwar von der Ordnung 12 im Falle der Abbildung 6 – man findet 6 Rotationen und 6 Spiegelungen, die das
Gitter invariant lassen – und von der Ordnung 8 im Falle des Gitters der Abbildung 7 (das also etwas weniger symmetrisch ist als
das erste). Die Frage nach den Symmetriegruppen von Gittern hat
in letzter Zeit großes Interesse erregt, unter anderem aus zahlen-
theoretischen Gründen. Untersucht man die Gitter im dreidimensionalen Raum, im vierdimensionalen Raum usw. vom Standpunkt

ihrer Symmetrieeigenschaften, so erscheint plötzlich im vierund-
zwanzigdimensionalen euklidischen Raum ein ganz besonderes Gitter mit sehr hoher Symmetrie. Es ist das sogenannte Gitter von
Leech (noch vor 30 Jahren unbekannt, was heute für viele Mathematiker kaum vorstellbar ist). Als J. Leech dieses Gitter entdeckte,
wußte er wohl nicht, daß es solche außergewöhnliche Symmetrieeigenschaften besaß: er war an einem ganz anderen Problem, nämlich dem Problem dichtester Packung von Sphären, interessiert.
Wenn man das eigentliche Gitter von Leech betrachtet, also die
von Leech selbst angegebene Konstruktion anschaut, so sieht dieses
Gitter überhaupt nicht besonders symmetrisch aus. In groben
Zügen läßt sich diese Konstruktion so beschreiben: man beginnt
mit einem schön symmetrischen, nämlich einem rechteckigen Gitter;
aus diesem nimmt man gewisse Punkte heraus und fügt statt dessen
andere Punkte hinzu. Beide Abänderungen sind ziemlich unsymme-
trisch und zerstören teilweise die Symmetrie des vorgegebenen Gitters. Es entstehen bei dem Prozeß aber neue Symmetrien, die allerdings gar nicht unmittelbar sichtbar sind: sie sind also »verborgene
Symmetrien« im obigen Sinn. Erst J.H. Conway bemerkte, daß das
Gitter von Leech eine riesengroße *Symmetriegruppe*, eine Gruppe
der Ordnung 8 315 553 513 086 720 000, besitzt. Ich kenne keine
explizite Konstruktion des Gitters von Leech, bei der die volle
Symmetrie auf einmal zu sehen ist: stets bleiben verborgene Symmetrien übrig, deren Auffindung recht mühsam ist.

Ein letztes berühmtes Beispiel ist das folgende. Man weiß, daß
es im Raum von einhundertsechsundneunzigtausendachthundert-
dreiundachtzig Dimensionen ein wunderschönes Gitter gibt, deren
Symmetriegruppe die Ordnung

808 017 424 794 512 875 886 459 904 961 710 757
005 754 368 000 000 000
$$= 2^{46} \cdot 3^{20} \cdot 5^9 \cdot 7^6 \cdot 11^2 \cdot 13^3 \cdot 17 \cdot 19 \cdot 23 \cdot 29 \cdot 31 \cdot 41 \cdot 47 \cdot 59 \cdot 71$$

hat. Dieses Gitter hat wohl niemand wirklich gesehen: wir wissen
bloß, daß es existiert, aber eine explizite Konstruktion fehlt noch.
Herstellen kann man allerdings die Symmetriegruppe des Gitters,
die sogenannte *Monstergruppe M* von R. Griess und B. Fischer.
Hier ist wieder das Herausfinden einer verborgenen Symmetrie ein
wesentlicher Schritt der Konstruktion. Die Gruppe *M* besitzt eine
gewisse Untergruppe der Ordnung

$$2^{46} \cdot 3^9 \cdot 5^4 \cdot 7^2 \cdot 11 \cdot 13 \cdot 23,$$

die man gut versteht. Um M zu erzeugen, konstruierte Griess im Raum der Dimension 196 883 ein Objekt, eine sog. Algebra, die diese kleinere Gruppe als (Teil seiner vollen) Symmetriegruppe hat, worauf er mit ziemlich großer Mühe eine zusätzliche, wahrhaft verborgene Symmetrie bestimmen konnte, die zusammen mit der schon bekannten Untergruppe die Gruppe M erzeugt. Der Verfasser zeigte nachher, daß M die *volle* Symmetriegruppe der Griess'schen Algebra ist.

Nun kann man sich natürlich fragen: warum interessiert man sich besonders für diese Monstergruppe? Ist dies mehr als ein schönes Spiel? Die Antwort hierzu ist, wie ich nachweisen möchte, entschieden positiv.

Seit Galois hat die Frage nach allen möglichen Symmetrietypen, also nach allen existierenden Gruppen (hier meine ich immer *endliche* Gruppen) einen klaren Sinn bekommen. Sie ist wohl eine natürliche, ja sogar grundlegende Frage, aber es stellt sich heraus, daß sie kein vernünftiges Problem darstellt, was ich kurz erklären möchte. Bekanntlich ist jede natürliche Zahl ein Produkt von *Primzahlen*, die also die »unzerlegbaren Atome« der Zahlentheorie sind. Genauso besitzt die (endliche) Gruppentheorie ihre »Atome«, die *einfache Gruppen* genannt werden: jede Gruppe ist in gewisser Weise eine Zusammensetzung solcher Atome. Aber während der Zusammensetzungsprozeß im Falle der Zahlen nichts anderes als die einfache Produktbildung ist, ist er in der Gruppentheorie (unter dem Namen »Erweiterung«) erheblich komplizierter und außerordentlich mannigfaltig: vorgegebene Atome (also einfache Gruppen) lassen sich unter Umständen auf mehrere Weise kombinieren, und wenn die Anzahl der betrachteten Bestandteile und ihre gegenseitige »Reaktivität« groß ist, kann die Gesamtheit der möglichen Kombinationen völlig unüberschaubar werden. Es liegt aber noch die Frage nahe, zumindest alle einfachen endlichen Gruppen aufzuzählen. Auch dieses Problem hätte man wohl vor 40 Jahren als unrealistisch angesehen, doch wurde es vor kurzem komplett gelöst. Das Ergebnis, dessen Beweis noch nicht vollständig aufgeschrieben ist und immerhin schon Tausende von Zeitschriftenseiten füllt (es ist das Resultat der von D. Gorenstein koordinierten Anstrengungen zahlreicher Spezialisten) ist erstaunlich. Erwartungsgemäß gibt es unendlich viele endliche einfache Gruppen: doch kann man sie alle ziemlich knapp und einheitlich beschreiben bis auf 26 Ausnahmen, die sich nicht in diesen schönen Rahmen einordnen lassen und die man *sporadische Gruppen* nennt. Hier spielt die oben erwähnte

Monstergruppe eine besondere Rolle: sie ist eine der zuletzt entdeckten einfachen und die größte unter den sporadischen Gruppen. Außerdem hat sie bemerkenswerte und noch ziemlich mysteriöse zahlentheoretische Eigenschaften. Diese Eigenschaften vollständig zu verstehen, ist heutzutage eine der faszinierenden Aufgaben der endlichen Gruppentheorie. Merkwürdigerweise spielt in einer neuen Arbeit von I. Frenkel, J. Lepowski und A. Meurman über dieses Thema die verborgene Symmetrie zwischen den zwei Problemen der Abbildung 4 eine wesentliche Rolle.

Zum Schluß will ich noch ein letztes schönes und wohlbekanntes symmetrisches Gebilde zeigen, nämlich ein *Ikosaeder*. Gerade wegen der am Anfang erwähnten psychologischen Wirkung der hohen Symmetrie erscheint der Ikosaeder mehr und mehr alltäglich in unserer werbungsorientierten Welt, aber das hier gezeigte Modell (Abbildung 8) ist besonders ehrwürdig, da es (angeblich) von Leonardo da Vinci stammt.

Es ist eine leichte Übung, alle Symmetrien des Ikosaeders zu bestimmen: man findet 60 Rotationen und die 60 Produkte dieser Rotationen mit der Zentralsymmetrie. Man soll sich nun im Raum von 196 883 Dimensionen einen Kristall vorstellen, der ein bißchen

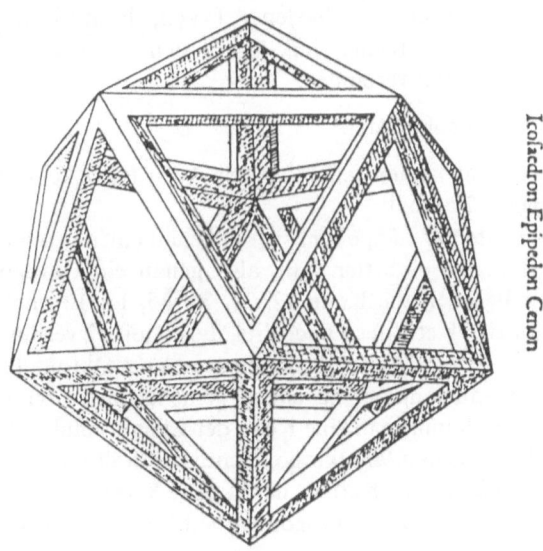

Icosaedron Epipedon Cænon

Abbildung 8   Icosaedron Planum Vacuum

so wie der Ikosaeder aussieht, der aber statt sechzig, 808 017 424
794 512 875 886 459 904 961 710 005 754 368 Milliarden Rotationssymmetrien besitzt. Diese Symmetrien bilden eben die Monstergruppe, von der gerade gesprochen wurde, und die der Leser
sich jetzt vorzustellen anfangen kann. Gleichzeitig gewinnt er vielleicht einen gewissen Eindruck von der Schönheit der Symmetrien
in der Mathematik.

NACHWORT

Bekanntlich hat Herr Dr. Götze eine Schwäche für die Oktogone!*
Er wird sich vielleicht freuen, 90 Oktogone – wohl etwas versteckt
und sehr unregelmäßig – in der Abbildung 3 entdecken zu können,
und zwar mit Hilfe der folgenden Hinweise:

- der auf dieser Abbildung dargestellte »Graph« (siehe oben)
  besitzt 30 (15 »weiße« und 15 »schwarze«) auf dem Randkreis
  liegende Eckpunkte;
- jede Kante verbindet einen weißen und einen schwarzen Eckpunkt;
- jeder Eckpunkt gehört zu genau drei Kanten;
- zwei Eckpunkte verschiedener Typen, die nicht Eckpunkte ein
  und derselben Kante sind, lassen sich durch genau eine Kette
  der Länge drei verbinden.

Daraus folgt sofort, daß

- der Graph keinen geschlossenen Weg (also kein Polygon) der
  Länge ≤ 7 enthält;
- jede Kette der Länge fünf sich in genau einen geschlossenen Weg
  der Länge 8 einbetten läßt, also genau ein Oktogon bestimmt
  (zum Beispiel: die Kette 12, $\overline{23}$, 34, $\overline{45}$, 15, $\overline{12}$ muß man durch
  die einzige Kette der Länge drei, die $\overline{12}$ mit 12 verbinden, nämlich
  mit $\overline{12}$, 46, $\overline{46}$, 12 vervollständigen, um ein Oktogon zu erhalten);
- zwei Kanten immer zu einem gemeinsamen Oktogon gehören
  (da sie sich immer in eine Kette der Länge 5 einbetten lassen);
- die Figur genau 90 Oktogone enthält (in der Tat ist die Anzahl
  der orientierten Ketten der Länge 5 gleich $30 \cdot 3 \cdot 2 \cdot 2 \cdot 2 \cdot 2$
  $= 90 \cdot 2^4$, und jedes Oktogon enthält $8 \cdot 2 = 2^4$ solche Ketten).

---

* H. Götze (1984): Castel del Monte, München: Prestel

André Weil

Sur quelques symétries dans l'Iliade

Qu'il y ait dans l'Iliade des symétries cachées, l'observation n'en est pas nouvelle. Il suffira ici de citer l'étude détaillée d'un grand nombre de telles symétries qui occupe tout un chapitre de l'ouvrage de C.H. Whitman, *Homer and the heroic tradition* (Harvard U. Press, 1958; v. Chap. XI, pp. 249–284; cf. aussi le Chap. V, et le dépliant en fin de volume). Fallait-il l'œil d'un mathématicien pour en apercevoir d'autres? Si c'était le cas, ce serait la justification du présent article.

Whitman intitule son chapitre « *The geometric structure of the Iliad* », mettant ainsi en rapport ces symétries avec les célèbres vases attiques du style dit «géométrique» dont ceux du Dipylon offrent le plus parfait modèle. Ceux-ci, et par exemple le grand vase funéraire du Musée National d'Athènes (v. fig. 1), présentent une composition basée sur un axe central autour duquel se disposent symétriquement les figures. Or c'est justement une telle symétrie qui frappe au premier coup d'œil jeté sur l'Iliade; et c'est ce que Whitman ne manque pas de souligner. Qu'on relise le premier vers[1]:

Chante, déesse, la colère d'Achille, le fils de Pélée

puis le dernier:

C'est ainsi qu'ils célèbrent les funérailles d'Hector
dompteur de cavales.

Comment ne pas sentir qu'ils se répondent l'un à l'autre comme un objet et son image dans un miroir, ou encore comme les deux piliers d'un portail? Et cette symétrie, comme l'observe Whitman, se prolonge aux livres premier et dernier tout entiers, l'un bâti sur un schéma a-b-c-d-e-f, l'autre sur le schéma inverse f-e-d-c-b-a. Par exemple la fin du livre I raconte la visite de Thétis chez Zeus en suppliante (I, 495–530) puis s'achève sur une assemblée des dieux

---

[1] Les traductions sont celles, un peu modifiées, de P. Mazon (Les Belles Lettres, 1955–1957). Il va sans dire que nulle traduction ne peut rendre la beauté indicible du texte homérique.

Figure 1. Vase attique (env. 760–750 av. J.-C.). Athènes, Musée Archéologique National, no. 804. Photo Hirmer Verlag München.

(I, 533–604), tandis que le livre XXIV, après un passage de transition, débute par une autre assemblée des dieux (XXIV, 23–76) suivie d'un dialogue entre Zeus et Thétis que Zeus a fait convoquer par Iris pour entendre ses ordres (XXIV, 77–120).

Whitman suggère d'opposer le couple Hector/Andromaque au couple Paris/Hélène; c'est là une observation littéraire, qui échappe à la structure «géométrique» de l'ensemble. Mais il est plus à propos ici de remarquer que, si le premier et le dernier livre mettent en opposition Achille et Hector, on peut en conclure que ce sont les couples Achille/Briséis et Hector/Andromaque qu'il y a lieu de mettre en relation l'un avec l'autre. Un point culminant du premier livre nous dit l'adieu silencieux de Briséis à Achille (I, 337–348):

> Patrocle ... fait sortir de la tente Briséis aux belles joues; il la leur donne à emmener ...; malgré elle la femme s'en va avec eux ...

à quoi répond au dernier livre le déchirant adieu d'Andromaque à Hector (XXIV, 723–740):

> Elle tient entre ses mains la tête d'Hector tueur d'hommes. «Epoux, tu quittes jeune le monde ... tu n'auras pas en mourant tendu vers moi tes mains, tu ne m'auras pas dit un mot plein de sens, dont je puisse éternellement me souvenir nuit et jour en pleurant ...»

Comment s'étonner après cela si, au célèbre discours d'Andromaque à Hector au livre VI répond au livre XIX le discours de Briséis, non pas directement à Achille (ce ne serait guère conforme à son silence du livre I) mais au cadavre de Patrocle? Dans le poème ils sont placés symétriquement, l'un à cinq livres de distance du début, l'autre à la même distance de la fin. Dans l'un Andromaque dit à Hector qu'elle n'a plus que lui au monde (VI, 414–430):

> Je n'ai plus mon père ni ma digne mère ... j'avais sept frères ... tous abattus par le divin Achille ... Hector, tu es pour moi père, digne mère, frère, et tu es le tendre compagnon de ma couche ...

De même Briséis (XIX, 287–300):

> Le mari auquel m'avaient donnée mon père et ma digne mère, je l'ai vu mis en pièces ... et trois frères, tous nés de la même mère que moi, ont atteint leur jour fatal ... mais ... tu m'assurais que tu ferais de moi l'épouse légitime du divin Achille ...

C'est là, peut-on dire, la «résolution» (au sens où l'on entend ce mot, je crois, en théorie musicale) du départ de Briséis au livre I; de même la lamentation d'Andromaque au livre XXIV est la «résolution» de son discours à Hector au livre VI.

Nous avons ainsi pour notre Iliade le schéma

Mais à ce schéma il manque encore un axe. A la vérité celui-ci pourrait être seulement imaginé, ou, comme dirait le mathématicien, virtuel. S'il était présent dans l'Iliade, ce devrait être au milieu du poème, donc dans l'un des livres XII et XIII.

Or le livre XII s'ouvre sur un extraordinaire passage prophétique, étranger à l'action, qui paraît tout désigné pour jouer ce rôle, sans que rien d'autre motive sa présence en ce point précis. C'est l'épisode de la destruction du mur des Achéens par Apollon et Poseidôn enfin réconciliés après la chute de Troie (XII, 8–34); Zeus lui-même y participe:

> … quand la ville de Priam, à la dixième année, eut été détruite, le dessein vint à l'esprit de Poseidôn et d'Apollon d'anéantir le mur, concentrant sur lui la fureur des fleuves … Apollon les détourna et en réunit les bouches; neuf jours il en dirigea le cours vers le mur, et Zeus fit pleuvoir continuellement afin de faire partir plus vite les murailles à la dérive. En personne, le trident à la main, l'Ebranleur de la Terre les guidait et sur leurs vagues en emmenait les fondations, troncs et pierres, qu'à grand'peine avaient mis en place les Achéens. Ainsi il nivela les bords de l'impétueux Hellespont. Puis il cacha sous les sables le vaste rivage; le mur était anéanti. Enfin il renversa de nouveau le cours des fleuves et leur fit à chacun retrouver le lit où avaient auparavant coulé ses belles eaux.

Ainsi tout a été vain, et nulle trace, sinon le poème, ne subsiste de la guerre de Troie. N'est-ce pas la philosophie de l'œuvre entière qui s'exprime ici, et ce passage ne domine-t-il pas l'Iliade dont il formule la conclusion «sous l'aspect de l'éternité»? On est tenté de dire qu'il est au poème ce que la Bhagavadgītā est au Mahābhārata.

Dans ces deux épisodes, s'agirait-il d'insertions faites après coup, la Gītā par un philosophe anonyme, la destruction du mur par

l'un des hypothétiques «arrangeurs» de Pisistrate? A l'égard de
la Gītā mon maître Sylvain Lévi avait coutume, paraît-il, de se
moquer de telles suggestions. «Autant vaudrait prétendre, disait-il,
qu'un noyau a été inséré après coup dans le fruit». Il est vrai aussi
que la Bhagavadgītā (le dialogue de Krishna et d'Arjuna) fait partie
de l'action du Mahābhārata et y a donc sa place marquée d'avance
à l'ouverture de la grande bataille de Kurukshetra où devait périr
l'humanité. Dans l'Iliade, au contraire, les occasions ne manque-
raient pas pour insérer l'épisode en question, ne serait-ce que lors
de la construction du mur (VII, 433–442) où sa destruction est
déjà annoncée, ou bien encore (comme l'aurait peut-être fait un
poète soucieux de logique) en épilogue à l'épopée. Si même il était
omis du livre XII, le lecteur moderne n'y soupçonnerait pas de
lacune. Qu'il se trouve au milieu même du poème témoigne du
sens aigu des symétries qui marque l'apogée du style géometrique.
La destruction du mur en occupe le centre, comme le cadavre du
mort occupe le centre du vase du Dipylon.

Don Zagier

Lösungen von Gleichungen in ganzen Zahlen*

Die Frage, wie man die Lösungen in ganzen oder Bruchzahlen einer unbestimmten Gleichung ausfindig macht, ist eine der ältesten, mit denen Mathematiker sich beschäftigt haben, gehört aber gleichzeitig zu den aktuellsten Gegenständen der mathematischen Forschung. In meinem heutigen Vortrag möchte ich Ihnen über beide Aspekte etwas erzählen.

Ich beginne mit einem Beispiel, das Ihnen allen von der Schule her bekannt ist, nämlich der Gleichung

(1) $$a^2 + b^2 = c^2,$$

welche die Beziehung zwischen den Katheten und der Hypotenuse eines rechtwinkligen Dreiecks ausdrückt (»Satz von Pythagoras«):

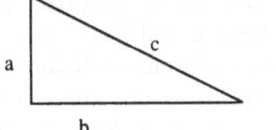 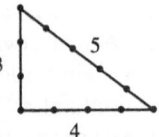

Uns geht es hier allerdings nicht um den geometrischen Inhalt dieser Gleichung, sondern darum, wie man ganzzahlige Lösungen findet, also Zahlentripel (a, b, c), die die Gleichung erfüllen. Da $3^2 + 4^2 = 9 + 16 = 25 = 5^2$ ist, hat man als einfachste Lösung das Tripel $(3, 4, 5)$, das schon allen antiken Völkern bekannt war[1] und z.B.

---

* Vortrag gehalten in Schweinfurt anläßlich der Verleihung des Carus-Preises, 27. 1. 84. Viele Anregungen zu diesem Vortrag, insbesondere über Diophants Werk und dessen Interpretation vom modernen Standpunkt, habe ich dem Büchlein *Diophant und diophantische Gleichungen* (Deutscher Verlag der Wissenschaften, Berlin 1974) von I.G. Bashmakova entnommen, das ich dem interessierten Leser sehr empfehlen möchte. Eine weitere Quelle war die von derselben Autorin kommentierte Übersetzung (ins Russische) von I.N. Veselovsky (Nauka Verlag, Moskau 1974).

[1] Zur Geschichte der pythagoreischen Gleichung (1) und ihrer ganzzahligen Lösungen s. Kapitel I von B.L. van der Waerden, *Geometry and Algebra in Ancient Civilisations* (Springer-Verlag, 1983).

von den Ägyptern in Form eines in regelmäßigen Abständen geknoteten Seils der Länge $3+4+5=12$ zur Konstruktion rechter Winkel in der Landvermessung gebraucht worden sein soll; ein anderes Tripel, (8, 15, 17), wurde u.a. von Plato angegeben. Die allgemeine Lösung der Gleichung (1) steht bei Euklid (Buch X, 29) – sie wurde auch in anderen Ländern entdeckt, z.B. in Indien im 7. Jahrhundert (Brahmegupta) – und hat die Form

$$(2) \qquad a=2pq, \quad b=p^2-q^2, \quad c=p^2+q^2,$$

wobei p und q beliebige positive Zahlen mit $p>q$ sind. (Eigentlich muß man auch Vielfache hiervon, also Tripel $(2hpq, h(p^2-q^2), h(p^2+q^2))$ mit irgendeinem positiven h, betrachten, aber solche proportionale Lösungen werden als im wesentlichen gleich angesehen.) Wir werden später sehen, wie man auf die Formeln (2) kommt.

Die Gleichungen (2) stellen lediglich ein Einzelergebnis dar. Die erste systematische Theorie von unbestimmten Gleichungen und ihren Lösungen wurde von dem genialen alexandrinischen Mathematiker Diophant (ca. 250 A.D.) entwickelt, zu dessen Ehre das ganze Gebiet heute die *diophantische Analysis* oder die Theorie der *diophantischen Gleichungen* heißt. Ich möchte an dieser Stelle auf Diophants Arbeit etwas näher eingehen, weil diese, wie wir sehen werden, die Keime aller späteren Entwickungen enthält.

Über Diophant selbst wissen wir sehr wenig, außer daß er, wie gesagt, in Alexandria lebte und anscheinend im Alter von 84 Jahren starb (sogar die Zeitangabe 250 A.D. könnte um bis zu 150 Jahre in beiden Richtungen falsch sein). Von seinen Werken sind die meisten verlorengegangen. Bis zum 16. Jahrhundert waren sie sogar alle verschwunden, und man wußte nur vage, was Diophant getan hatte; ein Manuskript von 6 der 13 Bücher der *Arithmetika* wurde erst 1570 von dem Astronom Johannes Müller aus Königsberg (bei Schweinfurt) dem sogenannten Regiomontanus, in Venedig entdeckt[2]. Vielleicht einmalig in der Geschichte der Wissenschaft ist die Tatsache, daß seine Schriften, als sie zu diesem Zeitpunkt – also über 1200 Jahre nach seinem Tode! – wieder auftauchten,

---

[2] Vor ein paar Jahren sind weitere vier Bücher aufgetaucht (s. J. Sesiano, *Books IV to VII of Diophantus' Arithmetica*, Springer-Verlag, 1982), was unter anderem eine Umnumerierung der bis dann bekannten Bücher mit sich brachte. In diesem Vortrag werden die neu entdeckten Bücher nicht berücksichtigt und die traditionelle Numerierung der Bücher verwendet.

keineswegs nur von historischem Interesse, sondern ganz im Gegenteil dem derzeitigen Wissensstand weit voraus waren und einen entscheidenden Anstoß für die Entwicklungen der nächsten ein- bis zweihundert Jahre gaben.

Diophant war der erste, der einen systematischen algebraischen Symbolismus einführte, so daß er eine Gleichung wie $3x^2 - 2x^2 = 4$ (was allerdings bei ihm so ausgesehen hätte: $K^Y \bar{\gamma} \Lsh \Delta^Y \beta \iota \overset{\circ}{M} \delta$) schreiben konnte; vor ihm konnte man eine solche Formel nicht einmal in Worten ausdrücken, da man die geometrisch aufgefaßten Kubik- und Quadratzahlen wegen der Inkompatibilität der Dimensionen nicht addieren konnte und ohnehin den Begriff einer Unbekannten, mit der operiert wird, noch nicht formuliert hatte. Darüber hinaus hat Diophant das System der natürlichen (d.h. positiven ganzen) Zahlen in zwei Richtungen erweitert – zunächst, indem er negative Zahlen zuließ und die Rechenregeln für diese (etwa: Minus mal Minus gleich Plus) genau formulierte, dann durch Hinzunahme der rationalen (d.h. Bruch-)Zahlen, mit denen er genauso umging wie mit ganzen Zahlen. Um zu sehen, wie fortschrittlich dies war, muß man bedenken, daß die rationalen Zahlen noch ganz unbekannt waren (Euklid betrachtete Verhältnisse von Längen, die aber nur miteinander verglichen werden konnten und nicht etwa addiert), und daß man in Europa tausend Jahre später die negativen Zahlen noch längst nicht als vollwertig anerkannte. Es ist übrigens ein in der Geschichte der Zahlentheorie immer wieder auftretendes Phänomen, daß die Erweiterung des Zahlsystems oder der Übergang zu neuartigen Zahlen notwendig wurde, um Probleme, die die gewöhnlichen ganzen Zahlen betreffen, besser zu verstehen. Jedenfalls war die Ausdehnung der Zahlen auf positive und negative ganze und Bruchzahlen bei Diophant sehr zweckmäßig, weil man jetzt alle vier Grundrechenarten unbeschränkt ausführen konnte und auch, weil viele Problemstellungen einfacher wurden. Teilt man z.B. Gleichung (1) durch c^2 und setzt $\dfrac{a}{c} = x$, $\dfrac{b}{c} = y$, so erhält man die nunmehr in rationalen Zahlen zu lösende Gleichung

$$(3) \qquad\qquad x^2 + y^2 = 1,$$

welche nur zwei Variable enthält.

Im ersten Buch der *Arithmetika* betrachtete Diophant lineare Gleichungen (also solche, in denen die Variablen nur zur ersten

Potenz auftreten); die Methoden hier waren zum Teil schon bekannt. Für Gleichungen von höherem Grad dagegen mußten ganz neue Verfahren entwickelt werden, die wir jetzt beschreiben.

Sehen wir uns zunächst die Methode an, die Diophant für quadratische Gleichungen (Gleichungen vom Grad 2) benutzte. Als Beispiel nehmen wir die pythagoreische Gleichung in der Gestalt (3). Wir setzen in dieser Gleichung

(4) $$y = 3x + 1$$

und erhalten

$$x^2 + (3x + 1)^2 = 1,$$
$$10x^2 + 6x + 1 = 1,$$
$$10x^2 + 6x = 0,$$
$$10x + 6 = 0,$$
$$x = -\tfrac{3}{5}, \quad y = 3x + 1 = -\tfrac{4}{5},$$

also (bis aufs Vorzeichen) die alte Lösung (3, 4, 5) von (1). Man sieht hier, daß die Wahl des Koeffizienten 3 in (4) völlig willkürlich war; man hätte genausogut irgendeine andere Zahl einsetzen können und würde so viele rationale Lösungen bekommen, wie man wollte. In Formeln ausgedrückt (was allerdings Diophant nicht gut konnte, weil sein Symbolismus nur eine Veränderliche auf einmal zuließ): mit dem Ansatz $y = kx + 1$ anstelle von (4) erhält man

$$(k^2 + 1)x^2 + 2kx + 1 = 1$$

und somit die allgemeine Lösung

(5) $$x = \frac{-2k}{k^2 + 1}, \quad y = \frac{1 - k^2}{1 + k^2}$$

von (3), die zu (2) äquivalent ist, wenn wir $k = -\dfrac{q}{p}$ setzen. Daß diese Lösung den freien Parameter k enthält, impliziert, daß man unendlich viele rationale Lösungen von (3) und somit unendlich viele ganzzahlige Lösungen von (1) hat. Das Verfahren funktioniert für jede quadratische Gleichung in zwei Veränderlichen x und y: ausgehend von einer bekannten Lösung (x_0, y_0) ersetzt man zunächst x durch $x + x_0$, setzt dann $y = kx + y_0$, und erhält für jeden rationalen Wert von k eine Lösung der Gleichung. Man findet in der *Arithmetika* vielfache Anwendungen.

Wir wenden uns jetzt dem – viel subtileren – Verfahren zu, das Diophant für Gleichungen dritten Grades verwendete. Diesmal nehmen wir als Beispiel die Aufgabe 24 des IV. Buchs: man zerlege eine gegebene Zahl – etwa 6 – in zwei Teile, so daß ihr Produkt gleich einer um ihre Wurzel verminderten Kubikzahl ist, also

$$(6) \qquad\qquad y(6-y) = x^3 - x.$$

Hier macht Diophant in Analogie zum quadratischen Fall zunächst den Ansatz

$$x = 2y - 1$$

mit dem willkürlich gewählten Koeffizienten 2 und erhält

$$6y - y^2 = 8y^3 - 12y^2 + 4y.$$

Er bemerkt, daß die neue Gleichung sofort rational lösbar wäre, wenn die beiden Koeffizienten von y, also die Zahlen 6 und 4, gleich wären, da man dann durch y^2 teilen und die entstehende lineare Gleichung lösen könnte. Die Zahl »6« kommt von der Aufgabenstellung, die Zahl »4« dagegen ist das Doppelte des Koeffizienten »2« in dem gewählten Ansatz. Dieser Koeffizient muß also durch 3 ersetzt werden, d.h. wir setzen $x = 3y - 1$ und finden

$$6y - y^2 = 27y^3 - 27y^2 + 6y,$$
$$y = \tfrac{26}{27}, \qquad x = \tfrac{17}{9},$$

die gewünschte Lösung. Diesmal ist der Parameter k im Ansatz $x = ky - 1$ nicht frei, und wir erhalten nur *eine* Lösung.

Wir können Diophants Erkenntnisse so zusammenfassen:

1. Quadratische Gleichungen lassen sich durch einen bestimmten linearen Ansatz auf vielerlei Weisen lösen. Solche Gleichungen werden heute als *rational* bezeichnet.

2. Bei kubischen (und gewissen quartischen) Gleichungen liefert ein ähnlicher Ansatz manchmal spezielle Lösungen. Gleichungen dieses Typs heißen heute *elliptisch*.

3. Gleichungen höheren Grades werden nicht betrachtet[3]. Diese nennt man heute *von höherem Geschlecht* (die Terminologie wird unten erläutert).

---

[3] Eine Ausnahme ist Aufgabe 18 des IV. Buchs, wo eine nichttriviale Lösung der Gleichung $y^2 = x^6 - 2b^2 x^3 + x + b^4$ (d.h. eine mit $x \neq 0$) durch den Ansatz $y = x^3 + b^2$ gefunden wird.

Von den drei Klassen haben die elliptischen die interessantesten Lösungen. Wir illustrieren dies anhand zweier historischer Beispiele.

FERMATS PROBLEM ÜBER RECHTWINKLIGE DREIECKE. In einem Brief an den Priester Mersenne stellte Fermat 1643 die folgende Aufgabe: man finde ein rechtwinkliges Dreieck mit ganzzahligen Seiten, so daß sowohl die Summe der Katheten als auch die Hypotenuse Quadrate sind. Fermat (1601–1665) war Mitbegründer der analytischen Geometrie (mit Descartes), der Wahrscheinlichkeitsrechnung (mit Pascal) und der Infinitesimalrechnung (mit Leibniz und Newton), seine größte Leistung war aber in der Zahlentheorie, wo er unmittelbar von Diophant inspiriert wurde, dessen Werk er als erster voll verstand und weiterentwickelte. Auch die genannte Frage stellte er erstmalig im Zusammenhang mit einer Aufgabe von Diophant (Nummer 22 aus Buch VI). An der von ihm gefundenen Lösung erkennt man, wie weit die Theorie in seinen Händen gekommen war:

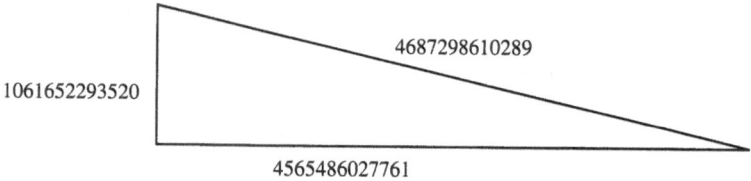

(Fermat konnte nachweisen, daß dies die *kleinste* Lösung ist!). Die zum Problem gehörige Gleichung findet man, indem man die Seiten des Dreiecks durch die Kathetensumme teilt; diese wird dann 1 und man sucht nunmehr ein rechtwinkliges Dreieck der Gestalt

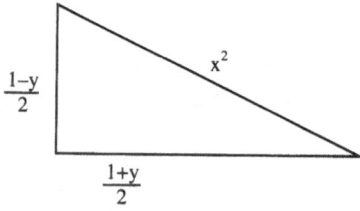

d.h. man muß die Gleichung

$$y^2 = 2x^4 - 1$$

mit x und y rational und der Nebenbedingung $-1 < y < 1$ lösen.

Daß eine so harmlos aussehende Gleichung eine so komplizierte Lösung haben kann, gibt einen Eindruck von der Schwierigkeit der Theorie der elliptischen Gleichungen.

KONGRUENTE ZAHLEN. Gegeben sei eine Zahl n; man soll entscheiden, ob n als Flächeninhalt eines rechtwinkligen Dreiecks mit rationalen Seitenlängen vorkommen kann:

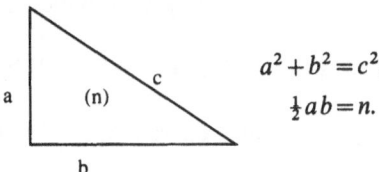

$$a^2 + b^2 = c^2,$$
$$\tfrac{1}{2}ab = n.$$

Eine äquivalente Aufgabe[4] ist, eine (rationale) Quadratzahl x zu finden, die sowohl um n vermindert als auch um n vermehrt eine Quadratzahl bleibt. Eine Zahl n, für die es ein Dreieck oder eine Zahl x mit den genannten Eigenschaften gibt, heißt klassisch »kongruent«. Das Problem steht bereits in einem über 1000 Jahre alten arabischen Manuskript und wurde von Leonardo Fibonacci von Pisa, einem Zeitgenossen Dantes, wieder aufgegriffen, der ihm ein ganzes Buch, das *liber quadratorum* (1225), widmete. Für $n = 5$ wird die Lösung des Problems durch das Dreieck

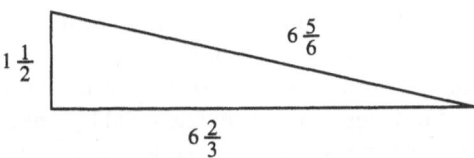

---

[4] Gilt $x = s^2$, $x - n = t^2$, $x + n = u^2$, so erfüllt das Dreieck mit den Seitenlängen

$$\frac{2xn}{stu} = \frac{2ns}{tu}, \quad \frac{x^2 - n^2}{stu} = \frac{tu}{s}, \quad \frac{x^2 + n^2}{stu}$$

die gegebene Bedingung; ist umgekehrt (a, b, c) ein pythagoreisches Tripel mit $\tfrac{1}{2}ab = n$, so hat man die Lösung $x = \left(\dfrac{c}{2}\right)^2$, $x \pm n = \left(\dfrac{a \pm b}{2}\right)^2$ der zweiten Aufgabe.

oder durch die Zahl $x = 11\frac{97}{144}$ gelöst:

$$11\tfrac{97}{144} = (3\tfrac{15}{12})^2,$$
$$6\tfrac{97}{144} = (2\tfrac{7}{12})^2,$$
$$16\tfrac{97}{144} = (4\tfrac{1}{12})^2.$$

Wenn wir aber $n = 157$ wählen, dann sehen wir noch deutlicher als beim vorigen Beispiel, wieviel in einer elliptischen Gleichung steckt, denn die kleinste Lösung x ist in diesem Fall

$$158\,\tfrac{1573895479089949747604806007758166260069136616365753818333091285006249883067347975098257208\underline{81}}{31771866588716253752986042920489352212245784987895186010677509621208919878703599127102993560\underline{0}}\,!$$

Für Gleichungen von höherem Geschlecht geben wir nur ein Beispiel:

(7) $x^n + y^n = 1$ (x, y rational)

bzw.

$a^n + b^n = c^n$ (a, b, c ganz)

mit $n > 3$ (für $n = 3$ ist (7) elliptisch). Fermat hat diese Gleichung am Rande seines Exemplars von Diophant angegeben und behauptet, er habe einen wahrlich wunderbaren Beweis für ihre Unlösbarkeit (für alle $n > 2$) gefunden, der Rand sei aber zu schmal, um ihn zu fassen. Dies ist der »letzte Satz von Fermat«, bis heute ungelöst und eins der berühmtesten und wichtigsten Probleme in der mathematischen Geschichte.

Ich möchte aber jetzt mit der Beschreibung der historischen Entwicklung fortfahren und zum schon erwähnten Thema zurückkehren, daß man oft in der Theorie der diophantischen Gleichungen durch eine Ausdehnung des benutzten Zahlsystems weiterkommt. Das erste Beispiel, der Übergang von den ganzen zu den rationalen Zahlen, haben wir schon gesehen; das zweite wäre der Übergang von den rationalen zu den *reellen* Zahlen, also Zahlen wie $\sqrt{2}$ oder π, die Streckenlängen darstellen können, aber nicht notwendigerweise Quotienten ganzer Zahlen sind. In der vorhin erwähnten analytischen Geometrie von Fermat und Descartes stellt man die Menge aller reellen Lösungen einer polynomialen Gleichung $f(x, y) = 0$ graphisch als Kurve in der Ebene dar; diese Kurve wäre im Fall der pythagoreischen Gleichung (3) ein Kreis

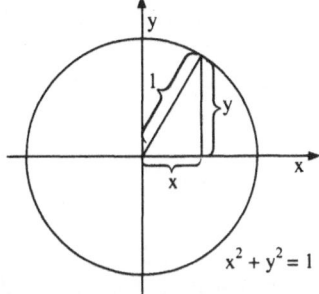

(allgemeiner, für irgendeine rationale Gleichung ein Kegelschnitt, also Ellipse, Hyperbel oder Parabel) und sähe für $y^2 = x^3 + 1$, eine typische elliptische Gleichung, so aus:

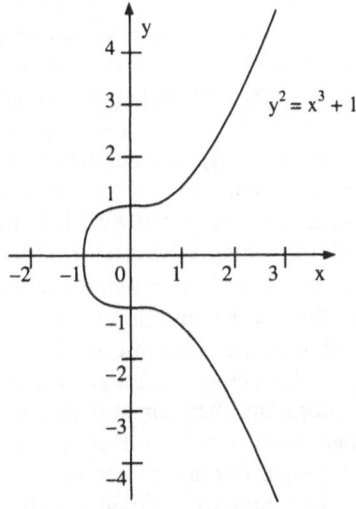

Wir werden ab jetzt Kurven und Gleichungen identifizieren, so daß wir von *rationalen Kurven, elliptischen Kurven* und *Kurven von höherem Geschlecht* sprechen und zwischen »Lösung einer Gleichung« und »Punkt auf einer Kurve« nicht unterscheiden werden. Der geometrische Standpunkt liefert neue Einsicht in den algebraischen Sachverhalt. Insbesondere können wir die vorher erläuterten Verfahren von Diophant jetzt veranschaulichen:

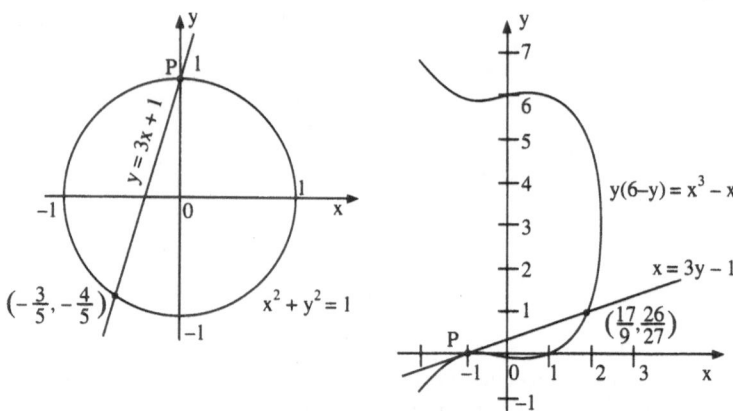

Im Fall einer rationalen Gleichung nehmen wir irgendeine bekannte Lösung und ziehen durch den entsprechenden Punkt P (hier $0, 1$)) auf der Kurve eine Gerade mit willkürlich gewählter Steigung k; der zweite Schnittpunkt der Geraden mit der Kurve ist dann eine neue Lösung. Bei einer elliptischen Kurve gehen wir ebenfalls von einem bekannten Punkt P (hier $(-1, 0)$) aus und zeichnen durch diesen Punkt die *Tangent*gerade zur Kurve; diese Gerade schneidet die Kurve in einem weiteren Punkt, der eine neue rationale Lösung liefert. Allgemein kann man von zwei bekannten Lösungen ausgehen und durch die entsprechenden Punkte P und Q der Kurve eine Gerade ziehen, deren dritter Schnittpunkt mit der Kurve wieder rationale Koordinaten haben und eine neue rationale Lösung darstellen wird; die Konstruktion mit der Tangenten ist der Grenzfall $P = Q$. (Dieses allgemeine Verfahren wurde häufig von Fermat benutzt, von Diophant aber nur in dem etwas entarteten Fall, daß P oder Q ein »unendlich ferner« Punkt ist.) Übrigens war Newton anscheinend der erste, der den geometrischen Inhalt von Diophants algebraischem Verfahren im rationalen Fall erkannte[5].

Der nächste Schritt in der Kette von Erweiterungen des Zahlsystems war der Übergang zu den *komplexen* Zahlen, also Zahlen der Gestalt $x + y\sqrt{-1}$ mit x und y reell. Ohne hierauf im Detail einzugehen, können wir sagen, daß die Lösungsmenge jetzt nicht

---

[5] *The mathematical papers of Isaac Newton*, Band IV, ed. D.T. Whiteside, Cambridge, 1971, S. 110.

mehr eine Kurve, sondern eine geschlossene *Fläche* ist, deren Geometrie (eigentlich: Topologie) unserer Einteilung von Gleichungen in 3 Typen zugrunde liegt: diese Fläche ist nämlich eine *Sphäre* (Kugeloberfläche) im rationalen Falle, ein *Torus* (Reifen) im elliptischen Falle und sonst eine *Fläche von höherem Geschlecht*

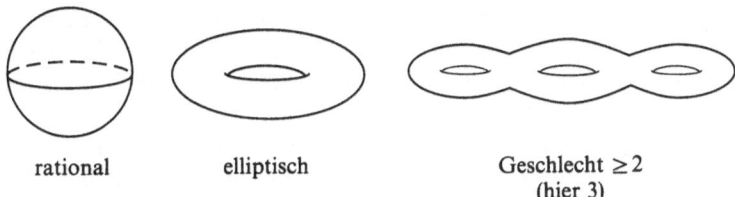

rational   elliptisch   Geschlecht ≥ 2
                                      (hier 3)

(das *Geschlecht* einer Kurve ist die Anzahl der Löcher oder Henkel, also 0 für die Sphäre, 1 für den Torus, und sonst ≥ 2). Der erste, der diophantische Gleichungen von diesem Standpunkt aus betrachtet hat, war Poincaré am Anfang dieses Jahrhunderts (die Flächen selbst waren schon 50 Jahre früher von Riemann studiert worden). Poincaré hat auch erkannt, daß die vorhin beschriebene Konstruktion einer dritten Lösung einer elliptischen Gleichung aus zwei bekannten als *Addition* aufgefaßt werden kann: mit anderen Worten, je zwei Punkten P und Q auf der Kurve mit rationalen Koordinaten kann man einen dritten Punkt $P+Q$ zuordnen[6], und die Operation $P, Q \rightarrow P+Q$ besitzt die Eigenschaften der gewöhnlichen Addition (in der Mathematik sagt man: die Menge der Punkte mit rationalen Koordinaten hat eine *Gruppenstruktur*). Er vermutete, daß jede elliptische Gleichung endlich viele Grundlösungen hat, welche bezüglich dieser Addition sämtliche anderen Lösungen erzeugen. Diese Vermutung wurde 1920 von Mordell bewiesen. Mordell hat wiederum seinerseits vermutet, daß Gleichungen von höherem Geschlecht stets nur endlich viele Lösungen haben. Das ergibt folgendes Bild. Die zunächst nur topologisch unterschiedenen

---

[6] Wenn die Kurve die Gestalt $y^2 = ax^3 + bx^2 + cx + d$ hat (jede elliptische Kurve kann so geschrieben werden), so ist $P+Q$ nicht der dritte Schnittpunkt der Geraden durch P und Q mit der Kurve, sondern dessen Spiegelung an der x-Achse. Das Nullelement 0 der Addition ist der unendlich ferne Punkt $y = \infty$, das Negative eines Punktes $P = (x, y)$ ist der gespiegelte Punkt $-p = (x, -y)$, und drei kollineare Punkte P, Q, R der Kurve erfüllen die Beziehung $P + Q + R = 0$.

Fälle von Gleichungen, deren komplexe Lösungen eine Sphäre, einen Torus oder eine mehrhenklige Fläche bilden, verhalten sich auch vom Standpunkt ihrer rationalen Lösungen grundverschieden; für eine Sphäre gibt es stets unendlich viele Lösungen [7], die parametrisch angegeben werden; für einen Torus kann die Lösungsmenge endlich oder unendlich sein, wird aber auch im zweiten Falle von einer endlichen Teilmenge mittels eines algebraischen Verfahrens erzeugt; für eine Fläche von höherem Geschlecht sind die Lösungen schwer zu finden, haben (anscheinend) keine besondere Struktur und es soll nur endlich viele geben.

Mordells Vermutung wurde 1983 von dem jungen deutschen Mathematiker Faltings bewiesen, eine Leistung, die als eine der wichtigsten Fortschritte der letzten Jahrzehnte angesehen wird. Sein Ergebnis impliziert insbesondere, daß die Fermatsche Gleichung (7) für gegebenes $n > 3$ nur endlich viele Lösungen hat (der Fall $n = 3$ wurde schon von Euler erledigt).

Durch Faltings' Resultat ist die Frage nach der Struktur der Lösungen im Falle Geschlecht > 1 weitgehend geklärt. Wenden wir uns wieder den elliptischen Kurven zu. Mordells Satz über die endliche Erzeugbarkeit läßt sich wie folgt präzise formulieren: Für eine elliptische Kurve kann man endlich viele Grundlösungen $P_1, \ldots, P_r$ finden, so daß überhaupt jede Lösung P geschrieben werden kann als

$$P = n_1 P_1 + \ldots + n_r P_r + Q$$

mit $n_1, \ldots, n_r$ ganze Zahlen und Q aus einem endlichen Lösungsvorrat. Die (minimal gewählte) Zahl r heißt *Rang* der elliptischen Kurve. Beispielsweise hat die Kurve

$$y^2 = x^3 - 432,$$

welche vermöge der Transformation $x = \dfrac{12c}{a+b}$, $y = 36\dfrac{a-b}{a+b}$ der Fer-

---

[7] Genauer gesagt: wenn es *eine* rationale Lösung gibt, gibt es unendlich viele, da man dann Diophants Methode anwenden kann. Es gibt aber auch Gleichungen, deren komplexe Lösungsmenge eine Sphäre ist, die aber überhaupt keine rationale Lösung haben (z.B. $x^2 + y^2 = 3$, ein Kreis vom Radius $\sqrt{3}$, oder die Gleichung $x^2 + y^2 = -1$, die nicht einmal reelle Lösungen besitzt). Diese Möglichkeit haben wir bei der Diskussion der Arbeit Diophants nicht erwähnt, weil er nur Gleichungen betrachtete, für die er mindestens eine Lösung kannte.

matschen Gleichung $a^3 + b^3 = c^3$ entspricht, den Rang 0, weil sie
nur die zwei Lösungen $x = 12$, $y = \pm 36$ besitzt. Die Gleichung

$$y^2 = 2x^4 + 1,$$

die im Zusammenhang mit Fermats Problem über Dreiecke vor-
kam, hat den Rang 1. Eine Grundlösung ist der Punkt P(13,239)
mit den Vielfachen $2\,\mathrm{P}(\frac{1525}{1343}, \frac{2750257}{1803649})$, $3\,\mathrm{P}(\frac{2165017}{2372159}, \frac{3503833734241}{5627138321281})$,
..., wobei erst $3\,\mathrm{P}$ die Zusatzbedingung $-1 < y < 1$ erfüllt und das
geometrische Problem löst. Ein Beispiel für eine Kurve höheren
Ranges ist die Gleichung

$$y^2 = 4x^3 - 28x + 25$$

mit dem Rang 3; der Leser kann ausprobieren, für wie viele kleine
ganze Zahlen x die linke Seite hier ein Quadrat wird. Man kennt
Beispiele von elliptischen Kurven mit Rang bis zu 14, weiß aber
nicht, ob es Kurven mit beliebig hohem Rang gibt.

Es stellt sich nun die Frage, wie man den Rang einer elliptischen
Kurve berechnet. Bis heute ist keine Antwort hierauf bekannt; es
gibt aber eine schöne Vermutung, die ich abschließend erklären
möchte. Hier ist noch einmal die entscheidende Idee der Übergang
zu einem anderen Zahlbereich, der aber diesmal nicht eine Vergrö-
ßerung, sondern eine Verkleinerung des Systems der ganzen Zahlen
darstellt: Man nimmt eine Primzahl p und arbeitet mit dem *endlichen* System, bestehend aus den möglichen Resten von ganzen Zahlen nach Division durch p. Diese endlich vielen Zahlen, die wir
mit k^* ($0 \le k \le p - 1$) bezeichnen werden, können genauso wie
gewöhnliche Zahlen addiert und multipliziert werden (man denke
an die Reste nach Division durch 10 – allerdings keine Primzahl
–, die man sich als Endziffer gewöhnlicher Zahlen vorstellen kann;
hier gibt es die 10 Möglichkeiten $0^*, 1^*, ..., 9^*$ und man hat
$8^* + 4^* = 2^*$, $7^* \cdot 9^* = 3^*$, ..., weil z.B. das Produkt zweier Zahlen
mit den Endziffern 7 und 9 stets die Endziffer 3 hat). Die Idee ist
jetzt, daß eine Gleichung, die viele Lösungen besitzt, also eine von
hohem Rang, ebenfalls viele Lösungen in solchen Resten besitzen
wird. Wir schreiben N_p für die Anzahl der Lösungen unserer Gleichung in Resten nach Division durch p. Für die früher erwähnte
Gleichung $y^2 = x^3 + 1$ und $p = 7$ ist z.B. $N_p = 11$. Man hat nämlich

x	0	1	2	3	4	5	6	7	8	9
x^3+1	1	2	9	28	65	126	217	344	513	730
Rest nach Division durch 7	1	2	2	0	2	0	0	1	2	2

,

also für die Reste

x^*	0*	1*	2*	3*	4*	5*	6*
$x^{*3}+1$	1*	2*	2*	0*	2*	0*	0*

und ähnlich

y^*	0*	1*	2*	3*	4*	5*	6*
y^{*2}	0*	1*	4*	2*	2*	4*	1*

.

Von den 49 möglichen Restepaaren (x^*, y^*) erfüllen also genau 11 unsere Gleichung:

$$x^* = 0^*, \qquad\qquad y^* = 1^* \text{ oder } 6^* \quad (2 \text{ Paare}),$$
$$x^* = 1^*, 2^* \text{ oder } 4^*, \quad y^* = 3^* \text{ oder } 4^*, \quad (6 \text{ Paare}),$$
$$x^* = 3^*, 5^* \text{ oder } 6^*, \quad y^* = 0^* \qquad\qquad (3 \text{ Paare}).$$

Die entsprechende Rechnung für andere Primzahlen ergibt die Tabelle

p	2	3	5	7	11	13	17	19	23	29	31
N_p	2	3	5	11	11	11	17	11	23	29	35

.

Die Zahlen N_p können in diesem Fall durch eine schöne (und tiefliegende) Formel ausgedrückt werden:

$$N_p = \begin{cases} p, & \text{falls } p=3 \text{ oder } p \text{ die Gestalt } 3n+2 \text{ hat,} \\ p \pm 2u, & \text{falls } p \text{ die Gestalt } 3n+1 \text{ hat,} \end{cases}$$

wobei im zweiten Fall die Primzahl p als $u^2 + 3v^2$ dargestellt wurde (dies ist für Primzahlen der Gestalt $3n+1$ immer auf genau eine Weise möglich!) und das Vorzeichen so gewählt werden muß, daß N_p nicht durch 3 teilbar ist. Aus der Tabelle oder der Formel sieht man, daß N_p immer recht nahe bei p liegt; in der Tat gilt

$$p - 2\sqrt{p} < N_p < p + 2\sqrt{p}$$

für alle p, und zwar nicht nur in diesem Beispiel, sondern für alle elliptischen Kurven (Hasse, 1933; die Verallgemeinerung – 1974 durch den belgischen Mathematiker Deligne – dieses schon schwierigen Satzes auf Gleichungen in mehreren Variablen gilt als eine der größten intellektuellen Leistungen unseres Jahrhunderts).

Da die Zahlen N_p ungefähr gleich p sind, sind die Verhältnisse $\dfrac{N_p}{p}$ nahe bei 1. Wir bilden daher das Produkt

$$(8) \qquad \frac{N_2}{2} \cdot \frac{N_3}{3} \cdot \frac{N_5}{5} \cdot \ldots \cdot \frac{N_p}{p}$$

(p groß) und erwarten – in Übereinstimmung mit der Idee, daß die Zahlen für Kurven von höherem Rang verhältnismäßig groß sein sollen –, daß das Produkt (8) mit wachsendem p um so schneller anwächst, je größer der Rang der Kurve ist. Dies wurde von Birch und Swinnerton-Dyer (1965) präzisiert: das Produkt (8) soll asymptotisch wie $(\log p)^g$ mit einer ganzen Zahl $g \geq 0$ anwachsen und es soll gelten [8]

$$(9) \qquad\qquad g = r.$$

Sehr viel der heutigen Forschung dreht sich um Versuche, diese Vermutung zu bestätigen. Die zwei bisher stärksten Teilergebnisse:

1. 1975 zeigten J. Coates und A. Wiles, daß man für eine große Klasse von elliptischen Kurven (die z.B. die Kurve $y^2 = x^3 + 1$ oben enthält) aus $g = 0$ auch $r = 0$ folgern kann, d.h. eine Gleichung, für die das Produkt (8) beschränkt bleibt, hat nur endlich viele rationale Lösungen.

---

[8] Die hier gegebene Formulierung der Vermutung von Birch und Swinnerton-Dyer ist die elementarste und zuerst gefundene. Meistens wird aber eine andere Version der Vermutung benutzt, die technischer, aber für Forschungszwecke geeigneter ist: das unendliche Produkt

$$L(s) = \prod_{p \text{ prim}} \frac{1}{1 + (N_p - p)/p^s + p/p^{2s}}$$

soll eine vernünftige Funktion von s sein, die eine Nullstelle von einer wohldefinierten Ordnung g bei $s = 1$ hat, und mit dieser Zahl g soll die Beziehung (9) gelten.

2. 1983 zeigten B. Gross und ich, daß man für eine noch umfassendere Klasse von elliptischen Kurven (die vermutlich sogar alle elliptischen Kurven enthält) aus $g = 1$ $r \geq 1$ folgern kann, d.h. eine Gleichung, für die das Produkt (8) genau logarithmisch wächst, besitzt immer unendlich viele Lösungen.[9]

Da anscheinend die meisten elliptischen Kurven $g = 0$ oder $g = 1$ haben, werden in der Praxis sehr viele Fälle durch diese beiden Resultate abgedeckt. Für das theoretische Verständnis stellen sie nur einen Anfang dar; 1700 Jahre nach Diophant sind wir immer noch weit davon entfernt, die rationalen Lösungen auch sehr einfacher polynomialer Gleichungen voll zu verstehen.

---

[9] (Zusatz bei der Korrektur) Inzwischen gibt es neue allgemeine Ergebnisse in Richtung der Vermutung von Birch und Swinnerton-Dyer. Insbesondere konnte Kolyvagin (1988) für die unter 2. erwähnte »noch umfassendere Klasse« zeugen, daß sowohl im Falle $g = 0$ wie auch im Falle $g = 1$ stets $r = g$ gilt.